Prentice Hall Advanced Reference Series

Computer Science

DUBES AND JAIN *Algorithms for Clustering Data*
SHERE *Software Engineering and Management*

Engineering

FERRY, AKERS, AND GREENEICH *Ultra Large Scale Integrated Microelectronics*
JOHNSON *Lectures on Adaptive Parameter Estimation*
MILUTINOVIC *Microprocessor System Design GaAs Technology*
WALRAND *Introduction to Queueing Networks*

Science

BINKLEY *The Pineal: Endocrine and Nonendocrine Function*
CAROZZI *Carbonate Depositional Systems*
EISEN *Mathematical Methods and Models in the Biological Sciences: Linear and One-Dimensional Theory*
EISEN *Mathematical Methods and Models in the Biological Sciences: Nonlinear and Multidimensional Theory*
FRASER *Event Stratigraphy*
WARREN *Evaporite Sedimentology*

Mathematical Methods and Models in the Biological Sciences

Linear and One-Dimensional Theory

MARTIN EISEN

Department of Mathematics
Temple University

PRENTICE HALL, *Englewood Cliffs, New Jersey* 07632

LIBRARY OF CONGRESS
Library of Congress Cataloging-in-Publication Data
Eisen, Martin M.
Mathematical methods and models in the biological sciences/
Martin Eisen.
p. cm. — (Prentice Hall advanced reference series)
Includes bibliographies and index.
Contents: v. 1. Linear and one-dimensional theory — v.
2. Nonlinear and multidimensional theory.
ISBN 0-13-561291-8 (v. 1). ISBN 0-13-561309-4 (v. 2)
1. Biology—Mathematics. 2. Biology—Mathematical models.
I. Title.
QH323.5.E37 1988
574′ .0724—dc19 88-4176
CIP

Editorial/production supervision: *Editing, Design & Production, Inc.*
Cover design: *Ben Santora*
Manufacturing buyer: *Mary Ann Gloriande*

Prentice Hall Advanced Reference Series

A Division of Simon & Schuster
Englewood Cliffs, New Jersey 07632

Printed in the United States of America
10 9 8 7 6 5 4 3 2 1

ISBN 0-13-561291-8

Prentice-Hall International (UK) Limited, *London*
Prentice-Hall of Australia Pty. Limited, *Sydney*
Prentice-Hall Canada Inc., *Toronto*
Prentice-Hall Hispanoamericana, S.A., *Mexico*
Prentice-Hall of India Private Limited, *New Delhi*
Prentice-Hall of Japan Inc., *Tokyo*
Simon & Schuster Asia Pte. Ltd., *Singapore*
Editora Prentice-Hall do Brasil, Ltda., *Rio de Janeiro*

To my father, Isaac, and my brother, Edward

Contents

PREFACE xi

1 INTRODUCTION TO CONTROL SYSTEMS 1

1.1 Introduction 1
1.2 Brief History of Control Theory 1
1.3 External Description of Systems (Input-Output Relations) and Block Diagrams 4
1.4 Mathematical Models of Systems 8
References 27

2 CLASSIFICATION OF CONTROL SYSTEMS 29

2.1 Introduction 29
2.2 Classification of Systems 29
2.3 Open and Closed Loops 40
2.4 Basic Units of Linear Systems Diagrams 50
References 55

3 ORDINARY DIFFERENCE EQUATIONS 57

3.1 Introduction 57
3.2 Definitions Concerning Difference Equations 57
3.3 Existence and Uniqueness of Solutions of Difference Equations 62
3.4 Linear Difference Equations 66
3.5 Linear Difference Equations with Constant Coefficients 71

3.6 Graphical Analysis of One-Loop Discrete Systems 80
3.7 Conclusion 91
References 91

4 ORDINARY DIFFERENTIAL EQUATIONS 92

4.1 Introduction 92
4.2 Definitions Concerning Differential Equations 92
4.3 Existence and Uniqueness of Solutions of Differential Equations 101
4.4 Linear Differential Equations 105
4.5 Linear Differential Equations with Constant Coefficients 108
4.6 Glucose Tolerance Test and the Indicator-Dilution Method 120
4.7 Conclusion 126
References 126

5 CONVOLUTION REPRESENTATION; PROPERTIES OF LINEAR SYSTEMS 128

5.1 Introduction 128
5.2 Convolution Representation of Linear Systems 128
5.3 Convolution Representation for Continuous Time-Invariant Linear Systems 135
5.4 Some Properties of Linear Systems 138
5.5 Response to Special Test Inputs 141
5.6 Interconnections of Linear Systems 144
References 146

6 THE *z*-TRANSFORM 147

6.1 Introduction 147
6.2 Definition of the *z*-Transform 147
6.3 Properties of the *z*-Transform 152
6.4 Table of *z*-Transform Pairs 157
6.5 Partial Fractions and Inverse *z*-Transforms 158
6.6 Direct Method of Finding the Inverse *z*-Transform 165
References 166

7 THE LAPLACE TRANSFORM 167

7.1 Introduction 167
7.2 Definition of the Laplace Transform 167
7.3 Properties of the Laplace Transform 170

7.4 Table of Laplace Transforms 175
7.5 Expansion into Partial Fractions 178
7.6 Laplace Transform Solution of Ordinary Differential Equations 184
References 188

8 TRANSFER FUNCTIONS AND THEIR PICTORIAL REPRESENTATION **189**

8.1 Introduction 189
8.2 The Transfer Function 189
8.3 Use of Impulse Inputs to Determine the Transfer Function 192
8.4 Block Diagrams 196
8.5 Signal Flow Graphics 201
8.6 Linearity and Nonlinearity 211
References 212

9 RESPONSE OF LINEAR SYSTEMS **214**

9.1 Introduction 214
9.2 Decomposition of the Output 214
9.3 Transient and Steady-State Response 217
9.4 Characteristics of Feedback Control Systems 223
9.5 Static Gain 233
References 239

10 STABILITY CRITERIA FOR LINEAR STATIONARY SYSTEMS **241**

10.1 Introduction 241
10.2 Direct Determination of the Zeroes of a Polynomial 243
10.3 Routh-Hurwitz Stability Criterion 254
10.4 Frequency Response Method; Nyquist Criterion 260
10.5 The Bode Plot 271
References 283

A METHOD OF LEAST SQUARES **285**

B THE DIRAC DELTA FUNCTION **287**

INDEX **291**

Preface

Biology is a primitive science in the quantitive sense as compared to physics. Few areas in biology are based on fundamental mathematical laws that can be used to predict the outcome of a large number of different processes, as in mechanics. Unifying laws may be more difficult to find in biology than in physics for the following reason. A biological process often involves hundreds of biochemical reactions, whereas a physical process frequently can be reduced to a few simple reactions. This inherent difficulty of biology may partially account for the qualitative nature of biology. However, the author feels that, even without the discovery of fundamental unifying laws, biology could become a more quantitative science by teaching biologists to think mathematically.

Many biologists have consulted the author to analyze their data. These data often were useless for calculating the desired parameters. The common error is the belief that "thinking mathematically" involves merely manipulating large masses of data to obtain a desired result. Actually, much more is involved, as shown by the following brief description of proper scientific procedure:

1. A hypothesis is formulated or some interesting result is observed.
2. A model is constructed to explain the hypothesis.
3. An experiment is performed to gather the relevant data.
4. The data are analyzed and smoothed.
5. The parameters of the model are evaluated using the data.
6. The model is solved and the predicted results are obtained.
7. Statistical tests are applied to determine if the predicted results are in close agreement with the experimental results. If they are not close enough, a new model is constructed and the whole procedure must be repeated.

Note that the model should be constructed before the experimental data are obtained. The biologist should seek the cooperation of a mathematician to help him or her to design an experiment that will yield data that can be used to evaluate the desired parameters.

This book is concerned with the most important step in the analysis of a scientific process, step 2, as well as the mathematics required in step 6. This is the type of mathematics that can be used to translate biological results into mathematical form. Of course, computers and other types of mathematics play important roles in steps 4 through 7. However, these procedures are used after the mathematical model has been formulated and involve simulating or numerically solving a given system of equations. This type of analysis can be performed by someone with no knowledge or interest in biology or any other science.

Even though complex biological systems are ultimately studied by simulation, the more difficult analytic approach should not be abandoned. The exclusive use of simulation cannot fully determine the behavior of a system for the following reasons. Any model of a biological system will contain several parameters, say 15. Often the numerical value of a parameter is uncertain or the behavior of the system is desired for several different parameter values. Thus the program must be run for several different parameter values; otherwise important features of the system might be overlooked. Suppose that 10 estimates of each of the 15 parameters are considered. This would require 10^{15} separate simulations. If each simulation took only 1 second, the total time required would be 10^{15} seconds—much greater than a scientist's lifetime. Many biological systems can be described by a system of differential equations that may become unstable if their form is varied slightly. Ideally, several different sets of governing equations should be tested; this is not possible using simulation. Finally, numerical methods and simulation are not perfect. Errors can arise from many sources. Thus computer programs can yield misleading results.

One of the groups to which this book is directed is biologists. Since biologists usually have little advanced mathematical training, the only prerequisite for reading this book is two or three terms of elementary calculus. All advanced mathematical concepts are explained and illustrated with examples. Further, each section contains exercises that require the techniques of that section for their solution. By working on these problems, the reader will acquire the skill necessary to apply these techniques to his or her own problems. Since the two volumes are self-contained, they can be used for self-study or as references.

Applied mathematicians should also be interested in this book. Although the book begins at an elementary level, it soon reaches control theory, which is not usually taught in mathematics departments. Even some of the elementary topics, such as difference equations, contain examples and results that are not ordinarily covered in a typical mathematics course. This type of mathematics is useful for analyzing many diverse systems other than biological ones. A mathematician will see that there are many challenging practical problems as well as theoretical ones due to the complexity of biological systems.

Many different biological topics are discussed in this book—for instance, optimal drug dosages, insect populations, predator-prey interactions, the glucose tolerance test, and protein synthesis. Each example includes an outline of the biology necessary to understand it. Some of the examples have been deliberately simplified in order to appeal to a larger audience. A person not interested in a certain area of biology is not likely to read an example several pages long. Furthermore, an example is not meant to be a biological thesis but rather an illustration of particular mathematical techniques. Once the mathematical methods are mastered, the reader should be able to apply them to more complicated systems.

This first volume considers problems that can be formulated in terms of a single independent variable and are linear. The second volume will treat nonlinear and multivariable problems.

Part of the book was used in a graduate course in the Department of Physiology and Biophysics at Temple University. The book was also used in a graduate course in mathematical biology in the Mathematics Department. Readers with a strong mathematical background could omit Chapters 3, 4, 6, and 7 without loss of continuity.

Portions of this book were written while the author was on a leave of absence with support from Temple's University.

1

Introduction to Control Systems

1.1 INTRODUCTION

The concept of control systems and their representation by block diagrams are introduced in this chapter. The above ideas are illustrated by means of examples from biology. The practical application of control theory requires a bridge between the real world and mathematics. This bridge is the process of modeling; hence, a section on modeling is included. To emphasize the important role played by biological ideas in the development of control theory, the chapter begins with a brief history of control theory. The development of optimal control theory is outlined in the introduction to Volume II, Chapter 5.

1.2 A BRIEF HISTORY OF CONTROL THEORY

Control theory arose in two fields, biology and engineering. At the time when biologists were first beginning to formulate some of the concepts of control theory in a descriptive and qualitative manner, engineers already had made considerable progress in developing and applying quantitative methods. However, these two lines of development did not mingle until the 1940s. Many modern books on control theory appear to be advanced treatises on mathematics. A student may lose sight of the fact that the motivating factors in the development of control theory were practical problems. To emphasize the role played by practical problems and to indicate the intertwining of biology and engineering, a brief history of control theory will be presented.

Some highlights of the early history of control theory are shown in Table 1.1. As seen in the table, the early history of control theory was characterized by the study of simple practical devices or organisms in an intuitive manner. Efforts to increase the accuracy of the control systems led to slower attenuations of the transient oscillations and even unstable systems. It then became necessary to develop a theory of automatic control.

TABLE 1.1 EARLY HISTORY OF CONTROL THEORY

Approximate date	Person	System
300–0 B.C.	Ktesibios	Water clock based on float regulator
	Philon	Oil lamp with fuel oil level maintained by float regulator
	Heron of Alexandria	*Pneumatica,* a book, described several forms of water-level regulation using floats
1202	Fibonacci [14]	Growth of rabbit populations
1600	Cornelis Drebbel	Temperature regulator
1680	Giovanni Borelli [7]	Mechanics of animal motion
1681	Dennis Papin	Pressure regulator for steam boilers
1765	I. Polzunov	Water float regulator for a boiler
1769	James Watt	Flyball governor for controlling the speed of a steam engine

In 1868 Maxwell [25] formulated a differential-equation model of governors. A few years later Vyshnegradskii [36] proposed a mathematical theory of regulators. The major emphasis in engineering for the next 140 years was on practical problems in the design of automatic regulating devices.

Possibly the first to recognize the prevalence of regulatory processes in biology was Claude Bernard. The following quotation from his book [5], published in 1878, reveals his use of modern concepts:

> It is the fixity of the "milieu interieur" which is the condition of free and independent life, . . . all the vital mechanisms however varied they may be, have only one object, that of preserving constant the conditions in the internal environment.

Seven years later Fredericq [16] also emphasized that living organisms have control systems that tend to minimize deleterious disturbances imposed by the environment:

> The living being is an agency of such sort that each disturbing influence induces by itself the calling forth of compensatory activity to neutralize or repair the disturbance. The higher in the scale of living beings, the more numerous, the more perfect and the more complicated do these regulatory agencies become. They tend to free the organism completely from the unfavorable influences and changes occurring in the environment.

These ideas were further extended by Cannon [10], who coined the word *homeostasis,* which he defined as "coordinated physiological reactions which maintain most of the steady states of the body."

The book by D'Arcy Thompson [35] on growth and form, published in 1917, marks the start of modern mathematical biology.

Meanwhile engineers, motivated by practical problems, slowly developed the theory of automatic regulating devices. In 1922, Minorsky's basic paper [28] on the theory of automatic steering of ships appeared. Another noteworthy paper was that of Hazen [20] on the theory of shaft-positioning types of servomechanisms. In the 1920s the development of radio and the subsequent rise of the electronics industry were the major factors in the further theoretical development of control theory in the United States. The problems of stability of amplifiers and the design of the feedback amplifier led to the basic work of Nyquist [29]. In 1933, Lee and Wiener published their investigations of systems of differential equations descriptive of many physical systems. These results were applied by Bode [6] and his co-workers at Bell Laboratories to the design of telephone systems and electronic feedback amplifiers. The frequency domain was used primarily to describe the operation of the feedback amplifiers.

Prior to World War II, control theory developed in a different manner in the USSR. Many eminent Soviet mathematicians were interested in the theoretical problems generated by practical problems in control. These mathematicians dominated the field of control theory. Hence, the Soviet theory tended to utilize a time-domain formulation based on differential equations.

During World War II the theory and practice of automatic control received a tremendous impetus from the necessity of designing and constructing automatic pilots, radar tracking systems, and automatic gun sighting and fire control systems. In general, before 1940, the design of control systems was an art involving trial and error. During the 1940s mathematical methods increased in number, and control engineering became an established branch of engineering.

The paper by Rosenbleuth et al. [32], appearing in 1943, was one of the first attempts to refine the work of Cannon. The book by Wiener [37] expressed the view that human beings were complicated control systems. He coined the term *cybernetics*.

Following and during World War II frequency-domain techniques based on the Laplace transform and complex frequency domain continued to dominate the theory of control. The advent of Sputnik and the space age imparted a new impetus to control engineering. Missiles and space probes required the design of highly complex and accurate control systems. The necessity to minimize the weight of satellites and control their trajectories very accurately gave birth to the important field of optimal control. Because of these requirements, the time-domain methods of Liapunov, Minorsky, and others became increasingly popular. One of the major reasons for this surge in popularity was the use of high-speed computers for control problems in the 1950s. Time-domain theorems could now be applied by utilizing numerical techniques that were not feasible to do by hand. At present both time-domain and frequency-domain approaches are utilized in the analysis and design of control systems.

The interdisciplinary exchanges between systems theory and biology have become increasingly apparent. The texts by Guyton et al. [19], Grodin [18], Yamamoto and Brobeck [38], Kalmus [22], and Bayliss [3] constitute significant contributions from biologists and physiologists to a growing borderland between biological and technological regulation. An engineering-oriented approach to this territory is taken in the texts by Milhorn [26] and Milsum [27]. Other relevant books are [8, 9, 11, 17, 21, 23, 31, 33, 34]. A perusal of the literature reveals that the physiologist is using engineering methods for the analysis of physiological control problems. On the other hand, the engineer and mathematician are investigating adaptive control problems that are similar to the control problems encountered in biology [4].

A more detailed history of some early regulators can be found in Mayr [24]. Adolph [1] describes the origins of physiological regulation. A brief outline of optimal control theory is contained in the introduction to Volume II, Chapter 5.

1.3 EXTERNAL DESCRIPTION OF SYSTEMS (INPUT–OUTPUT RELATIONS) AND BLOCK DIAGRAMS

Webster's Dictionary defines a *system* as "a group of diverse units so combined by nature or art as to form an integral whole, and to function, operate, or move in unison, often in obedience to some form of command." The elements of the system function together, thereby giving rise to the behavior of the whole. The operation of the system depends not only on the operation of the individual elements but also on the nature of their interconnection. For example, the structures and processes associated with digestion can be called the digestive system. Its functioning depends not only on the nature and behavior of the components, such as the stomach, certain parts of the circulatory system, and the chemical and neural control loops, but also on the precise manner of their interconnection.

In referring to a system there is a real or imaginary boundary that separates the real world into two parts. The part of the real world under study is within the boundary and is called the *system*. The rest of the real world is referred to as the *environment*. By means of this division, two types of systems can be defined. An *open system* is a system that is affected by its environment. A *closed system* is one that is isolated from its environment—that is, none of the environment's variables affects the system. In reality, all biological systems are open. Therefore, any representation that regards a system as being closed is an approximation to the truth. Such approximations may be valid in certain situations, such as when examining effects that occur over a short period of time.

Systems theory is the study of the behavior of systems subject to certain conditions or inputs. Mathematics is the natural language of systems theory, since it is concerned with abstract properties rather than the physical form of the constituent parts. Hence, the positioning of a missile launcher or the distribution of iodine in humans can be studied by similar systems techniques.

A *control system* is a system whose purpose is to regulate or control the flow of hormones, energy, information, or other quantities in some desired way. *Control theory* is concerned with all questions related to the design and analysis of control systems.

To utilize mathematics in the design and analysis of systems, it is necessary to formulate a mathematical model of a system. An informal representation of a system will be presented, since the purpose of this book is to concentrate on applications of systems theory to biology.

The following notation will be used to describe systems:

$$t \text{ (time)} = \text{a real number or a discrete variable.}$$

u (input variables or signals) = *controls* chosen by the operator or mechanism controlling the system or perturbations in the system.

y (output variables or signals) describe either:

1. The behavior of the system and the outputs correspond to the observations that can be made on the system, or
2. What is produced by the system.

Note that both the input and output variables can be vectors. However, a knowledge of vectors is not required to understand the essential concept of a system presented below. The reader who is not familiar with vectors should just regard each input or output variable as a single real variable. Vectors will be introduced in Volume II, Chapter 1.

The concept of a *(dynamical)* system is a *(dynamic)* transformation of inputs into outputs. The term dynamic (as opposed to *static* or *memoryless*) is used because past input values $[u(t), t_0 \leq s < t]$ as well as the present value $u(t)$ determine the output $y(t)$. In other words,

$$y(t) = f(u[t_0, t], t) \tag{1}$$

where f is a function that describes the output of the system at time t for an input function u whose past and present values are denoted by

$$u[t_0, t] = \{u(s) \,|\, t_0 \leq s < t\}$$

This direct description in terms of the input–output relation (1) is called the *external description** of the system.

A system can be represented by a block diagram as shown in Fig. 1.1. In cybernetic terminology a block diagram is called a *black box*. This nomenclature is intended to emphasize the transformation of u into y rather than the corresponding mechanism. The omission of operational details may be deliberate or because of ignorance. Some examples of systems appear in Table 1.2.

TABLE 1.2 EXAMPLES OF SYSTEMS

System	Input	Output
Respiratory	CO_2	Respiratory rate
Telephone	Microphone vibration	Earphone vibration
Boat	Position of rudder	Heading
Heart	Flow through coronary artery	Cardiac output
Automobile	1. Accelerator	Speed
	2. Brake	

Inputs of a system can vary with the experiment. For example, an engineer may be interested only in studying the effects of varying a single input of a multiple-input

* Other methods of describing a system will be studied in the later chapters; for example, there is also an *internal description* of a system.

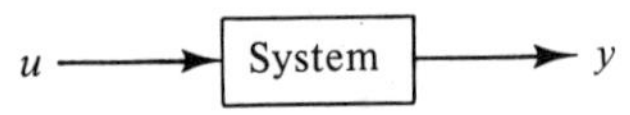

(a) Single input and output

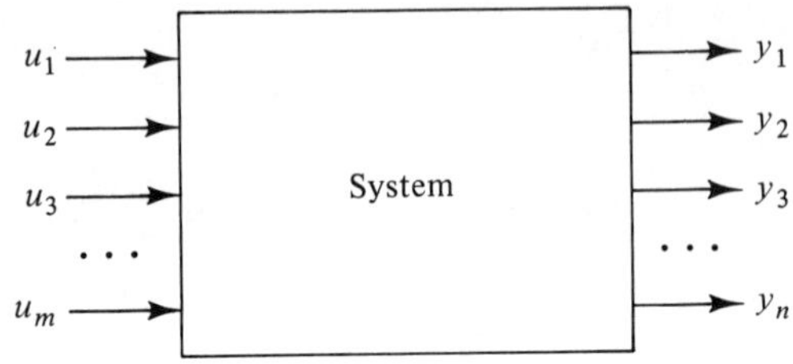

(b) Multiple inputs and outputs

Figure 1.1 Block diagram of a system.

system. The behavior of a system that has no input can also be studied. Conversely, an experimenter may want to increase the number of inputs. For example, the effects of O_2 and CO_2 on the respiratory rate may be desired (see Table 1.2).

Frequently, complex systems are decomposed into subsystems by representing each subsystem by a block diagram. The blocks are joined by directed line segments, which represent the flow of signals between the blocks. The output signals of some subsystems may be merged to produce an input for another subsystem. The device that performs this service is called a *junction*. A junction with two inputs is diagrammed in Fig 1.2.

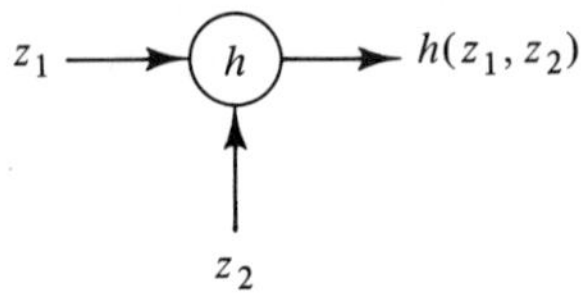

Figure 1.2 Two-input junction.

For example, if $h(z_1, z_2) = z_1 - z_2$, the two-input junction is diagrammed as

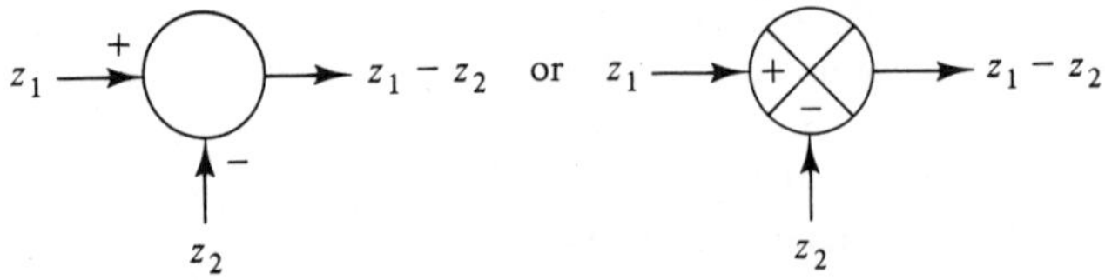

An example of a block diagram for a complex system appears in Example 1.

Example 1 (Human Circulatory System)

A block diagram of the human circulatory system is shown in Fig. 1.3. The two major veins carrying blood back to the heart from the upper and lower parts of the body combine near the right atrium of the heart. The venous blood enters the right atrium and flows into the right ventricle of the heart. From here it is squeezed through a sequence of arteries–capillaries–veins (the *pulmonary circulation*) to the left atrium. Then the blood flows to the left ventricle, where it is pumped out of the heart into the aorta. The ascending aorta has arterial branches that carry the blood to the head and arms, whereas the descending aorta supplies the rest of the body. The blood travels through a branching network of smaller and smaller vessels until it finally reaches the very numerous capillaries, where exchange of nutrients, O_2, and CO_2 occurs. Leaving the capillaries, the blood flows from a converging branched network of vessels into successively larger veins, which finally lead back to the right atrium. The loop from the left ventricle to the right atrium is termed the *systemic circulation*.

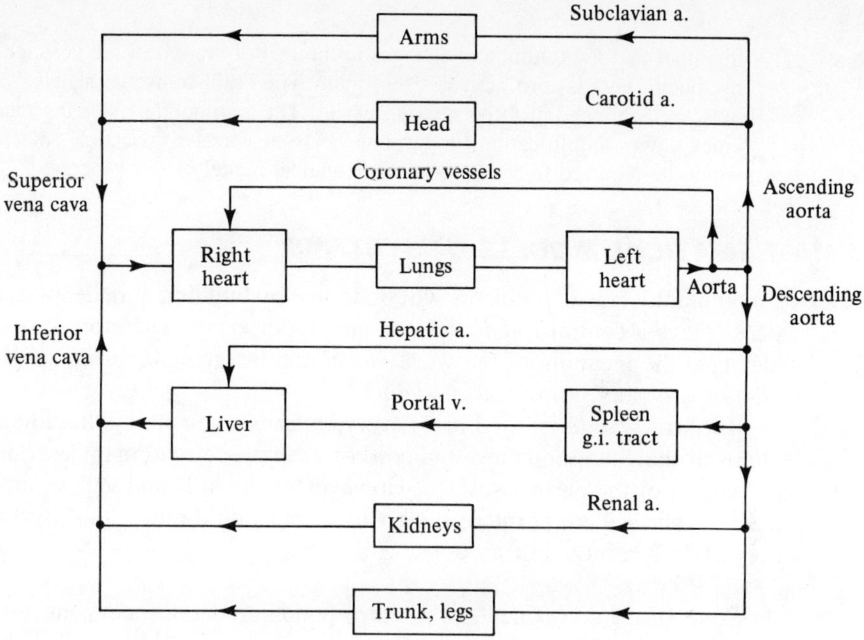

Figure 1.3 The human circulatory system.

A unique feature of the systemic circulation is the portion known as the *portal circulation*. This consists of the hepatic artery, which goes directly into the liver, and the portal vein, which carries blood from the digestive organs, the spleen, the pancreas, and the gallbladder to the liver. This system can protect the body. It permits the liver to detoxify any noxious substances that might have been picked up in the gastrointestinal tract before they can enter the rest of the circulatory system.

Example 2 (Driving an Automobile)

Consider the system driver plus automobile. The objective of the driver is to keep his automobile in the center of a chosen lane on the road. The input to this system is the desired path of the car, and the output is the actual path of the car. This system can be decomposed into subsystems as illustrated in Fig. 1.4. An error in the difference between

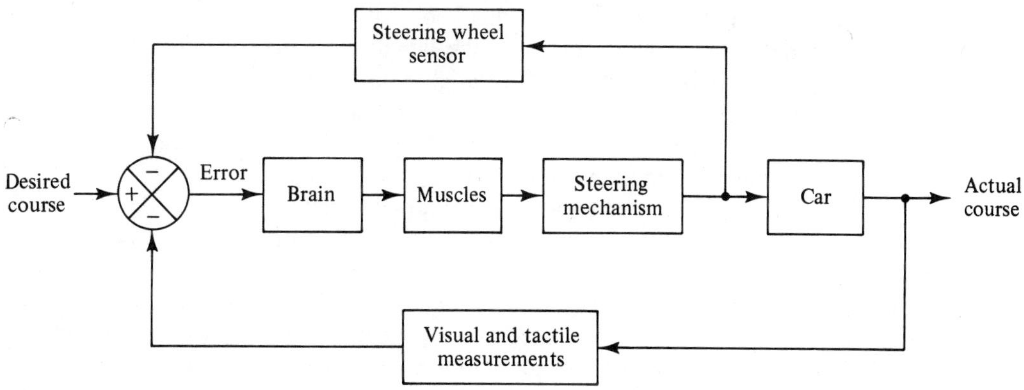

Figure 1.4 Automobile steering control system.

the input and the output and changes in the feel of the wheel are detected visually and by touch. This is turn activates the brain. The brain transmits signals to the driver's muscles, which control the steering wheel. The automobile's steering mechanism provides power amplification for turning the front wheels. Of course, much more detail would be required to construct a mathematical model of this process.

1.4 MATHEMATICAL MODELS OF SYSTEMS

Throughout this book, problems will be solved by building a model of a system. This process plays a central role in science and engineering. Therefore it is important to understand the meaning of the word *model* and the steps involved in constructing a model of a system, known as *modeling*.

The word *model* is used in its everyday sense: something that imitates relevant features of the situation being studied. For example, a road map describes highways on a portion of the earth's surface. However, if the hills and valleys of a portion of the earth's surface are of interest, a contour map would mimic this aspect. Formally, an object M is a model of an object S if:

1. Some of the components of M correspond in a one-to-one manner with some of the components of S.
2. For at least some relationships, the relation between the components of M is analogous to that between the corresponding components of S.

For instance, a dressmaker's mannequin is a model of a human being. Certain features of a mannequin correspond to certain features of a human being. Some relations between the corresponding features are the same in both. However, not all human features are imitated by the mannequin.

An important point is that the definition of a model does not require that the model M be an exact duplicate of S. Condition 1 does not require every component of M to correspond to a component of S or vice versa. Nor does condition 2 require that every relationship between components of S be mirrored by an analogous relationship in the model. Thus one of the first tasks in constructing a model is to decide which characteristics of the object or system are to be represented in the model.

The only models considered in this book will be mathematical models. Mathematical models mimic reality by using the language of mathematics. These models consist of collections of equations. The variables in these equations are associated with the physical entities of the system being studied so that relations are preserved. In other words, a relation between the variables in the mathematical expression is analogous to the relation between the corresponding physical entities.

Frequently the same phenomenon can be described by many different models. Some exhibit nearly perfect correspondence with all the physical parameters of the process being modeled. Others exhibit complete ignorance of the underlying mechanism. Which of these are correct? The criterion used by an applied mathematician is a practical one. If the results predicted by a model are sufficiently close to the observed data, the model is suitable. Hence, two models that give different results can both be satisfactory if they are both close to the observed data when biological variability and experimental errors are taken into account.

Models of systems can be constructed by two different methods. Experimental modeling, or modeling by synthesis, is the selection of mathematical relationships that seem to fit observed input–output data. Analytical modeling consists of a systematic application of basic physical laws to system components and the interconnection of these components. This book will emphasize analytical modeling.

An outline of the analytical approach to modeling is depicted in Fig. 1.5. Mechanically following the steps in Fig. 1.5 may not lead to a suitable model. Modeling often requires ingenuity coupled with a deep knowledge of the particular subject being studied as well as general physical or chemical laws. Constructing a suitable model is often the hardest step in solving a problem. The steps outlined in Fig. 1.5 are discussed in the following paragraphs.

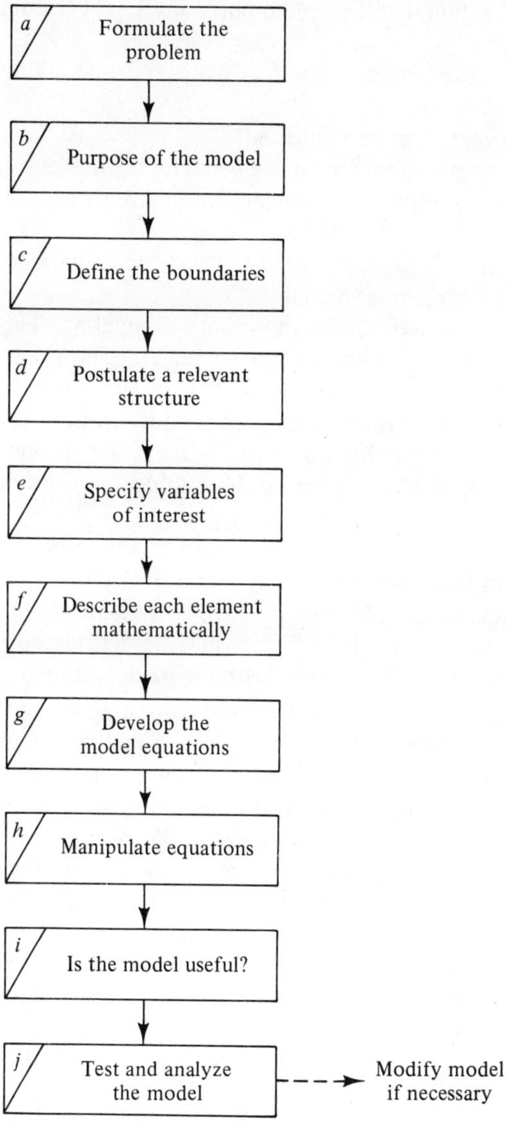

Figure 1.5 Steps in modeling.

(a) What you wish to know must be clearly specified. Defining the problem can be difficult. Frequently, the insight gained in the process of model building leads to a re-formulation of the problem. Care must be exercised in not defining the problem too narrowly or too rigidly thereby becoming blind to serendipitous insights.

(b) The intended purpose of the model must be clearly specified. The parts of the problem that the model is intended to clarify or analyze should be delimited.

(c) The boundary of the system is a real or imagined separation of part of the real world under study (the *system*) and the rest of the real world (the *environment*). A requirement for the selection of the boundary is that all signals crossing the boundary should be one-way interactions. The system output should not affect the environment to the extent that it would modify any signals from the environment to the system.

To aid in defining the boundary the real world must be divided into three parts:

(1) Factors whose effects can be neglected.

(2) Factors that affect the system but whose behavior the model is not designed to study. These are called independent variables, input, or exogenous variables.

(3) Factors the system is designed to study. These are called dependent variables, output, or endogenous variables.

Categories (1)–(3) are extremely important in modeling. The model will not be satisfactory if the wrong things are neglected. The other extreme of including too much can result in a model that is hopelessly complex and requires an enormous amount of data. Sometimes a compromise is required. Things are neglected even though they may be important, because they cannot be handled readily. The modeler hopes that neglecting these things will not invalidate the conclusions.

It may not be possible to carry out step (c) initially, since it may not be clear which factors can be neglected. It may be necessary to return to this step after the model is completed. A common practice is to begin with a crude model and rough data estimates to see which factors may be considered in the model and how accurately the exogenous variables must be determined.

(d) The actual system is reduced to an ideal system consisting of interconnections of simple, idealized elements that can be described mathematically. A block diagram is usually drawn for a complicated system. The resulting structure should mimic the behavior of the parts of the real system to be studied.

The above procedure is not straightforward. There is no single model of a complex system that is appropriate for all purposes. A finer level of decomposition often leads to simpler components but to a more complicated interaction network. The problem of how to lump parts of the system taxes the investigator's knowledge of the system being studied. The component subsystems of the idealized system do not necessarily correspond to physically identifiable components of the real system being studied.

(e) To develop a mathematical model, variables must be assigned to all attributes of interest. If a quantity of interest does not exist and so cannot be labeled, step (d) must be modified to include it.

(f) Translate the physical laws describing each model element into mathematical terms. The resulting mathematical relations may be algebraic, difference, differential, integral, linear or nonlinear, constant or time-varying. The type of relations are the basis of classifying systems as described in the next chapter.

(g) After a system has been reduced to an interconnection of idealized elements, with each element described mathematically, equations must be developed to describe the interconnection effects.

(h) The equations developed in step (g) must be manipulated into a form in which they can be solved or convey some qualitative information.

(i) Can all parameters in the model be determined from the available data so that the model can be used to make the desired predictions? If the answer is no, then the model must be re-formulated [step (c)] and perhaps even the problem [step (a)]. Sometimes new experiments may have to be performed to gather additional data. Note that "useful" does not mean reasonable or accurate; these properties are considered in step (j).

(j) Use the model to make predictions that can be checked against data or intuition. It is not advisable to rely entirely on intuition, because it may well be wrong. However, for certain problems experience in the field being modeled is essential, since no data may be obtainable or a relevant experiment cannot be performed. For example, how can a model of bacteriological warfare be tested? Begin with easy predictions; don't waste time on involved calculations with a model that may be no good.

If the predictions are unacceptable and there are no mathematical errors, return to step (c) or step (a). If the predictions do not have the desired accuracy, try to understand why, since this may uncover implicit or false assumptions. Once the model has been checked to your satisfaction, seek the solution to the motivating problem.

If necessary, revise the model by beginning at step (c) or even step (a). It may be necessary to revise the physical system being studied by using the results of new experiments that have been suggested by the failure of the previous model. Do not simply revise the mathematical model and force it to fit the data. A frequent pitfall of this last approach is to believe that the biological system must actually contain components that correspond to the various parts of the mathematical model because the computer behaves as a real biological system. Modeling is an iterative process. Steps (a) through (j) may have to be carried out many times before an acceptable model is constructed. The following examples illustrate the modeling process.

Example 3 (Binary Fission)

Cells of higher plants and animals and some one-celled creatures reproduce asexually. The parent cell divides into two daughter cells, which are similar to the parent cell. This process is called *binary fission*.

(a) The following experiment is performed on a fixed number of cells that reproduce by binary fission. The cells are grown on a suitable medium with an abundant food supply and without crowding. The number of cells at times $t_1, \ldots, t_n$ is recorded. The size of a population of the same type of cells grown under similar conditions is to be predicted given the initial number of cells without further experimentation.

(b) The purpose of the model is to obtain a quantitative formula that gives the population size at time t when the population size is known initially at $t = 0$.

(c) The given system of cells is confined to a certain area and does not suffer from overcrowding or lack of food. There is no immigration or emigration of cells. The population size changes because of the birth or death of cells.

(d) The only purpose of the mathematical model is to mimic the change in size of a real population of cells. Therefore, an ideal cell will only model the processes of birth and death. Other features of real cells such as size or metabolism are not modeled, because they are nonessential for determining population size.

The real population of cells is replaced by an ideal population of identical cells. Each ideal cell behaves like an average physical cell as far as birth or death is concerned. Over a sufficiently small time interval, each ideal cell reproduces a constant fraction of itself per unit of time. Simultaneously, another constant fraction of each ideal cell is destroyed per unit of time. Note that the ideal population size can be a fraction, while the real population size is a nonnegative integer. The population size is assumed to be so large that the percentage error between the real and ideal population sizes is small.

The simple model proposed above is not unique; there are many other models. For example, the fractional reproduction rate may vary with time. The crucial test to determine whether a model is suitable is to determine if its predictions agree with the experimental data.

(e) The following notation is introduced in order to translate the model described in step (d) into mathematical language:

t = a continuous variable denoting time $(0 \le t < \infty)$.
$N(t)$ = the size of the ideal population of cells at time t.
β = the fraction of an ideal cell that is produced by each ideal cell per unit of time. This fraction is a constant on a sufficiently small time interval.
δ = the fraction of each ideal cell that is being destroyed per unit of time and is a constant on a sufficiently small time interval.

(f) The increase in population size in a small interval of time Δt is

$$\beta N\,\Delta t + 0(\Delta t)$$

where $0(\Delta t)$ is some function of Δt such that

$$\lim_{\Delta t \to 0} \frac{0(\Delta t)}{\Delta t} = 0$$

The decrease in population size in a small interval of time Δt is

$$-\delta N\,\Delta t + 0(\Delta t)$$

Hence, the net change in population size in the time interval t to $t + \Delta t$ is

$$N(t + \Delta t) - N(t) = (\beta - \delta)N(t)\Delta t + 0(\Delta t) \tag{2}$$

(g) Dividing both sides of (2) by Δt and taking limits as $\Delta t \to 0$ gives

$$\frac{dN}{dt} = (\beta - \delta)N \tag{3}$$

(h) Equation (3) can be solved by separating the variables,

$$\frac{dN}{N} = (\beta - \delta)\,dt$$

and integrating from $t = 0$ to $t = t_1$:

$$\int_0^{t_1} \frac{dN}{N} = (\beta - \delta)\int_0^{t_1} dt$$

This gives

$$\ln \frac{N(t_1)}{N(0)} = (\beta - \delta)t_1 \tag{4}$$

where ln denotes the natural logarithm function. Exponentiating both sides of (4) and using the fact that

$$e^{\ln x} = x$$

yields

$$N(t) = N(0)e^{(\beta-\delta)t} \tag{5}$$

where the subscript on the t has been dropped.

(i) The model contains two parameters, β and δ. These parameters cannot both be determined from the experimental data determined in step (a). Fortunately, the *fractional* (or specific) growth rate,

$$k = \beta - \delta = \frac{1}{N}\frac{dN}{dt}$$

can be determined from the given data. Equation (5) can be rewritten in the form

$$N(t) = N(0)e^{kt} \tag{6}$$

using the new notation. As shown by (6), the new model is useful for predicting the size of the population at an arbitrary time t, given the initial population size. In order to determine both the parameters β and δ, a new experiment must be designed to determine one of these parameters.

(j) Does the theoretical growth curve (6) describe the experimental growth curve, and if so, what is the constant k? To answer this question, it is convenient to take natural logarithms of both sides of (6). The resulting relationship between $\ln N$ and t is

$$\ln N = \ln N(0) + kt \tag{7}$$

The graph of (7), when $\ln N$ is plotted as a function of t, is a straight line with slope k and intercept $\ln N(0)$. Therefore, if such a plot is made with the experimental data and the resulting curve is a straight line, within experimental error, then the growth of the population is described by (7). Furthermore, the growth rate

$$k = \frac{\ln N(t_2) - \ln N(t_1)}{t_2 - t_1} \tag{8}$$

from the usual analytic formula for the slope of straight line, where $N(t_1)$ and $N(t_2)$ are the ordinates on the straight line determined from the experimental data.

To complete the discussion of this example, the problem of fitting a straight line to the experimental data must be resolved. The difficulty arises from the fact that even though the relationship between two variables is suspected to be linear, the observations may not all fall on a straight line because of experimental error. A method (least squares) for determining the straight line that "best" fits the data is presented in Appendix A. This book will not discuss procedures for fitting observations to an assumed functional form and for determining the goodness of fit. These aspects of model testing can be found in books on numerical analysis [2, 12, 15] and statistics [13].

Cell populations that are described by (6) are said to be in *log phase*. The reason for this terminology is the relation (7). This type of growth is exhibited by isolated colonies of bacteria in the early growth stage. However, the colony of bacteria does not grow indefinitely. Possible reasons for this are lack of space, lack of oxygen, disappearance of nutrients, the appearance of toxic substances, or changes in the ion concentration in the medium, such as the pH. Hence, for longer time intervals, the model described in Example 3 is not valid. Generalizations of this model in which the population size levels off will be discussed later.

The parameters β and δ in (5) can be defined in terms of two biologically important time periods. The mean generation time, T_g, is the mean time required for a cell to divide. Since cell deaths are not considered in finding the mean generation time, $\delta = 0$. The initial population doubles in a time interval T_g; hence,

$$N(T_g) = 2N(0)$$

Substituting this relation into (5) with $\delta = 0$ gives $2N(0) = N(0)e^{\beta T_g}$, or

$$\beta = \frac{\ln 2}{T_g} \tag{9}$$

The *doubling time, T_d*, is the length of time required for the population to double if $\beta - \delta > 0$. By analogous reasoning to that used in deriving (9), it follows that

$$\beta - \delta = \frac{\ln 2}{T_d} \tag{10}$$

If, in addition to recording the population size, the mean generation time or the mean doubling time is found experimentally, then the parameters β and δ can be determined. For example, the division time of a large number of cells can be found by time-lapse photography. The average of these results is an estimate of the mean generation time. The parameter β can then be calculated from (9). The growth rate $k = \beta - \delta$ is the slope of the growth curve. The doubling time is just the abscissa corresponding to the ordinate $2N(0)$ of the smoothed growth curve. Since β, $\beta - \delta$, and T_d are known, δ can be determined from (10).

Modeling steps (a)–(j) will not be discussed explicitly in the remainder of the book. This will shorten the presentation of each example, thereby allowing a greater variety of topics to be presented without unduly lengthening the book. However, the reader should mentally separate steps (a)–(j) for the remaining examples in this section.

The population model of Example 3 is not applicable to the study of human populations with no immigration or emigration. The reason is that the birth rate and death rate are dependent on the specific age group under study. The next example takes this fact into account. The model can also be applied to other species with age-dependent birth and death rates.

Example 4 (Growth of Human Populations)

The following variables are required to formulate a mathematical model of human population growth without immigration or emigration:

$x_i(k)$ = number of people who are i years old at the beginning of year k

b_i = number of children produced yearly by a typical person of age i years (it is assumed that the children are born at the beginning of the year)

s_i = fraction of people aged i who typically live through that year ($0 \le s_i \le 1$)

$y(k)$ = population size at the beginning of year k

N = oldest observed age

Since children who are age 1 at the beginning of year $k + 1$ must have been born to a parent aged i ($i = 1, 2, \ldots, N$) in year k, it follows that

$$x_1(k + 1) = b_1x_1(k) + b_2x_2(k) + \cdots + b_Nx_N(k) \tag{11}$$

Individuals who are older than 1 at the beginning of a year are persons who have survived from the previous year and so have aged one year. Hence,

$$x_{i+1}(k + 1) = s_ix_i(k) \qquad i = 1, 2, \ldots, N - 1 \tag{12}$$

It is apparent that the total population at the beginning of the kth year is

$$y(k) = x_1(k) + x_2(k) + \cdots + x_N(k) \tag{13}$$

Equations (11) and (12) are called *recursive equations,* * since they can be solved recursively as follows. Suppose that the initial values $x_i(0)$ for $i = 1, 2, \ldots, N$ are given. Setting $k = 0$ in (11) yields the value of $x_1(1)$ on terms of the given initial values. Letting $k = 0$ in (12) gives

$$x_{i+1}(1) = s_ix_i(0) \qquad i = 1, 2, \ldots, N - 1$$

Since $x_i(1)$ for $i = 1, 2, \ldots, N$ are known, setting $k = 2$ in (11) and (12) determines $x_i(2)$. By repeating this procedure, $x_i(k)$ can be found for any finite value of k. The population size can then be determined from (13). A numerical example appears in Exercise 15 at the end of this section.

In general, the survival and offspring coefficients will vary as a function of the time parameter k. For example, improvements in the medical care, famine, and war are factors that will create changes in s_i, the birth rate coefficients b_i will tend to change in a manner reflecting the current attitudes of society. Hence, the model of population growth will be time-varying in nature. More work will be involved in obtaining statistical estimates of b_i and s_i. However, the numerical solution of the model equations is no more difficult than in the situation where these coefficients are constants.

The next example is concerned with a drug dosage problem. The dosages for many drugs are not calculated scientifically but are determined by trial and error. If the desired effect is not obtained, the dosage is increased. On the other hand, if adverse symptoms appear, the dosage is decreased.

The mathematical description of drug quantity and activity changes within the body is the concern of pharmacokinetics, a relatively new science. The mathematical models are based on physiologic concepts of absorption, distribution, and elimination

* They are also called difference equations. The reason for this nomenclature and some analytic methods for their solution will be discussed in Chapter 3.

(metabolism and excretion). Even though complex chemical and physiological processes are involved, the overall processes such as drug absorption and elimination are often described by first-order kinetics (the rate of change of a quantity is proportional to the amount of that quantity). This last assumption allows simple mathematical models to be used in many situations where a few rate-limiting processes govern the overall drug transport or metabolism.

Example 5 (Determination of Effective and Safe Drug Dosages)

A constant dose of a drug capable of raising its concentration in the blood to C_0 is administered at fixed time intervals of length t_0. Suppose that the drug is known to be ineffective below a concentration C_L and harmful above some higher concentration C_H. Can values of t_0 and C_0 be found that will produce safe concentrations (not above C_H) that are effective (above C_L)?

A model to answer this last question will be constructed based on the following two assumptions:

1. The decrease of concentration of the drug in the bloodstream is proportional to the concentration.
2. The fixed dosage of drug diffuses so rapidly through the blood that its concentration in the blood immediately after its administration increases by C_0.

Assumption 1 is supported by clinical evidence for many drugs. Assumption 2 is a reasonable approximation if the drug is administered intravenously but may be invalid for some drugs that are taken orally.

In order to translate the model into mathematical form, let $C(t)$ denote the concentration of the drug in the blood at time t. Assumption 1 becomes

$$\frac{dC}{dt} = -kC \tag{14}$$

where $k > 0$ is known as the *elimination constant*. The solution of (14) is

$$C(t) = C(0)e^{-kt} \qquad (0 \le t < \infty) \tag{15}$$

The effects of repeating equal doses are shown in Fig. 1.6. As soon as the initial dose is administered at $t = 0$, the drug concentration rises to C_0 by assumption 2. By assump-

Figure 1.6 Effects of repeated equal doses of a drug.

tion 1 the concentration decreases exponentially, according to (15), until the residual concentration R_1 is reached at $t = t_0$. At this time the second dose is administered and the graph rises vertically to reach concentration $C_1 = R_1 + C_0$ according to assumption 2. This sequence of events is repeated as shown in Fig. 1.6. In general,

$$C_{i-1}e^{-kt_0} = R_i \qquad i = 1, 2, 3, \ldots \tag{16}$$

$$R_i + C_0 = C_i \qquad i = 1, 2, 3, \ldots \tag{17}$$

From (16) and (17) it follows that

$$C_i = C_0 + C_{i-1}e^{-kt_0} \qquad i = 1, 2, 3, \ldots \tag{18}$$

Some quantitative results will now be deduced from the model equations (16)–(18). If the time, t_0, between successive doses is sufficiently long, then from (16) with $i = 1$, R_1 will be close to zero. Hence, from (17), with $i = 1$, $C_1 = C_0$ for all practical purposes. Repeating the preceding argument for $i = 2, 3, \ldots$, shows that $R_i = 0$ and $C_i = C_0$. In other words, the peak concentration of the drug is C_0 at the time of its administration, and the concentration drops to zero between successive drug doses.

Next the situation in which t_0 is small will be investigated. Setting $i = 1$ in (18) yields

$$C_1 = C_0 + C_0e^{-kt}$$

Once again from (18) with $i = 2$,

$$\begin{aligned} C_2 &= C_0 + C_1e^{-kt_0} = C_0 + (C_0 + C_0e^{-kt_0})e^{-kt_0} \\ &= C_0(1 + e^{-kt_0} + e^{-2kt_0}) \end{aligned}$$

By induction it can be shown that

$$\begin{aligned} C_n &= C_0(1 + e^{-kt_0} + e^{-2kt_0} + \cdots + e^{-nkt_0}) \\ &= C_0\left(\frac{1 - e^{-(n+1)kt_0}}{1 - e^{-kt_0}}\right) \end{aligned} \tag{19}$$

where the last inequality results from the formula for summing a geometric series. From (17) and (19) it can be deduced that

$$\begin{aligned} R_n &= C_0(e^{-kt_0} + e^{-2kt_0} + \cdots + e^{-(n+1)kt_0}) \\ &= C_0\left(\frac{1 - e^{-(n+1)kt_0}}{1 - e^{-kt_0}}\right)e^{-kt_0} \end{aligned} \tag{20}$$

Equations (19) and (20) show that both C_n and R_n are increasing sequences. Since

$$\lim_{n\to\infty} e^{-(n+1)kt_0} = 0$$

it follows from (19) and (20) that the limiting concentration

$$C = \frac{C_0}{1 - e^{-kt_0}} \tag{21}$$

and the limiting residual concentration

$$R = \frac{C_0e^{-kt_0}}{1 - e^{-kt_0}} \tag{22}$$

Taking limits of both sides in Eq. (17) as $i \to \infty$ yields

$$R + C_0 = C \tag{23}$$

The graph of the drug concentration when the interval between doses is short is shown in Fig. 1.7. The first part of the graph is similar to that in Fig. 1.1. However, as $n \to \infty$, C_n and R_n approach their limiting values C and R. From (21) and (22), $Ce^{-kt_0} = R$; while from (23), $C = C_0 + R$. Hence, for sufficiently large t, the graph of the concentration will oscillate between $C_0 + R$ and R.

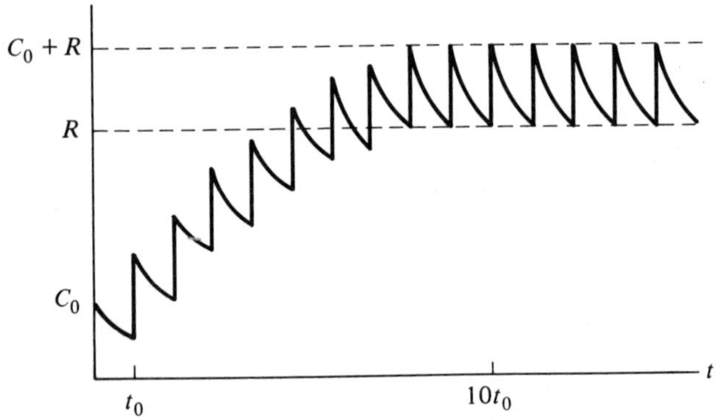

Figure 1.7 Concentration for a short interval between doses.

Figure 1.7 suggests how to find values of C_0 and t_0 that will produce a concentration $C(t)$ that is safe (not above C_H) but still efficient (not below C_L). An effective level

$$C_H = C_0 + R \tag{24}$$

can be achieved immediately by a loading dose. The drug concentration in the blood will fall to

$$R = C_L \tag{25}$$

in a time interval t_0. Thus, the loading dose should be followed every t_0 hours by a dose that raises the concentration by C_0, where

$$C_0 = C_H - C_L \tag{26}$$

is determined from (25) and (24). When the values of R and C_0, defined by (25) and (26), are substituted in (22), it is found that

$$C_L = \frac{(C_H - C_L)e^{-kt_0}}{1 - e^{-kt_0}}$$

Solving this last equation for e^{kt_0} gives

$$e^{kt_0} = \frac{C_H}{C_L} \tag{27}$$

Taking the logarithm of both sides of (27) yields

$$t_0 = \frac{1}{k} \ln \frac{C_H}{C_L} \tag{28}$$

The final example is concerned with quantifying the time required for the central nervous system to respond to a stimulus*. For example, a driver is sitting in his car with his foot on the brake pedal waiting for a red light to change. As soon as the light changes to green he removes his foot from the brake pedal and places it on the accelerator. The interval between the time he first sees the green light and the time he begins to withdraw his foot from the brake pedal is called his reaction time. Experience suggest that the reaction time is probably a decreasing function of the strength of the stimulus. In the following example, a mathematical model to predict this dependence will be constructed.

Example 6 (Reaction Time)

In order to formulate a mathematical model of reaction time, the effect of a stimulus on a nerve will be described. Since a nerve is composed of a large number of nerve cells (neurons), the effect of a stimulus on a neuron will be considered first. A neuron has the ability to respond to a stimulus by generating an electrical impulse (the action potential) and transmitting this impulse or signal through the cell to neighboring cells. However, the action potential is generated by the neuron only if the strength (intensity) of the stimulus exceeds a certain threshold level peculiar to that neuron. The magnitude of the response of a neuron is independent of the strength of the stimulus, once the threshold level has been exceeded (the all-or-none law). Based on the preceding discussion, a threshold stimulus acting on a nerve is defined to be a stimulus that is just strong enough to elicit an action potential in the most excitable neuron† in the nerve.

Next the sequence of events that occurs when an above-threshold stimulus is applied to a portion of the body will be described. The effect of the stimulus is transmitted to the brain along an afferent nerve. The intensity of excitation E in the afferent pathway depends on the number of neurons in the afferent pathway excited by the stimulus. Hence, $E = E(S)$ is an increasing function of S. Furthermore, if S is constant, E does not change with time, because both the structure of the afferent pathway and the behavior of its neurons are time-invariant. The exact nature of the function E will be described below. The excitation in the afferent pathway induces a response within the brain, which builds up with the passage of time provided that the stimulus continues. Even though the exact physiological nature of this response is not known, its magnitude, $\epsilon(t)$, can be determined under certain assumptions described below. If the *excitatory factor*, $\epsilon(t)$, exceeds a certain level l_E, the efferent neuron leading from the brain to an effector organ is stimulated. This results in a response to the stimulus.

The discussion in the preceding paragraph can be translated into a mathematical form that shows how the reaction time depends on S. The reaction time T is

$$T = t_A + t_B + t_E \tag{29}$$

where t_A = the time required for the signal to be transmitted in the afferent pathway from the point of application of the stimulus to the brain
t_B = the time required for $\epsilon(t)$ to grow from 0 to the threshold level l_E
t_E = the time required for the response to be transmitted from the brain to the effector organ

The transmission times t_A and t_E depend only on the structure of the afferent and efferent

* A stimulus is an environmental change capable of producing a reaction in living tissue without harming it.

† The neuron with the smallest threshold.

pathways and so are independent of the intensity of the stimulus. However, t_B is dependent on the stimulus intensity S. Thus, from (29), T depends on S only because of the dependence of t_B on S through the function ϵ. The mathematical form of this dependence can be found as follows. Let the first instant that the excitation produced by the stimulus reaches the brain be taken as $t = 0$; then

$$\epsilon(t_B) = l_E \tag{30}$$

Hence, the dependence of t_B on S must be determined indirectly through the relation (30). This requires the form of the function ϵ or, more precisely, its derivative.

In order to find the excitatory factor ϵ, the following two assumptions will be made:

1. The rate of increase of ϵ is proportional to E.
2. The rate of decay of ϵ is proportional to ϵ.

Mathematically these two assumptions state that the rate of change of ϵ is

$$\epsilon'(t) = k_1 E - k_2 \epsilon(t) \tag{31}$$

where k_1 and k_2 are positive constants. Assumption 1 is plausible since the rate of change of ϵ should vary directly as the number of excited neurons in the afferent pathway. However, the number of excited neurons in the afferent pathway is proportional to E. The basis of assumption 2 is the hypothesis that ϵ is proportional to the number of excited neurons (or neuron cycles) or the amount of some substance in the brain. Each portion of the substance (or each cycle) has the same chance of being destroyed (or returning to the quiescent state). Then it is plausible that the loss per second in ϵ is proportional to the amount of neuron cycles or substance present.

By means of (31), t_B can be expressed in terms of l_E and E. Thus, from (29), T can be expressed in terms of l_E and E and so in terms of S, since E is a function of S, which will be determined later. Equation (31) can be rewritten as

$$1 = \frac{\epsilon'(t)}{k_1 E - k_2 \epsilon(t)}$$

and then integrated

$$\int_0^{t_B} 1\, dt = \int_0^{t_B} \frac{\epsilon'\, dt}{k_1 E - k_2 \epsilon(t)} = -\frac{1}{k_2} \int_0^{t_B} \frac{d[k_1 E - k_2 \epsilon(t)]}{k_1 E - k_2 \epsilon(t)}$$

Hence,

$$t_B = \frac{1}{k_2} \ln\left(\frac{k_1 E}{k_1 E - k_2 \epsilon(t_B)}\right) = \frac{1}{k_2} \ln\left(\frac{k_1 E}{k_1 E - k_2 l_E}\right)$$

from (30) and the fact that $\epsilon(0) = 0$.

Setting

$$T_0 = t_A + T_E$$

it follows from (29) that

$$T = T_0 + \frac{1}{k_2} \ln\left(\frac{k_1 E}{k_1 E - k_2 l_E}\right) \tag{32}$$

Finally, E must be determined as a function of S. Unfortunately, this cannot be done on the basis of direct observation or theoretically. The following approach will be adopted. The shape of this graph of E will be predicted based on a known biological fact. Then a simple function whose graph has this shape will be postulated for E. Lastly, the assumed function of E will be substituted into (32) and the resulting values of T will be checked against experimental data for various values of S. Three different functions will be used depending on the strength of the stimuli.

The functional form of E will be investigated for weak stimuli. The neurons in the afferent pathway have a threshold level l_A below which no excitation occurs. In other words, if $S < l_A$, $E(S) = 0$. Hence, E should depend on the amount by which S exceeds l_A or $S - l_A$. The simplest type of dependence of $S - l_A$ is direct proportionality:

$$E(S) = k(S - l_A) \qquad S > l_A \tag{33}$$

where k is a positive constant. Substituting the value of E from (33) into (32) yields

$$T = T_0 + \frac{1}{k_2} \ln\left(\frac{S - l_A}{S - l_A - C}\right) \qquad S > l_A \tag{34}$$

where $C = k_2 l_E / k_1 k$.

Next the form of E for moderate stimuli will be considered. If S increases, then, from (33), it follows that E increases. However, as the stimulus grows stronger, the number of neurons that are not excited decreases. This suggests that fewer additional neurons are excited per unit increase in S. In other words, $E'(S)$ is positive but decreasing; that is, $E''(S) < 0$. A simple function satisfying the above restrictions and $E(l_A) = 0$ is

$$E(S) = k_3 \ln \frac{S}{l_A} \qquad (S > l_A) \tag{35}$$

where k is a positive constant. Replacing E in (32) by its value from (35) gives

$$T = T_0 + \frac{1}{k_2} \ln\left(\frac{\ln S - \ln l_A}{\ln S - \ln l_A - C}\right) \tag{36}$$

where $C = k_2 l_E / k_1 k$.

If the stimulus is sufficiently strong, Eq. (35) is not satisfactory. There is an upper bound on the excitation corresponding to all neurons in the afferent pathway being excited. However, from (35), $\lim_{s\to\infty} E(S) = \infty$. A function E such that

$$\lim_{S\to\infty} E(S) = E^*$$

is

$$E(S) = k \ln\left(\frac{S}{l_A + Se^{-E^*/k}}\right) \tag{37}$$

where k and E^* are positive constants.

The theoretical derivation of the reaction times and the proposed functions for E have been checked empirically. Experimental data for reaction times obtained by varying S for auditory and visual stimuli were fitted using (36). The constants were chosen so that the theoretical curve fit the experimental data as well as possible. Similarly, experimental reaction times for various gustatory (taste) stimuli were also fitted using (34). Good agreement between the experimental data and the theoretical reaction times for low (34) and moderate (30) stimuli was obtained. The details and graphs can be found in Rashevsky [30], pages 39–41.

The examples in this section show that modeling requires some effort. Why not study the real world instead? The following principal uses of models should justify the energy and time required for their formulation.

1. Costly and unnecessary experiments can be avoided. For instance, the optimal size and composition of a group of subjects can be determined for certain clinical trials of drugs.
2. Undesirable experiments need not be performed. For example, the effect of hunting on the size of a population of animals can be predicted.
3. Modeling can be used as a substitute for experiments that cannot be performed. For example, the ages of certain biological specimens can be calculated by radioactive dating.
4. Mathematical models are often used only for descriptive purposes. Here quantitative relationships are given in terms of equations for conciseness. This results in ease of analysis and handling of data. For example, if one variable is proportionally related to another, then a linear equation relating the two is more concise and easier to handle than graphical or verbal descriptions.
5. In a complex system with many interacting relations, a mathematical expression of the relations is required in order to analyze the system.
6. Models are used predictively to determine how a system would respond to a stimulus or to a change in the system.
7. A model can be used to gain insight and understanding of a system by describing the ways in which different features of system behavior and structure depend on each other.
8. A model can be used to test certain hypotheses about a system. If the theoretical results predicted by the model do not agree with the experimental results, then the hypotheses are rejected. Sometimes new experiments are suggested by the process.
9. A mathematical model may be used for the purpose of calculating, from observations on accessible variables, values of variables that are inaccessible to measurement. Related to this is the use of models to diagnose, from observable behavior, structural causes of an organism malfunction.

EXERCISES

1. The number of sex-related mutations in *Drosophila melanogaster* increases almost linearly with the X-ray dosage for dosages not exceeding 6 kR (kiloroentgen). For dose 0 no mutation is observed. With a dose of 3 kR the mutation rate is 8.4%. Let x denote the dose in kiloroentgen and y the percentage mutation rate. Find the equation satisfied by y and x.

2. Assume that a single cell is spherical in shape.

(a) Look up or derive formulas for the volume and surface area of a sphere. Show that if the cell grows sufficiently large (small), the ratio of its volume to its surface area becomes arbitrarily large (small).

(b) Assume that the metabolic capacity of a cell is proportional to its volume and the various physical and chemical properties of the cell components remain unaltered while the cell grows. Give a plausible argument showing that the radius of the cell must have a lower and an upper critical limit in order for the cell to survive.

3. A small piece of an artery or vein may be modeled as a cylinder of cross-sectional radius R. Like any liquid, the blood has an inner friction called the viscosity. The viscosity is denoted by η and measured in poise* ($\text{cm}^{-1}\ \text{g sec}^{-1}$). If the velocity, v, of flow does not exceed a certain critical value, the flow is laminar; that is, all particles of the liquid move parallel to the tube and

$$v = \frac{P}{4\eta l}(R^2 - r^2)$$

where l denotes the length (cm), P is the pressure difference between the two ends of the tube ($\text{dyne/cm}^2 = \text{cm}^{-1}\ \text{g sec}^{-2}$), and r is the distance of any point of the liquid from the central axis.

(a) Show that the velocity of flow increases regularly from zero at the wall to $PR^2/4\eta l$ at the center.

(b) Arterial blood of viscosity $\eta = 0.027$ poise is flowing through an arteriole of length $l = 2$ cm and radius $R = 8 \times 10^{-3}$ cm. The pressure difference between the two ends of the arteriole is $P = 4 \times 10^3\ \text{dyne/cm}^{-2}$ (= 3 mm of mercury). Verify that

$$v = 1.185 - (1.55 \times 10^4)r^2$$

4. Consider the trunk of a quadruped to be a flexible beam supported at the ends by legs. A result from elasticity theory states that if a beam of length l, vertical thickness t, and cross-sectional area A is subjected to a uniform load L while its end points are held fixed, the maximum deflection

$$\delta \propto \frac{Ll^3}{t^2 A}$$

It is reasonable to assume that there exists some physically determined upper limit to δ/l, the relative sagging, above which the animal's trunk will be deformed and probably touch the ground.

(a) Assume that L is due mainly to the weight of the trunk and so is roughly proportional to lA. Show that $l/t^{2/3}$ has an upper limit.

(b) Determine $l/t^{2/3}$ for the following animals:

Animal	l, cm	t, cm
Ermine *(Mustela erminea)*	12	4
Dachshund	35	12
Indian tiger *(Panthera tigris)*	90	45
Llama *(Lama glama)*	122	73
Indian elephant *(Elephas maximus)*	153	135

* This word is an abbreviation of the name Poiseuille. Jean Louis Poiseuille (1799–1869) discovered the law of laminar flow.

(c) Most other quadrupeds have $l/t^{2/3}$ smaller than the values in (b). Note that the shoulder–hip length of a dachshund is about three times its vertical thickness. Do you think it is possible to have a giant dachshund with

$$l = 350 \text{ cm} \quad \text{and} \quad t = 120 \text{ cm?}$$

5. (Continuation of Exercise 4) For many animals δ/l is a constant and the mass m of the trunk is proportional to lA. Under these assumptions, show that:
(a) Larger animals have relatively thicker trunks, or $t \propto l^{2/3}$
(b) $l \propto m^{1/4}$
(c) $t \propto m^{3/8}$

6. Weber's law states that just noticeable difference (jnd) in signal intensity (psychological intensity S) is proportional to the intensity of the signal (physical intensity F). Let ΔF be the increase in a signal of magnitude F that just produces a noticeable difference. Then Weber's law states that the ratio

$$r = \frac{\Delta F}{F}$$

is a constant* that does not depend on F when F is neither too small nor too large.
(a) Let F_0 be a fixed value of F. Prove that the nearest noticeably higher stimulus is

$$F_1 = F_0 q \qquad (q = 1 + r)$$

(b) Show that noticeably different stimuli follow each other in a geometric sequence:

$$F_0,\ F_1 = F_0 q,\ F_2 = F_0 q^2,\ F_3 = F_0 q^3,\ \ldots$$

(c) Experiments show that if a person holds a weight in his hand, then 63 g can be discriminated from 60 g and 84 g from 80 g but that the interval cannot be reduced. Find r. What will be the jnd for 20 g?
(d) Suppose that animals are being tested for their response to doses of different levels. Should the sequence 10 mg, 20 mg, 30 mg, 40 mg, . . . , or the sequence 10 mg, 20 mg, 40 mg, 80 mg, . . . , be used?

7. (Continuation of Exercise 6) Fechner assumed that all jnd's were psychologically equal for a given type of stimulus. This assumption can be used to derive the Weber-Fechner law relating psychological intensity S_1, measured in jnd's, to physical intensity F as follows.
(a) Suppose that F_i is mapped on S_i. Show that Weber's law states that if S_1 and S_2 differ by 1 jnd, then $\log F_1$ an $\log F_2$ differ by some constant k.
(b) Prove that if S_1 and S_2 differ by an integral number N of jnd's, then $\log F_1$ and $\log F_2$ differ by kN.
(c) Conclude that $S = a \log F + b$.† Determine the constants a, b, and k.
(d) Use the Weber-Fechner law to show that if $F_1/F_2 = F_3/F_4$, then $S_1 - S_2 = S_3 - S_4$.

8. Stevens discovered that equal ratios of physical intensity (see Exercise 6 for notation) correspond to equal ratios of psychological intensity; that is,

$$\frac{F_1}{F_2} = \frac{F_3}{F_4} \quad \text{if and only if} \quad \frac{S_1}{S_2} = \frac{S_3}{S_4}$$

* The sensitivity of human senses is illustrated by the following values of r: visual brightness (0.2), tone (0.1), smell for rubber (0.125); taste for saline solution (0.25).

† Examples are loudness in decibels, $L = 10 \log(I/I_0)$, where $I_0 = 10^{-12}$ W/m²; and apparent magnitude of stars, $m = c - 2.5 \log I$, where I is the light intensity and c is a constant depending on the units of I.

(a) Let $s = \log S$ and $f = \log F$. Show that ds/df is a constant by verifying that the slope of the secant joining (s_1, f_1) and (s_2, f_2) is equal to the slope of the secant joining (s_3, f_3) and (s_4, f_4). Then let $f_2 \to f_1$ and $f_4 \to f_3$ so that the slopes of these secants remain the same.

(b) Describe the function $S = g(F)$ in Stevens' law.

9. Let $c(t)$ be the concentration of a certain solute in a cell at time t and let c_m be the concentration of this solute in the surrounding medium. According to Fick's law of diffusion, the rate of change of concentration

$$\frac{dc}{dt} = k(c_m - c)$$

where k is a constant depending on the size of the cell and the properties of the cell membrane. If $c(0) = c_0$, verify that when $c_0 < c_m$,

$$c(t) = c_m + (c_m - c_0)e^{-kt}$$

What will the concentration inside the cell eventually approach?

10. Let x be the amount of radiation, measured in rads, to which a population of cancer cells is subjected. Let y denote the number of surviving cancer cells. Suppose that the change, Δy, in the number of surviving cancer cells when x is increased to $x + \Delta x$ is proportional to both y and Δx.

(a) By letting $\Delta x \to 0$, show that

$$\frac{dy}{dx} = -ky$$

where k is the constant of proportionality.

(b) If, when $x = 0$, $y = y_0$, the initial number of cancer cells in the patient, show that

$$y = y_0 e^{-kx}$$

(c) A dose of 0.1 rad is sufficient to kill 50% of the cells of a certain type of cancer. How much radiation must be given to kill 99% of the cells?

11. The half-life of a radioactive substance, denoted $T_{1/2}$, is the time interval in which 50% of the radioactive atoms decay. Radioactive thorium is used to date corals and shells. Its rate of decay is $dy/dt = -9.2 \times 10^6$ years when t is measured in years. What is the half-life of radioactive thorium?

12. Consider a vertical beam entering the water with intensity I_0 at the surface. The Beer-Lambert law* gives the reduced intensity

$$I = I_0 e^{-kx}$$

at a depth x in meters. The parameter $k > 0$ is called the absorption coefficient and depends on the purity of water and the wavelength of the beam.

(a) What is reduction in intensity of light at a depth of 1 m?

(b) Why does plant life exist in lakes only in the top layer, which is about 10 m deep?

13. In a recently discovered sample there were 3 g of thorium (^{232}Th). Thorium decays to lead-208 (^{208}Pb) by the reaction

$$^{232}\text{Th} \longrightarrow {}^{208}\text{Pb} + 6(4^4\text{He})$$

* This law is applicable to any homogeneous transparent substance such as glass, liquids, and thin layers viewed under the microscope, and other electromagnetic waves such as X-rays and γ-rays.

It was determined that 0.0376 g of lead-208 was produced by the disintegration of the original thorium in the sample. Given that the half-life of thorium is 13.9 billion years, derive the age of this sample. (*Hint:* The lead-208 is the product of the decay of $(232/208) \times 0.0376$ g of thorium.)

14. Living plants (from ingestion of carbon dioxide) and animals (from eating plants) have a certain equilibrium concentration of radioactive ^{14}C while living. When an organism dies, its concentration of ^{14}C decreases because of disintegration of the ^{14}C present. Assume that the original rate of disintegration of ^{14}C in a sample such as charcoal is the same as the rate measured today. The radioactivity decay law states that $dN/dt = -\lambda N$, where $N(t)$ is the amount of ^{14}C in a sample at time t and λ is the decay constant.

(a) Find an expression for the rate of disintegration, $R(t) = \lambda N(t)$.

(b) Show that $R(t)/R(0) = e^{-\lambda t}$ and so

$$t = \frac{1}{\lambda} \ln \frac{R(0)}{R(t)}$$

(c) The half-life of ^{14}C is 5568 years. A sample of charcoal gave an average count of 0.97 disintegrations per minute per gram. Living wood gave 6.68 disintegrations. Estimate the age of the charcoal.

15. Using the notation of Example 4, the size of a population, y, is described by

$$x_1(k+1) = \tfrac{3}{4}x_2(k) + \tfrac{1}{4}x_3(k)$$

$$x_{i+1}(k+1) = \tfrac{1}{2}x_i(k) \qquad i = 1, 2$$

$$y(k) = x_1(k) + x_2(k) + x_3(k)$$

for $k = 1, 2, 3, \ldots$. Given that $x_1(0) = x_2(0) = x_3(0) = 10$, verify that

$$x_1(3) = \tfrac{35}{8}, \quad x_2(3) = \tfrac{5}{2}, \quad x_3(3) = \tfrac{5}{2}, \quad y(3) = 9\tfrac{3}{8}$$

16. Consider laminar flow of a liquid through a tube of cross-sectional radius R as described in Exercise 3. Then the velocity of all particles of the fluid that are a distance r from the central axis is

$$v = kP(R^2 - r^2)$$

where k is a constant.

(a) Show that the flow (volume per second) through a cross section is

$$F = cPR^4 \text{ cm}^3/\text{sec}$$

(*Hint:* The flow due to a thin shell of thickness dr and distance r from the center is $v(2\pi r)\,dr$, since v is approximately constant if dr is small. Hence $F = 2\pi\int_0^R vr\,dr$.)

(b) The average velocity of flow, V, is defined by equating the flow when each particle moves with velocity V to the real flow F; that is, $F = V\pi R^2$. Verify that

$$V = aPR^2$$

where $a = c/\pi$.

17. The results in Exercise 16 can be used to study the human cough. X-rays reveal that the radius of the tracheal tube contracts to about two-thirds its radius during a cough. Show that such a constriction produces the maximum of the average velocity, V, of the air in the trachea under the assumption of laminar air flow and perfect elasticity of the trachea. Perfect elasticity means that

$$R_0 - R = bP$$

where $b > 0$ is a constant, R_0 is the normal radius of the trachea when one is not coughing, R is the contracted radius of the trachea during a cough, and P is the difference in pressure between the pressure in the lungs and the atmosphere during a cough. The perfect elasticity assumption is valid for small changes in pressure P satisfying

$$0 \leq P \leq \frac{R_0}{2b}$$

If larger values of P occur, the tracheal wall stiffens.* (*Hint:* Use the given expression for P to prove that the last inequality is equivalent to $R_0/2 \leq R \leq R_0$ and also to express V, given in Exercise 16(b), as a function of R. Find the value of R that satisfies the deduced restrictions, which gives the absolute maximum of V.)

REFERENCES

1. Adolph, E. F. *Origins of Physiological Regulations*. Academic Press, New York, 1968.
2. Atkinson, D. E. *An Introduction to Numerical Analysis*. Wiley, New York, 1978.
3. Bayliss, L. E. *Living Control Systems*. English Universities Press, London, 1966.
4. Bellman, R. *Adaptive Control Processes: A Guided Tour*. Princeton University Press, Princeton, NJ, 1961.
5. Bernard, C. *Les Phenomenes de la Vie*. Paris, 1878.
6. Bode, H. W. Feedback amplifier design. *Bell System Tech. J. 19,* 42, 1940.
7. Borelli, G. *De motu animalium (On animal motion),* Vol. I. Bernabo, Rome, 1680.
8. Brown, J. H. U., and Grann, D. *Engineering Principles in Physiology,* Vols. 1–3. Academic Press, New York, 1973.
9. Brown, J. H. U., Jacobs, J. E., and Stark, L. (eds.). *Biomedical Engineering*. F. A. Davis, Philadelphia, 1971.
10. Cannon, W. B. Organization for physiological homeostasis. *Physiol. Rev. 9,* 399, 1929.
11. Clynes, M., and Milsum, J. H. (eds.). *Biomedical Engineering Systems*. McGraw-Hill, New York, 1976.
12. Conte, S. D., and deBoor, C. *Elementary Numerical Analysis. An Algorithmic Approach*. McGraw-Hill, New York, 1980.
13. Daniel, C., Wood, F. S., and Gorman, J. W. *Fitting Equations to Data Computer Analysis of Multifactor Data*. Wiley, New York, 1980.
14. Fibonacci, Leonardo of Pisa. *Liber abaci (Book of the Counting Board)*. 1202.
15. Forsythe, G. E., Malcolm, M. A., and Moler, C. B. *Computer Methods for Mathematical Computations*. Prentice-Hall, Englewood Cliffs, NJ, 1977.
16. Fredericq, L. Influences de milieu ambiant sur la composition du sang des animaux aquatiques, *Arch. de Zool. Exper. et Gen. 3,* 35, 1885.
17. Goodwin, B. C. *Temporal Organization in Cells, A Dynamic Theory of Cellular Control Processes*. Academic Press, New York, 1963.
18. Grodin, F. S. *Control Theory and Biological Systems*. Columbia University Press, New York, 1963.
19. Guyton, A. C., Taylor, A. E., and Granger, H. J. *Circulatory Physiology. I. Cardiac Output and Its Regulation*. Saunders, Philadelphia, 1973. *II. Dynamics and Control of the Body Fluids*. Saunders, Philadelphia, 1975.
20. Hazen, H. L. Theory of servomechanisms. *J. Franklin Inst. 218,* 279, 1934.
21. Iberall, A. S., and Guyton, A. C. (eds.). *Regulation and Control in Physiological Systems*. Instrument Society of America, Pittsburgh, 1973.
22. Kalmus, H. *Regulation and Control in Living Systems*. Wiley, New York, 1966.

*This is fortunate, since if the trachea contracted too much, a person would suffocate.

23. Kline, J. *Biological Foundations of Biomedical Engineering*. Little, Brown, Boston, 1976.
24. Mayr, O. *The Origins of Feedback Control*. M.I.T. Press, Cambridge, MA, 1970.
25. Maxwell, J. C. On governors. *Proc. Roy. Soc. London 16,* 1868, in *Selected Papers on Mathematical Trends in Control Theory,* Dover, New York, 1964, pp. 270–283.
26. Milhorn, H. T. *Applications of Control Theory to Physiological Systems*. Saunders, Philadelphia, 1966.
27. Milsum, J. H. *Biological Control System Analysis*. McGraw-Hill, New York, 1966.
28. Minorsky, N. Directional stability of automatically steered bodies. *J. Am. Soc. Naval Eng. 34,* 200, 1922.
29. Nyquist, H. Regeneration theory. *Bell System Tech. J. 11,* 126–147, 1932.
30. Rashevsky, N. *Mathematical Biophysics,* Vol. 2. Dover, New York, 1960.
31. Riggs, D. S. *Control Theory and Physiological Feedback Mechanisms*. Williams & Wilkins, Baltimore, 1970.
32. Rosenbleuth, A., Wiener, N., and Biglow, J. H. Behavior, purpose, and teleology. *Phil. Sci. 10,* 18–24, 1943.
33. Schwan, H. P. (ed.). *Biological Engineering*. McGraw-Hill, New York, 1969.
34. Talbot, S. A., and Gessner, U. *Systems Physiology*. Wiley, New York, 1973.
35. Thompson, D'Arcy W. *On Growth and Form*. Cambridge University Press, Cambridge, 1917.
36. Vyshnegradskii, I. A. On controllers of direct action. *Izv., SPB Tekhnolog. Inst.,* 1977.
37. Wiener, N. *Cybernetics*. M.I.T. Press, Cambridge, MA, Rev. ed., Wiley, New York, 1961.
38. Yamamoto, W. S., and Brobeck, J. R. (eds.). *Physiological Controls and Regulations*. Saunders, Philadelphia, 1965.

2

Classification of Control Systems

2.1 INTRODUCTION

In this chapter the major types of control systems are defined. The concepts of open and closed loops are discussed. Positive and negative feedback are defined, and their importance in biology is illustrated. Finally, basic units for drawing block diagrams of linear systems are introduced in order to obtain a pictorial representation of the equations that describe a system.

2.2 CLASSIFICATION OF SYSTEMS

Models of systems are classified according to the types of equations used to describe them. The tree diagram in Fig. 2.1 illustrates the major classifications. Dashed lines in Fig. 2.1 indicate the existence of subdivisions similar to the ones shown at the right

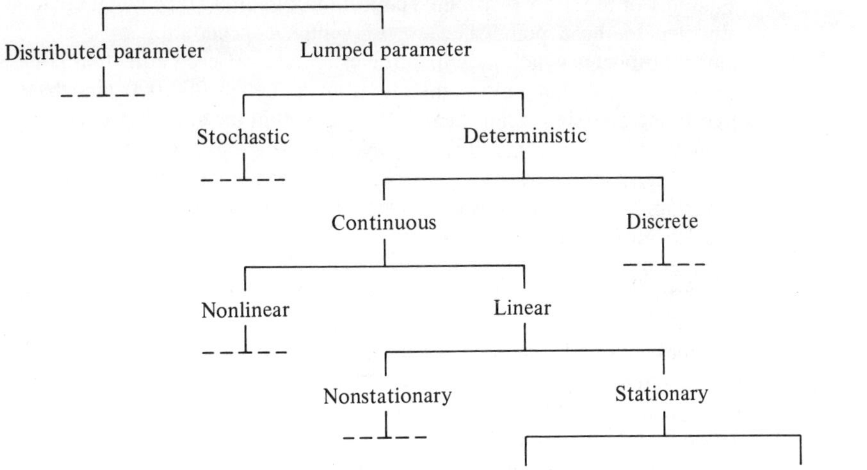

Figure 2.1 Major classes of system equations.

on the same level. Most systems are described by combinations of these classes, such as lumped parameter, stochastic, and nonlinear systems. A description of the various classes follows.

A. Distributed-Parameter versus Lumped-Parameter Systems

In a *distributed-parameter system* the elements of the system are distributed in space so that the variables they influence must be treated as differing not only from time to time but also from point to point. These systems must be analyzed by solving partial differential or difference equations, since the dependent variables depend on both time and space.

In a *lumped-parameter system* the elements are small enough (or the objectives of our analysis are limited enough) so that each such element can be treated as if it were concentrated ("lumped") at one particular point in the system. The position of the point can change with time. Thus, such systems can be analyzed using ordinary differential or difference equations. For example, to study the acceleration of a car, the car can be considered to be a point. However, to determine the effect of rear- and front-end collisions on a car, the car cannot be considered to be a point.

All of the mathematical examples up to this point have been lumped-parameter systems. The following example is a distributed-parameter system.

Example 1 (Aggregation of Slime Mold Amoeba)

A mathematical model of the behavior of the slime mold amoeba *(Dictyostelium discoideum)* will be formulated after its life cycle is discussed. This one-celled organism is about 10^{-3} cm in diameter and moves by extending contractile portions of itself (pseudopods). The natural habitat of these amoebas is soil or dung. They feed on bacteria by engulfing them and multiply by mitosis when the food supply is plentiful. If the food supply becomes exhausted, there is an interphase period of random and feeble movement over the area available to them. After a few hours the amoebas begin to aggregate into a number of regularly distributed collections of points. It is thought that the amoebas are attracted to these points because of a high concentration of cyclic 3′,5′-adenosine monophosphate, which they themselves secrete. At each attraction point a multicelluar slug is formed that ranges in size from a few to 200,000 amoebas. The formerly free-living amoebas retain their cell walls within the slug; however, the slug moves as a unit. The slug eventually stops and erects a stalk, on top of which is a roundish container of spores. During the spore stage, each amoeba is dormant within a protective covering. When conditions are favorable, an amoeba emerges from its spore and the life cycle begins again.

A simple mathematical model that determines the time of onset of aggregation and the spacing of the aggregation centers will be described. Generalizations of this model are contained in the exercises in [23] and the cited references. Since the distance between amoebas is small compared to the average distance between aggregation centers, a continuum model will be used. Let the aggregation occur in the xy plane. For simplicity it is assumed that no quantities change with y; hence, only x variation will be considered. The following notation will be used:

$a(x, t)$ = number of amoebas per unit area at the position x and time t.
$J(x, t)$ = net rate per unit length at which amoebas cross the line $x = x_1$ at time t (flux

density). The flux density is positive if there are more amoebas crossing in the direction of increasing x than in the opposite direction.

$Q(x, t)$ = net rate per unit area at which amoebas are being generated at time t. The generation of amoebas means the excess of births over deaths.

An equation connecting a, J, and Q will be derived from the following general balance principle. Consider the amoebas located in a rectangle of unit width contained between the lines $x = x_0$ and $x = x_0 + \Delta x$. A general balance principle for this rectangle states:

$$\begin{pmatrix}\text{rate of change of}\\ \text{number of amoebas}\end{pmatrix} = \begin{pmatrix}\text{rate at which amoebas}\\ \text{flow across boundary}\end{pmatrix} + \begin{pmatrix}\text{rate of generation of}\\ \text{amoebas in region}\end{pmatrix}$$

Translating the above balance law into mathematical notation gives

$$\frac{\partial}{\partial t}\int_{x_0}^{x_0+\Delta x} a(x, t)\,dx = J(x_0, t) - J(x_0 + \Delta x, t) + \int_{x_0}^{x_0+\Delta x} Q(x, t)\,dx \tag{1}$$

Dividing both sides of Eq. (1) by Δx and letting $\Delta x \to 0$ yields

$$\frac{\partial a(x_0, t)}{\partial t} = -\frac{\partial J(x_0, t)}{\partial x} + Q(x_0, t) \tag{2}$$

where x_0 is arbitrary.

The generation of amoebas, Q, can be considered to be zero for the following reasons. The reproduction of amoebas can be ignored, since there is little reproduction in the absence of food. Deaths can also be ignored, since there are few during the time interval of interest.

A formula for the flux, J, will be obtained under the assumption that

$$J = J_r + J_c$$

where J_r and J_c are the fluxes caused by differences in amoeba and concentration, respectively. The magnitude of the random flux at x depends on $\partial a/\partial x$, the concentration gradient of the number of amoebas at x. The simplest dependence is

$$J_r(x, t) = -\mu\frac{\partial a(x, t)}{\partial x} \tag{3}$$

where μ is a positive constant. The negative sign is required since, by convention, J_r must be positive for amoebas moving to the right from a region of high concentration to a region of low concentration ($\partial a/\partial x < 0$). The flux J_c is due to chemotaxis, which is the directional motion induced by variations in chemical concentration. Let $\rho(x, t)$ be the density of the attractant. Assume that J_c is proportional to the attractant gradient $\partial\rho/\partial x$ for small values of the gradient. For a given gradient, if the amoeba density is twice as large, the net flux should be twice as large. Thus, the proportionality factor should be a multiple of a. A reasonable assumption is that

$$J_c = \chi a\frac{\partial\rho}{\partial x} \tag{4}$$

where χ is a constant that measures the strength of chemotaxis. Note that in contrast to (3), there is no negative sign in (4) because amoebas tend to move toward attractant concentrations.

Replacing the total flux J in (2) by the sum of the random contribution J_r, given by (3), and the chemotactic contribution J_c, given by (4), yields

$$\frac{\partial a}{\partial t} = \frac{\partial}{\partial x}\left(\mu \frac{\partial a}{\partial x} - \chi a \frac{\partial \rho}{\partial x}\right) \tag{5}$$

Finally, an equation for the attractant density will be derived. This will be of the general form (2):

$$\frac{\partial \rho}{\partial t} = -\frac{\partial J_a}{\partial x} + Q_a$$

where the subscript a refers to attractant. The flux is assumed to be proportional to gradient, as in (3):

$$J_a = -D\frac{\partial \rho}{\partial x}$$

where D is a positive constant. The generation term Q_a has a positive contribution fa as a result of the secretion of attractant by the amoebas. Here $f > 0$ is the rate of secretion per unit amoeba density. The rate of decay in attractant activity is assumed to be proportional to the amount of attractant present, with decay constant $k > 0$. Hence, $Q_a = fa - k\rho$ and the desired equation for $\partial\rho/\partial t$ is

$$\frac{\partial \rho}{\partial t} = fa - k\rho + D\frac{\partial^2 \rho}{\partial x^2} \tag{6}$$

Partial differential equations (5) and (6), for two unknown functions $a(x, t)$ and $\rho(x, t)$, are a mathematical model describing the aggregation of slime mold amoebas. The system of amoebas is a distributed-parameter system. Readers interested in the analysis of this system are referred to [23]. Only lumped-parameter systems will be considered in the remainder of this book.

B. Stochastic versus Deterministic Systems

A *deterministic system* is one whose response to a specified input applied under specified conditions is completely predictable from physical laws. Therefore the response is precisely reproducible time after time. All of the mathematical examples up to this point have been deterministic systems.

By contrast, it is not possible to predict which output will occur in a purely *stochastic system*. However, a set of all possible outputs for each given input may be known. Probability theory is required to make predictions about the output. It may be argued that if enough were known about the system, it would no longer be non-deterministic. In general, this information is not available. As an example of a stochastic system, consider microscopic particles immersed in a liquid. Each particle exhibits a random motion, called Brownian motion, due to collisions with the molecules of the liquid. The exact position of the particle at time t is unpredictable.

To emphasize the difference between deterministic and stochastic systems, a stochastic version of the deterministic model of population growth in Example 3 of Chapter 1 will be formulated. An actual population of living organisms is subject to random fluctuations in the birth and death rates. Its size at any time t cannot be predicted. A more realistic description of population growth could be obtained from Eq. (3) in Chapter 1 by considering β and δ to be random variables.

Only deterministic systems will be considered in the remainder of this book.

C. Continuous versus Discrete Systems

A *continuous system* is a system in which all the essential variables are continuous in nature and the time that the system operates is some interval (or intervals) of the real numbers. A continuous system can be described by differential equations. Examples 3, 5, 6, and 7 in Chapter 1 are continuous systems.

A *discrete system* is a system in which all essential variables are discrete and the time that the system operates is a finite or countably infinite subset of the real numbers. A discrete system can be described by difference equations. Example 4 in Chapter 1 is an example of a discrete system.

A system that contains both continuous and discrete variables is known as a *hybrid system*.

In many applications information must be extracted from a continuous output signal with the aid of a digital computer. The use of a computer program to automatically screen for heart disease by analyzing a continuous-time electrocardiogram signal is but one example. It is therefore necessary to convert such a signal into a format consistent with digital computation. This can be done in two ways. The continuous signal can be converted into a discrete signal (a sequence of numbers) by the process of sampling. Alternatively, the problem can be avoided by replacing the continuous system by its discrete analog, which of course has a discrete output signal.

A model for sampling a continuous function, $x(t)$, is shown in Fig. 2.2. The switch closes instantaneously at $t = t_k$ and is open otherwise. Therefore, the sampler transforms the continuous signal into the discrete signal $x(t_0)$, $x(t_1)$, $x(t_2)$, . . . , which is a sequence of numbers appearing at the sampling instants t_k and having values equal to the amplitudes of the continuous signal at the sampling instants. It is necessary to choose the sampling instants t_k to be sufficiently close together so that little or no information is lost compared to that contained in the continuous signal.

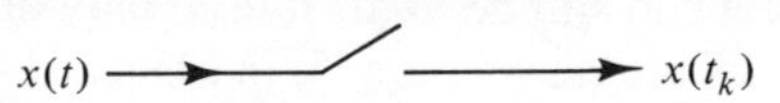

Figure 2.2 Model of the sampling process.

Often, the time interval between successive samples is the same, say, τ. Then the sampling instants $t_k = k\tau$, $k = 0, 1, 2, \ldots$, and the process is called *uniform sampling*.

The discrete analog of a continuous system, described by differential equations, is obtained by replacing each dependent variable by uniform sampling with $t_k = k\Delta$. The derivatives are replaced by difference quotients that can be made to approximate them by choosing Δ sufficiently small. Thus, the system of differential equations become difference equations that represent the corresponding discrete system. To illustrate this process, consider the differential equation (3) in Chapter 1 representing the growth of a population of cells. The derivative dN/dt is approximately

$$\frac{N(k\Delta) - N[(k-1)\Delta]}{\Delta}$$

Thus, (3) in Chapter 1 is replaced by the difference equation

$$N(k\Delta) = [1 + (\beta - \delta)\Delta]N[(k-1)\Delta] \qquad k = 1, 2, 3, \ldots \tag{7}$$

The solution of (7) is

$$N(k\Delta) = [1 + (\beta - \delta)\Delta]^k N(0) \tag{8}$$

as is shown in Exercise 1(a) at the end of this section. Letting $k \to \infty$ and $\Delta \to 0$ in such a way that $k\Delta \to t$, which is finite, it can be shown, as in Exercise 1(b), that

$$N(k\Delta) \quad \longrightarrow \quad N(t) = e^{(\beta-\delta)t}N(0)$$

in agreement with the solution (5) in Chapter 1 of the corresponding differential equation.

A discrete model of a system can often by formulated directly without first obtaining a continuous model and then transforming it into a discrete model. For example, consider cells that grow by binary fission. Assume that the time from birth to fission is a constant T and that there are $N(0)$ newborn cells at $t = 0$. Under the hypothesis that there are no deaths, the number of cells N at time $t = kT$ satisfies the difference equation

$$N(kT) = 2N((k-1)T) \qquad k = 1, 2, 3, \ldots$$

The solution of this difference equation is

$$N(kT) = 2^k N(0) \tag{9}$$

Since a system can be modeled both by difference and differential equations, which of these approaches should be used? The discrete model for cell population size seems more realistic than the continuous model of Example 3 in Chapter 1, since the number of cells can only increase by an integer. However, to the naked eye a tumor appears to increase continuously—no sudden jumps in size are observed. Thus, the continuous model seems more appropriate than the discrete model. Accordingly, models should not necessarily be judged by the criterion of realism but mainly on the basis of a high degree of relevance to experimental data. Even if the system seems discrete, a continuous model can still be used. It need only fit closely for the discrete set of observations. For example, the solution (5) in Chapter 1 of the continuous model in Example 3 in Chapter 1 coincides with the solution (9) of the discrete model at $t = kT$ provided that $\beta = \ln 2/T$ and $\delta = 0$ (since there are no deaths). Conversely, a continuous system can be approximated by a discrete system. Therefore, the type of model that is used is frequently a matter of convenience. When a continuous model can be solved analytically and provides a good fit to the data, it should be used. On the other hand, if an analytic solution is difficult to find and so ultimately digital processing is necessary, a discrete model may be preferable.

D. Nonlinear versus Linear Systems

Suppose that the admissible inputs u_i $(i = 1, 2)$ to a system produce outputs y_i. Then the system is *linear* if, and only if, the input $a_1u_1 + a_2u_2$ produces the output $a_1y_1 + a_2y_2$ for any real numbers a_1 and a_2. A system that is not linear is said to be *nonlinear*. Defining

$$a_1 u_1[t_0, t] + a_2 u_2[t_0, t] = \{a_1 u_1(s) + a_2 u_2(s) \mid t_0 \le s \le t\}$$

the criterion for linearity when applied to (1) in Chapter 1 becomes

$$f(a_1 u_1[t_0, t] + a_2 u_2[t_0, t], t) = a_1 f(u_1[t_0, t], t) + a_2 f(u_2[t_0, t], t) \qquad (10)$$

By induction, the linearity property continues to hold for a finite number, say n, of input functions:

$$f\left(\sum_{i=1}^{n} a_i u_i[t_0, t], t\right) = \sum_{i=1}^{n} a_i f(u_i[t_0, t], t) \qquad (11)$$

However, the linearity property for an infinite number of inputs is

$$f\left\{\int a(s) u_s[t_0, t]\, ds, t\right\} = \int a(s) f(u_s[t_0, t], t)\, ds \qquad (12)$$

where $a(s)$ are continuous functions, does not follow from (11) unless f satisfies appropriate conditions, such as continuity.

Linear systems are important for the following reasons. Historically, the first examples of the external description of systems were linear systems. Linear systems provide the background for the modern state space approach to system analysis. There are many general theorems that can be used to study the behavior of a linear system. Many nonlinear systems can be approximated by linear systems for certain values of their variables. The resulting linear systems can be analyzed by powerful linear systems techniques. There are many applications of linear systems theory to science and engineering. Diverse applications to physiology and biology appear in [3–5, 13, 16, 20, 27, 28].

The model of population growth described in Example 3 in Chapter 1 provides an example of a linear system. Suppose that cells enter the system at a rate of $u(t)$ cells per second. Then taking this input into account, (3) in Chapter 1 becomes

$$\frac{dN}{dt} = kN + u \qquad (13)$$

where $k = \beta - \delta$. Consider two different inputs u_1 and u_2 and let N_1 and N_2 be the corresponding outputs; that is,

$$\frac{dN_1}{dt} = kN_1 + u_1 \qquad (14)$$

$$\frac{dN_2}{dt} = kN_2 + u_2 \qquad (15)$$

To find the output corresponding to $a_1 u_1 + a_2 u_2$, multiply both sides of (14) by a_1 and (15) by a_2 and add the resulting equations. This procedure yields

$$\frac{d(a_1 N_1 + a_2 N_2)}{dt} = k(a_1 N_1 + a_2 N_2) + (a_1 u_1 + a_2 u_2)$$

It follows that $a_1 N_1 + a_2 N_2$ is the output corresponding to the input $a_1 u_1 + a_2 u_2$. Hence, the system described by (13) is linear.

The following equivalent definition of a linear system is provided for readers familiar with the terminology of differential and difference equations. A system described by differential or difference equations is linear provided that the describing equations are linear.

The next example is a nonlinear model of population growth.

Example 2 (Limited Population Growth)

A more realistic model of population growth will be formulated in this example. This model will account for the fact that a colony of cells or bacteria contained in a nutrient medium will not grow indefinitely. Some reasons for this are lack of space, lack of oxygen, decrease of nutrients, the appearance of toxic wastes, and changes in the ion concentration in the medium, especially pH.

As in Example 3 of Chapter 1, the equation for the rate of change in the size, N, of the population contains the term kN, which represents the excess of the birth rate over the death rate. In addition, a term that limits the growth will be introduced under the assumption that the main limiting factor in growth is the appearance of toxic substances. If the toxic material diffuses freely throughout the intercellular medium, then a given cell detects the cumulative toxic effect of all N cells. Thus the toxic effect on a given cell is proportional to N. The toxic effect on all N cells is N times the effect on one cell and hence proportional to $N \cdot N$. Hence, the rate equation contains the term $-\beta N^2$, where $\beta > 0$. The above results are summarized by the differential equation

$$\frac{dN}{dt} = kN - \beta N^2 \qquad k > 0,\ \beta > 0 \tag{16}$$

The reader can verify that this system is nonlinear because of the presence of the N^2 term.

Finally, it will be shown directly from the form of the differential equations (16) that N cannot increase indefinitely. The right-hand side of (16) can be written in the form $N(k - \beta N)$ and so is negative for $N > k/\beta$. Hence, $dN/dt < 0$ for $N > k/\beta$ and so N must decrease. A population whose size is governed by the differential equation (16) is said to satisfy the *logistic law* of growth. The graph of N is shown in Fig. 2.3. The details for sketching the graph appear in Exercise 2. Whether a population actually satisfies the logistic growth law can be determined only by comparing the results predicted from the theory with those obtained from experiments.

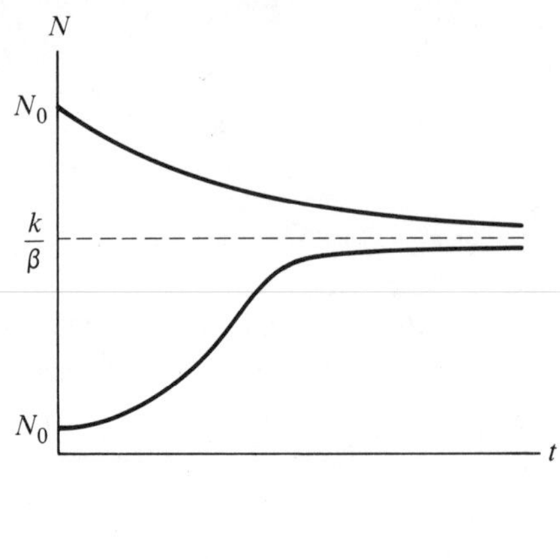

Figure 2.3 The logistic growth law.

E. Stationary versus Nonstationary Systems

The system is *stationary* or *time-invariant* if the form of the equations characterizing the system do not vary with time when the inputs are constant; otherwise the system is *nonstationary* or *time-varying*.

The same terminology will be used to describe difference and differential equations in Chapters 3 and 4. According to this terminology a system is stationary or nonstationary if, and only if, the equations characterizing the system are stationary or nonstationary.

All of the mathematical examples up to this point have been stationary systems. If in (3) of Chapter 1 the specific growth rate $k = \beta - \delta$ was a function of t, then the system represented by the resulting differential equations would be nonstationary. Similarly, the system corresponding to (16) in which either k or β depended on t would be nonstationary. The system represented by (11), (12), and (13) of Chapter 1 would be nonstationary if any of the parameters b_i or s_i depended on t.

F. Homogeneous versus Nonhomogeneous Systems

If there are no external inputs and the system behavior is determined entirely by its initial conditions, then the system is called *homogeneous* or *unforced;* otherwise the system is *nonhomogeneous*.

The same terminology will be introduced for difference and differential equations in the next two chapters. Using this terminology a system is homogeneous or nonhomogeneous if, and only if, the equations describing the system are homogeneous or nonhomogeneous.

The homogeneous systems described by (3) of Chapter 1, (5), (6), and (16), become nonhomogeneous if input terms are added. For instance, the nonhomogeneous version of (3) in Chapter 1 is (13).

EXERCISES

1. (a) Consider the difference equation (7). Verify that

$$N(\Delta) = [1 + (\beta - \delta)\Delta]N(0)$$

$$N(2\Delta) = [1 + (\beta - \delta)\Delta]N(\Delta) = [1 + (\beta - \delta)\Delta]^2 N(0)$$

$$N(3\Delta) = [1 + (\beta - \delta)\Delta]N(2\Delta) = [1 + (\beta - \delta)\Delta]^3 N(0)$$

and finally that (8) is valid.

(b) Assume that $\Delta \to 0$ and $k \to \infty$ in such a way that $k\Delta \to t$. Using the fact that $\lim[1 + (\beta - \delta)\Delta]^{1/\Delta} = e^{\beta-\delta}$, show that under the stated assumptions,

$$\lim_{\substack{\Delta \to 0 \\ k \to \infty}} [1 + (\beta - \delta)\Delta]^k = \lim[1 + (\beta - \delta)\Delta]^{(1/\Delta)(k\Delta)} = e^{(\beta-\delta)t}$$

2. Solve the differential equation (16) as follows.

(a) Verify that (16) can be written in the form

$$\int_0^t \frac{dN}{kN - \beta N^2} = \int_0^t dt$$

(b) Using the partial fractions, show that

$$\int_0^t \frac{dN}{kN - \beta N^2} = \frac{1}{k}\left(\int_0^t \frac{dN}{N} + \beta \int_0^t \frac{dN}{k - \beta N}\right)$$

$$= \frac{1}{k}\left[\ln\left(\frac{N}{k - \beta N}\right) - \ln\left(\frac{N_0}{k - \beta N_0}\right)\right]$$

where $N_0 = N(0)$.

(c) Evaluate the integrals appearing in (a) to show that

$$\ln\left(\frac{N}{k - \beta N}\right) - \ln\left(\frac{N_0}{k - \beta N_0}\right) = kt$$

or

$$\frac{N}{k - \beta N} = \frac{N_0}{k - \beta N_0} e^{kt}$$

(d) Prove that

$$N = \frac{N_0 k/\beta}{N_0 + (k/\beta - N_0)e^{-\beta t}}$$

3. A fair die is rolled. Let $y(k)$ be the sum of the number of spots on the uppermost face of the die that appears in the first k rolls. Let $u(k)$ be the number of spots on the uppermost face of the die on the kth roll. Show that

$$y(k) = y(k - 1) + u(k) \qquad k = 1, 2, 3, \ldots$$

4. Consider the system of bees foraging for nectar in rape flowers. The rape flower has four nectaries (nectar-containing tissues). If the population of bees in the field is high, flowers may be visited frequently. Hence, a bee may find a nectary that is empty or only partly filled. The nectaries are filled within 30 min of being emptied. When a bee finds an empty nectary, it may abandon the flower or try another nectary in the same flower. Is this system stochastic or deterministic? Why?

5. Fill in as many correct adjectives as possible in the appropriate blank spaces in the table on p. 39.
(*Hint*: c.f. Exercise 3 before filling in column 2.)

6. In a town of 1000 people there are 2 people with influenza. The spread of this disease follows the differential equation

$$\frac{dy}{dt} = 0.001y(1000 - y)$$

where y is the number of people infected at time t (measured in weeks).

(a) Is the influenza epidemic system linear or nonlinear?

(b) Find y as a function of t and sketch its graph.

(c) How long will it take for three-quarters of the population to be infected?

7. If y is the amount of the substrate present at time t that is being transformed by an enzyme, then according to the Michaelis-Menton equations the rate of transformation of y is

$$\frac{dy}{dt} = -\frac{My}{K + y}$$

System representation	Distributed (D) or lumped (L)	Deterministic (D) or stochastic (S)	Discrete (D) or continuous (C)	Linear (L) or nonlinear (N)	Homogeneous (H) or nonhomogeneous (N)
$\left(\frac{\partial^2 y}{\partial t^2}\right)^2 + \frac{\partial^2 y}{\partial x^2} = u(x, t)$					
$y(k) - 2y(k-1) = u(k)$					
$\frac{d^2 y}{dt^2} - 3y\frac{dy}{dt} + 5y = 0$					
$y(k, x) - 2y(k-1, x-1) = u(k, x)$					
$t^2 \frac{dy^2}{dt^2} + 2\frac{dy}{dt} = 4t^3$					
$ky(k)y(k-1) = u(k)$					

where M and K are constants.

(a) Classify the above system as thoroughly as possible.

(b) Find an implicit equation expressing y as a function of t.

8. The global population of blue whales grows according to the logistic model with $k/\beta = 200{,}000$ and $\beta = 6 \times 10^{-7}$ (time is measured in years). In 1972 there were 6000 blue whales. Assuming that the present ban on harvesting these whales is maintained, how long will it take for the population to recover to 100,000?

9. A body whose temperature is T above its surroundings loses heat by conduction and convection. Newton's law of cooling states that its temperature falls according to the differential equation $dT/dt = -kT$, where k is a constant.

(a) Classify this system as thoroughly as possible.

(b) The body is 100°C hotter than its surroundings at $t = 0$, and at $t = 15$ min this temperature difference has fallen to 60°C. How long will it take the body to reach a temperature 10°C above that of its surroundings?

2.3 OPEN AND CLOSED LOOPS

The concepts of open and closed loops will be introduced qualitatively by means of Goodwin's simplified model of the control of protein synthesis [12] in a cell. The corresponding mathematical description will be given in Examples 4 and 5 of Chapter 9. Let $u(t)$ be the concentration of messenger RNA (mRNA) at time t. This mRNA directs the synthesis of an enzyme whose concentration is denoted by $n(t)$ [40]. The enzyme is a catalyst for a reaction that produces a protein product whose concentration is denoted by $y(t)$.

The example above is an *open-loop control system*. The control u is altered by changes in the rate of production of mRNA as dictated by the nucleus. In this simple model these alterations in u are determined at any time based on a goal for the protein production system, such as $y > 0$, and all available prior knowledge about the system. The input, u, is not influenced by the output, y, of the system at time t. A block diagram of an open-loop system is shown in Fig. 2.4.

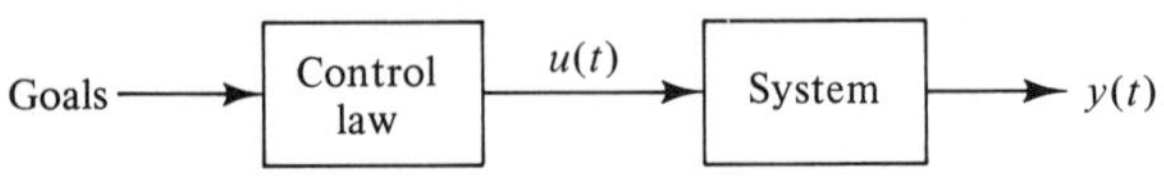

Figure 2.4 An open-loop control system.

If the behavior of an open-loop system is not completely understood or if unexpected disturbances act on it, then there may be considerable, unpredictable variations in the output. Open-loop systems are not used when accuracy is required.

The simple model of protein synthesis described above will be modified. Since nature is economical, the assumption that the synthesis of y is unaffected by its concentration will be changed. If the concentration of y becomes too high, then its production is slowed down. On the other hand, safeguards are provided so that if the concentration of y becomes too low, then its production is increased. With these modifications the resulting protein synthesis model is an example of a *closed-loop* or *feedback* control system.

In a closed-loop system, the control u is modified in some way by information about the behavior of the system output. Figure 2.5 is a block diagram for a closed-loop system. Closed-loop systems can often be designed to cope with lack of knowledge of system behavior, inaccurate components, and unexpected disturbances. With all their many advantages, feedback systems have a very serious disadvantage: Closed-loop systems may inadvertently develop unstable oscillations.

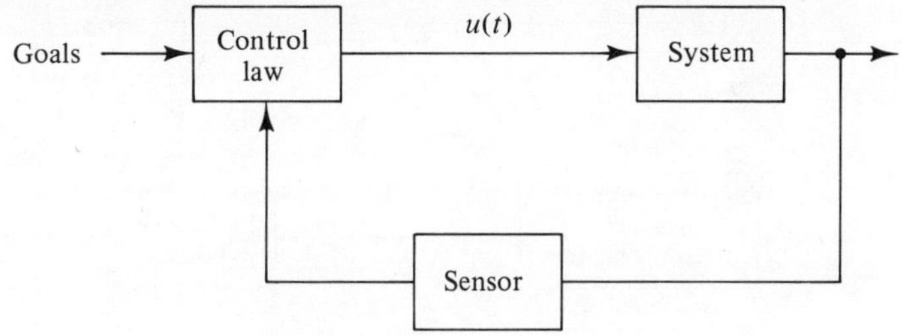

Figure 2.5 A closed-loop system.

Feedback is one of the most important phenomena occurring in living organisms. It is the basis of maintaining regulatory processes within the organism, such as thermoregulation. The core temperature of the human body in normal circumstances is accurately controlled by fine and coarse control processes that involve feedback.

The coarse control of core temperature is probably performed by the shivering and sweating mechanisms. Temperature information is fed back from the skin via the nervous system to the thermoregulatory center in the hypothalamus. If the information fed back indicates that the temperature is too low, signals are transmitted from the thermoregulatory center to the muscles to increase oxidation (by shivering when extremely cold) and hence supply heat to the body. The reverse happens if the skin temperature is too high. In addition, sweating occurs.

The fine control of core temperature occurs by adjusting the blood flow rate (and hence heat flow rate) from the core to the surface of the body, consequently regulating the heat loss from the surface of the body to the environment. A simplified block diagram of this process is shown in Fig. 2.6. Thermoreceptors sense the core and surface temperatures of the body and code these temperatures as nervous impulses. These signals are sent to the hypothalamus where, according to the "set-point theory" [35], they are compared with the "set points" or desired temperatures. It is assumed that the central nervous system receives signals that represent the differences between the set point and the actual temperature. If necessary, the blood flow through the core and the surface of the body is adjusted by means of appropriate sympathetic and parasympathetic stimulation of the blood vessels. The changes in blood flow result in temperature changes of the core and surface of the body in the direction of the set-point temperature.

Four types of single closed-loop control have been observed in biological systems.

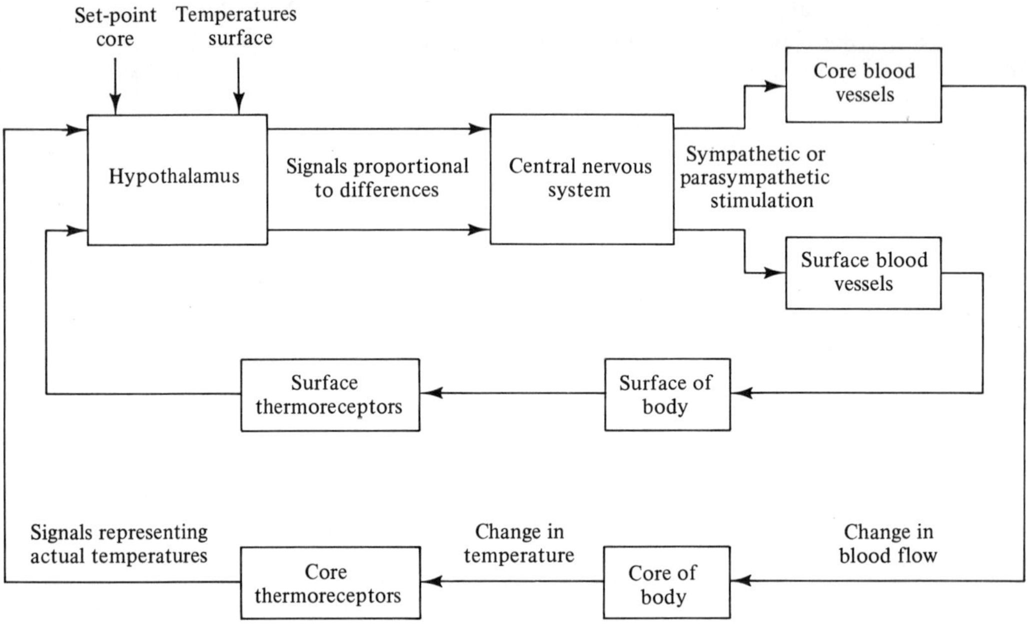

Figure 2.6 Temperature control by means of blood flow.

A. Negative Feedback

Negative feedback exists when an increase (decrease) in the output leads to a decrease (increase) in the output. A typical single-loop negative feedback system is illustrated in Fig. 2.7. The feedback signal y inhibits the production of u_1 from u_0 as indicated by the symbol ⊖ in Fig. 2.7. Hence an increase in y decreases the net rate of y production. A decrease in y increases the production of u_1 from u_0 and eventually causes an increase in the synthesis of y. Hence, the total effect is that the system tends to return the output to its previous level.

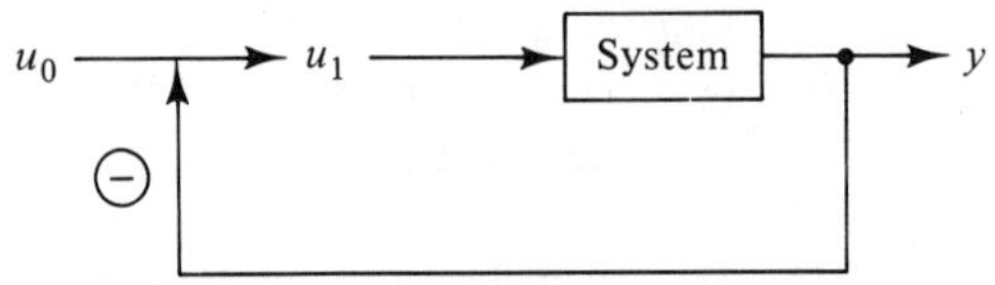

Figure 2.7 Negative feedback control.

Negative feedback systems are used for self-regulation in biology. In a typical biochemical process the system consists of a series of chemical reactions

$$x_2 \longrightarrow x_3 \longrightarrow \cdots x_{n-1}$$

where x_2 is produced from u_1 and x_{n-1} produces y. The regulatory enzyme that catalyzes the reaction $u_0 \rightarrow u_1$ is inhibited by the terminal metabolite, y, of the pathway. The first experimentally confirmed example of negative feedback control of a metabolic process was the inhibition of threonine deaminase by isoleucine [38] in 1956. By 1963 it was realized that single-loop feedback control is the most common form of biological regulation [29].

The predominance of single-loop negative feedback systems can be explained by the work of Savageau [33]. In unbranched pathways, he examined several possible patterns of negative feedback control that had the same responsiveness to changes in initial substrate concentration. He discovered that the single-loop system was the least sensitive to parameter variation. It was also shown that a compromise between input responsiveness and desensitization must be made, but whatever mix of the two was chosen, single-loop negative feedback assures the maximum of both. Therefore it is plausible that the process of natural selection would favor single-loop end-product inhibition.

B. Positive Feedback

A single-loop positive feedback system appears in Fig. 2.8. An increase in y enhances the production of u_1 from u_0, as indicated by the sympol ⊕ in Fig. 2.8, and so eventually causes an increase in the production of y. A decrease in y decreases the production of u_1 from u_0 and leads to a further decrease in y. Therefore, positive feedback is unlikely to be useful in biological regulation, since it causes its output to depart unidirectionally from a desired level.

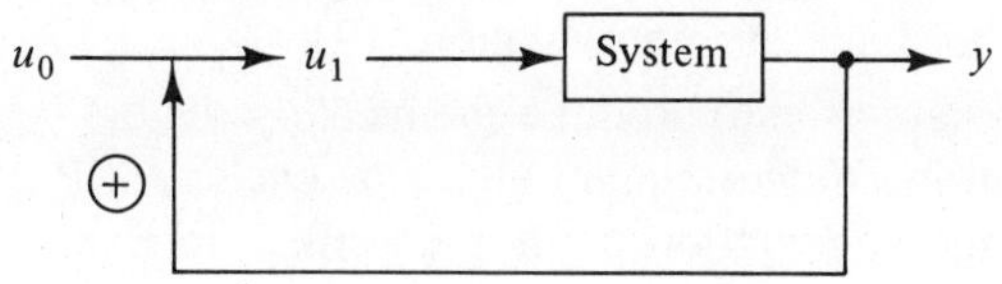

Figure 2.8 Positive feedback control.

Positive feedback has at least three functions.

1. It occurs in pathological conditions when normal homeostatic regulation has broken down. A positive feedback loop whose operation tends to intensify the pathophysiological changes responsible for a particular disease is known as a *vicious circle*.

   The vicious circle present in *acute gouty arthritis* is shown in Fig. 2.9. An increase in the concentration C_p of uric acid in the plasma leads to an increase in the concentration C_s of uric acid in the synovial fluid. This increase favors the formation of monosodium urate crystals in the synovial fluid of the joint [34]. The crystals cause a local inflammatory reaction, possibly by activating Hageman factor, which may result in the formation of "kinin-like inflammatory mediators" [21]. The acute inflammatory response includes the invasion of the

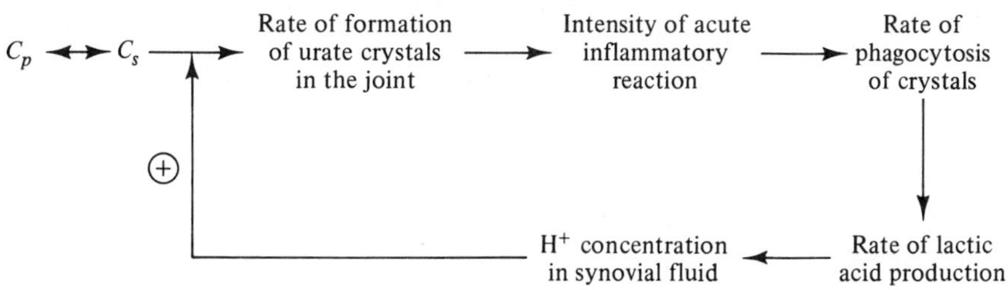

Figure 2.9 Acute gouty arthritis.

joint fluid by leukocytes that phagocytose many of the crystals. The increased metabolic activity of the leukocytes due to phagocytosis results in increased lactic acid production. The lactic acid increases the hydrogen ion concentration of the synovial fluid, which decreases the solubility of monosodium urate. Thus, the rate of crystallization is increased. This completes the vicious circle. Hemorrhagic shock is another example in which vicious circles play a vital role (see Fig. 2.12).

2. Positive feedback loops are used in the construction of chemical on–off switches, especially when a rapid all-or-none switching response is required. The generation of an action potential in neurons is caused by a positive feedback loop. A small initial membrane depolarization results in an increase in sodium permeability, which allows an inward sodium flux. This additional sodium increases the depolarization of the membrane. Eventually, the nerve fiber becomes positive and sodium influx ceases (saturation of response). Active pumping moves sodium out of the nerve cell. Saturation of response is common in biological systems and explains why the response (output) generated by a positive feedback loop is finite. Two other examples in which positive feedback provides a rapid-response switch is blood clotting [6, 10, 30] and the sudden release of luteinizing hormone during ovulation [19].
3. Positive feedback loops are used to produce chemical oscillators—for example, cyclic AMP oscillation in *Dictyostelium* [11]. This cyclic AMP oscillation in slime molds controls the aggregation process described in Example 1. Oscillations of the glycolytic pathway are also due to positive feedback [2]. The function of these oscillations is not clear. A possible explanation is that the purpose of the positive feedback is to produce a chemical switch. The oscillations may be a switch defect. Switch "chatter" in the form of oscillations is common in physical control systems.

C. Negative Feedforward

Negative feedforward control exists when a sufficiently high (low) input leads to an eventual decrease (increase) in the output. A single-loop negative feedforward system is shown in Fig. 2.10.

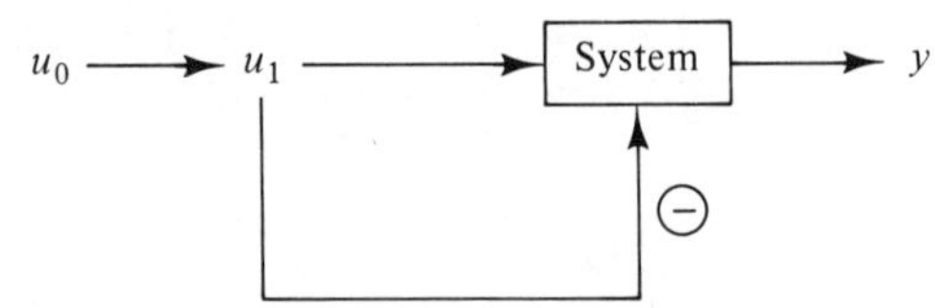

Figure 2.10 Negative feedforward control.

The inhibition of an enzyme by a high concentration of its substrate is an illustration of negative feedback control. This inhibition prevents the unnecessary synthesis of intermediate compounds. Table 2.1 contains some particular examples. Kinetic equations for this type of inhibition appear in [17] and [9].

TABLE 2.1 INHIBITION OF AN ENZYME BY A HIGH SUBSTRATE CONCENTRATION

Enzyme	Substrate	References
Liver carboxylesterase	Ethyl butyrate	[31]
Xanthine oxidase	Hypoxanthine	[8]
Snake venom L-amino acid oxidase	Leucine	[7]

D. Positive Feedforward

In positive feedforward control the input is fed forward into a certain part of the system and adjusts the system so that sufficiently high (low) inputs eventually lead to an increase (decrease) in the output. Figure 2.11 illustrates a single-loop positive feedforward control system.

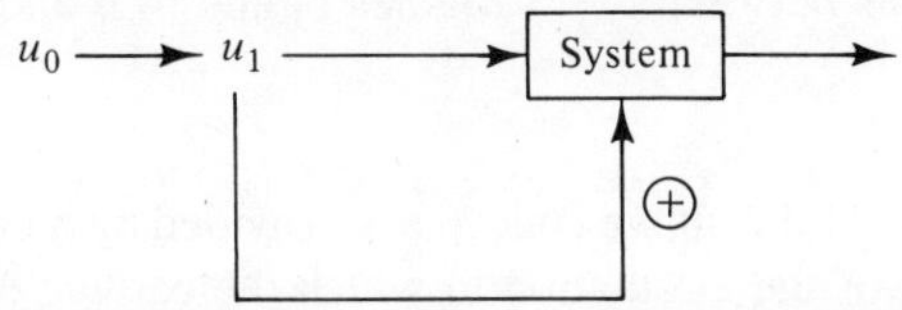

Figure 2.11 Positive feedforward control.

In biochemical systems, positive feedforward control adjusts enzyme activity to the precursor concentration. Examples are the activation of fumarase hydrathase [25] and phosphodiesterase by cyclic AMP [39]. Some other examples of positive feedforward control are cortisol secretion after hemorrhage [1, p. 528], cardiovascular and thermoregulatory response to exercise [5, p. 242], and in autonomic involuntary sensory motor systems [36, p. 222].

Positive feedforward control can be used to its full advantage when combined with prediction. Information about a measured or predicted disturbance can be fed forward, effecting necessary adjustments before a feedback controller would have time to operate. This type of control is used in human-machine coordination [5]. In biology, circadian rhythms are combined with positive feedforward control to control physiological processes. Hoffman [18] states:

> In general it is assumed that the adaptive significance of such endogenous rhythms lies in the fact that by proper phase adjustment, the organism is prepared to meet the challenge of cyclic changes in the environment; the organism can prepare in advance while a stimulus-response system might not allow sufficient time for the appropriate morphological, physiological and behaviorial changes.

For example, the movement of petals in plants is slow. If the orientation of the petals to the sun did not begin until sunrise, then several hours of sunlight would be lost. The circadian oscillator predicts dawn, and a feedforward control loop begins the petal movement prior to sunrise.

Most biological control networks are multiple-loop systems composed of the four basic types of control loops described above. A convenient way to show the relationship in such complicated systems is by means of a symbol–arrow graph. Let v_1 and v_2 be any two variables that can affect each other. The following notation will be used:

$v_1 \longrightarrow v_2$ means that a primary increase ($\uparrow$) or decrease ($\downarrow$) in v_1 causes a secondary $\uparrow$ or $\downarrow$, respectively, in v_2. In other words, v_1 and v_2 change in the same direction.

$v_1 \dashrightarrow v_2$ means that a primary $\uparrow$ or $\downarrow$ in v_1 causes a secondary $\downarrow$ or $\uparrow$, respectively, in v_2. In other words, v_1 and v_2 change in the opposite direction.

The two types of feedback can be defined in terms of this notation.

1. *Negative feedback* exists between two variables v_1 and v_2 if and only if $v_1 \begin{smallmatrix}\longrightarrow \\ \dashleftarrow\end{smallmatrix} v_2$ or $v_1 \begin{smallmatrix}\dashrightarrow \\ \longleftarrow\end{smallmatrix} v_2$.
2. *Positive feedback* exists between two variables v_1 and v_2 if and only if $v_1 \leftrightarrows v_2$ or $v_1 \begin{smallmatrix}\dashrightarrow \\ \dashleftarrow\end{smallmatrix} v_2$.

An interesting example of the above concepts is provided by hemorrhagic shock. Circulatory shock is a state of the circulation in which the cardiac output is too low to supply the normal nutritional needs of the body's tissues. Guyton and Crowell [15] noted the following results when the arterial pressure of dogs was reduced rapidly by acute hemorrhage. The animals whose pressure was only slightly reduced recovered rapidly, whereas those whose pressure was greatly reduced died rapidly. However, at intermediate pressure levels it was impossible to determine for many hours whether the animal would live or die.

Guyton and Crowell [15] explained these results in terms of feedback. They identified six separate types of feedback that further depress the cardiac output in shock. In addition, they stated four types of negative feedback mechanisms that help to oppose the progression of shock. Three of these processes are shown in the symbol-and-arrow diagram in Fig. 2.12.

Loops 1 and 3 are positive feedback loops. Decreased cardiac output leads to decreased arterial pressure in both these loops. In loop 1, decreased coronary blood flow results in decreased myocardial oxygenation, thereby producing heart failure. Heart failure still further decreases the cardiac output. In loop 3 the decrease in blood flow leads to an increase in capillary permeability due to the decrease in oxygenation of the peripheral tissues. Consequently, blood volume, venous return, and cardiac output all decrease.

Loop 2 is a negative feedback loop. A decrease in capillary pressure in shock causes fluid to be absorbed into the circulatory system. Hence, the blood volume and the cardiac output both increase. This raises the capillary pressure back to normal.

The experimental results can now be explained. Above a critical pressure level the positive feedback mechanisms of deterioration took charge and caused death.

Some other biological examples of multiple-loop systems are models for repression of seven enzymes by acetate in *Pseudomonas fluorescens* in the linear mandelate-acetate pathway [24, p. 463], a division of the autonomic nervous system [37],

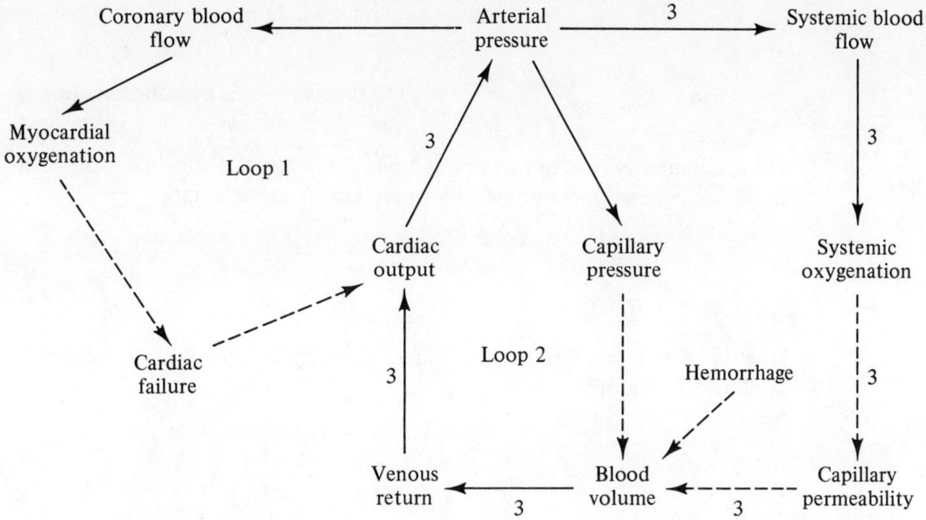

Figure 2.12 Symbol–arrow graph of some factors in hemorrhagic shock.

granulocyte production [22], cortisol secretion [3, p. 213] and [1, p. 528], and the nerve membrane [36, p. 82]. Other examples appear in the exercises. The biological background can be found in [14,38]. Many of the examples in this section were used by Rapp [32] to illustrate feedback and feedforward control in biology.

Mathematical difficulties have limited theoretical investigation of multiple-loop systems in biology. Some work has been done on particular configurations [26]. Multiple-loop systems are studied by simulation using electrical or hydrodynamic analogs. Another approach is to solve the system equations numerically using digital or analog computers.

EXERCISES

1. Describe the operation of the following system for regulating electrolyte concentration. The symbol [ion] denotes the concentration of the ion in the extracellular fluid.

(a)

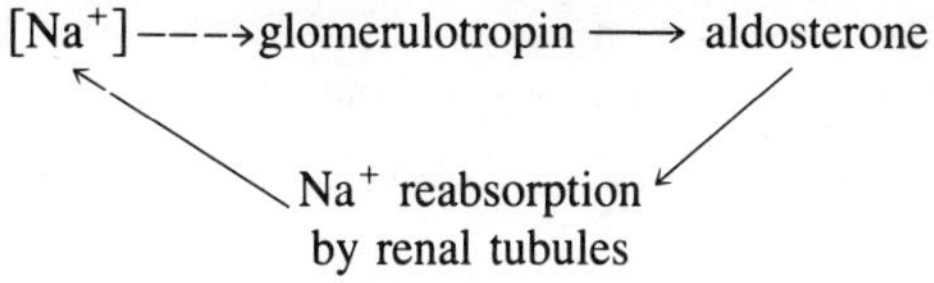

(b)

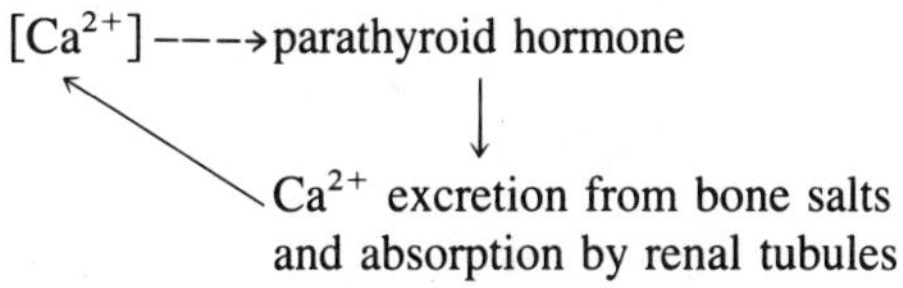

2. Discuss the control of extracellular fluid osmolarity (Os) as described by the following symbol–arrow graph:

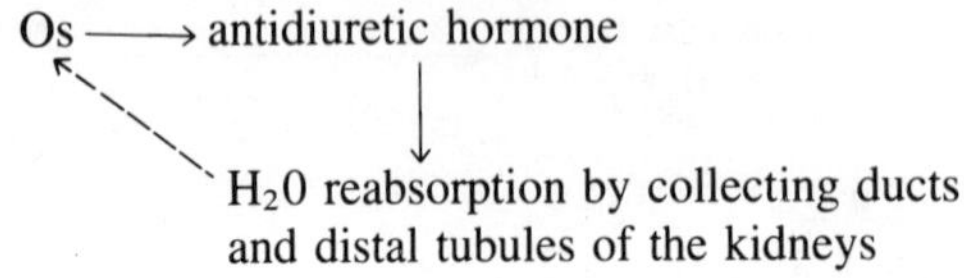

3. The respiratory system can act in 1 to 3 min to regulate the acid–base balance of your body. Explain its operation from the following symbol–arrow graph.

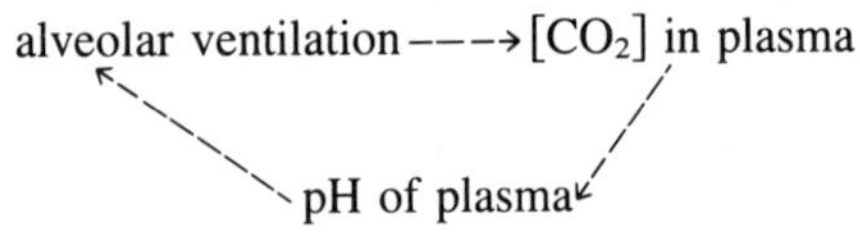

4. Discuss the regulation of red blood cell (r.b.c.) production using the following symbol–arrow graph:

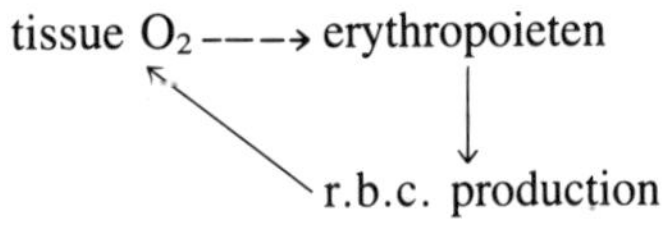

5. Discuss the regulation of arterial pressure by the following simplified symbol–arrow graphs for the following three control systems.

(a) Almost immediate changes in arterial pressure (a.p.) can occur due to changes in capillary pressure (c.p.) as follows:

c.p. $\rightleftarrows$ fluid leak

(b) The long-term regulation of a.p. is regulated by the release of renin by the kidneys as shown below:

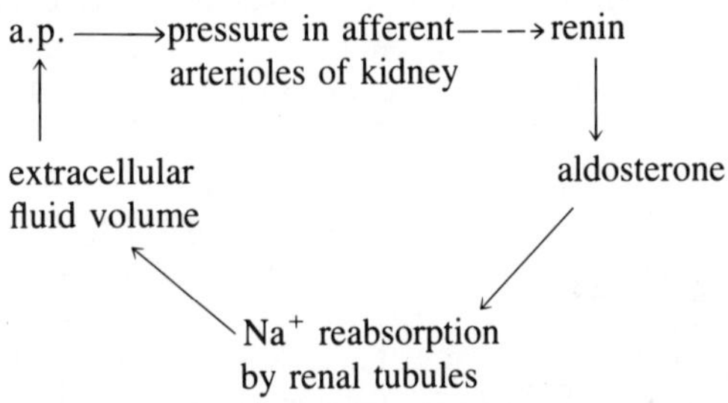

(c) The nervous system, as shown below, takes about 90 min to change a.p. through transducers in the carotid sinus and in thoracic pressure receptors:

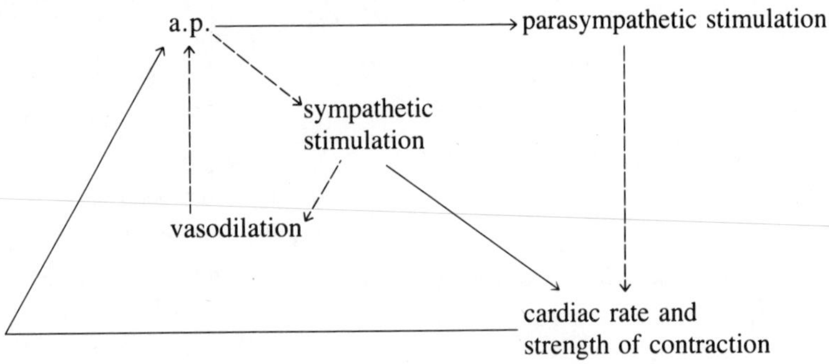

6. Decreasing arterial pressure (a.p.) to as low as 60 mm Hg does not decrease blood flow to the brain because of local autoregulatory mechanisms that cause the blood vessels to dilate when the blood pressure falls. The CNS ischemic reflex acts as an emergency a.p. control system that acts rapidly to prevent a further decrease in a.p. when the blood flow to the brain decreases dangerously close to lethal level. Discuss the symbol–arrow graph for the reflex:

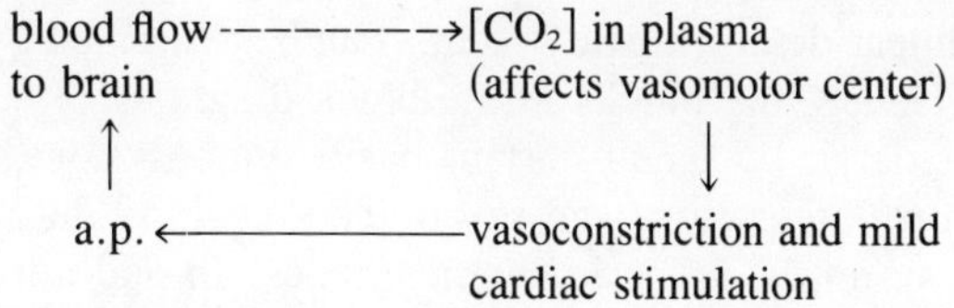

7. The following simplified symbol–arrow graph indicates some control systems regulating the concentration of glucose in the blood, [Gl]:

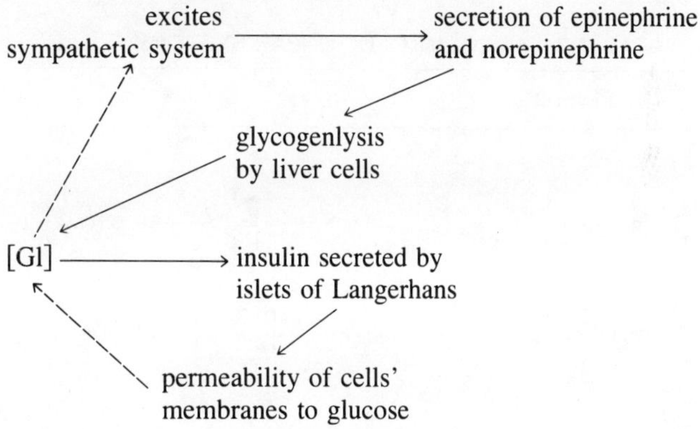

Describe the regulation of blood glucose concentration.

8. **(a)** Draw a symbol–arrow graph for acute gouty arthritis (c.f. Fig. 2.9).
   (b) Phenylbutazone may decrease the intensity of the acute inflammatory reaction. What is the sequence of events leading to relief of pain in the joints when this drug is administered?
   (c) Colchicine may decrease the metabolic activity of the leukocytes and the rate at which they produce lactic acid. Answer (b) for colchicine.
9. The following symbol–arrow graph indicates some control circuits that control the release of glucocorticoids by the adrenal cortex. The symbols ACTH and CRF denote the hormone adrenocorticotropin and its releasing factor, respectively.

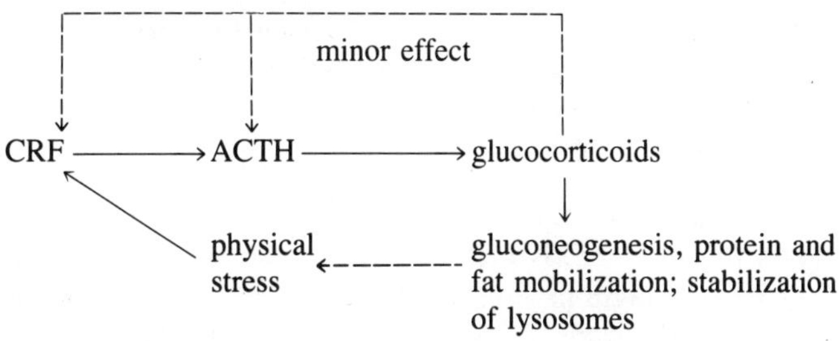

Describe the operation of this control system.

10. Verify that an even number of dashed arrows in a symbol–arrow graph of a control loop gives positive feedback, while an odd number of dashed arrows gives negative feedback.

2.4 BASIC UNITS OF LINEAR SYSTEM DIAGRAMS

Most linear deterministic systems can be described by differential or difference equations. Hence, the blocks in the block diagrams of such systems can be described in greater detail by using certain basic symbols for the operation of addition, multiplication, integration, and so on. The resulting diagram gives a visual interpretation of the system's dynamic characteristics. In addition, it reveals possible methods to engineers for implementing the equations governing the system. For a discrete system the basic block diagram also serves as a flow chart for writing a program to describe the behavior of the system. The six basic units appear in Table 2.2.

TABLE 2.2 BASIC UNITS OF LINEAR SYSTEMS

Element	Symbol	Input–output relation
1. Integrator	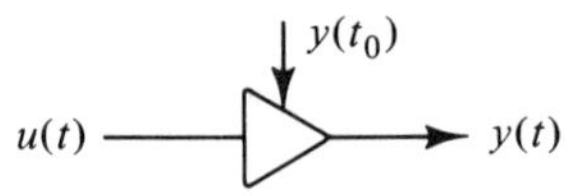	$y(t) = y(t_0) + \int_{t_0}^{t} u(s)\, ds$
2. Delay[a]	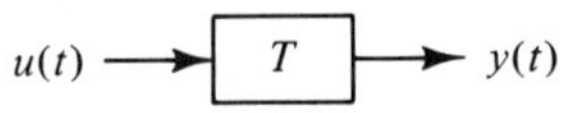	$y(t) = u(t - T)$
3. Summing junction	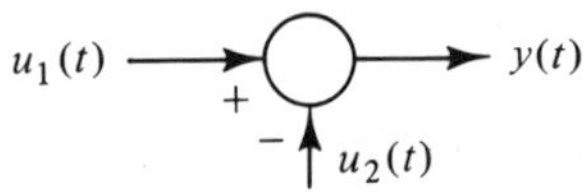	$y(t) = u_1(t) - u_2(t)$
4. Gain change	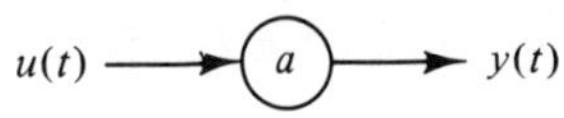	$y(t) = au(t)$
5. Multiplier	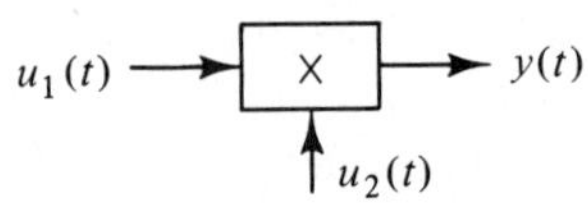	$y(t) = u_1(t)u_2(t)$
6. Single-valued function generator	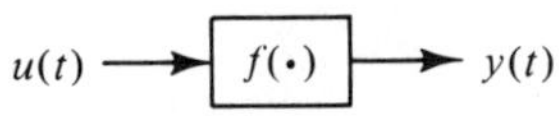	$y(t) = f[u(t)]$

[a] The symbol z^{-T} is often used instead of T.

Another convention frequently used in a system diagram is a *splitting point* as shown in Fig. 2.13. Notice that the signal, $u(t)$, follows each pathway unchanged.

Block diagrams that contain only the above basic units are frequently called *simulation diagrams*. These diagrams can be used to simulate the behavior of the

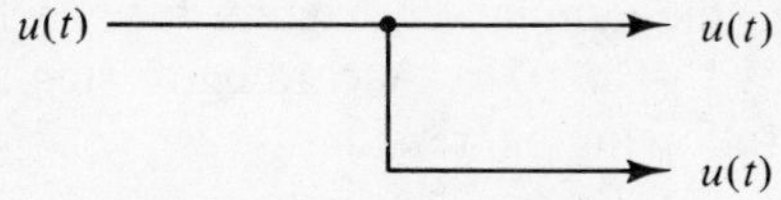

Figure 2.13 A splitting point.

system they represent. They can serve as flow diagrams for programming a digital or analog computer* for discrete or continuous systems, respectively. Moreover, an electrical system can be constructed from such simulation diagrams using standard electronic components. Frequently, biological systems are so complicated that a complete analytic study is impossible and the above simulation methods are required.

Note that a differentiating element,

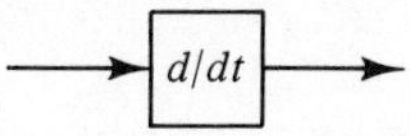

is not included. The reason for this convention arises from practical considerations. Reliable electronic differentiating devices are more expensive and difficult to construct than integrators.

Drawing simulation diagrams involving differential equations may require a great deal of ingenuity in complicated systems. A systematic approach that is easy to use with constant-coefficient systems is as follows:

1. Solve each differential equation for its highest-order derivative.
2. Integrate each equation as many times as the order of the highest derivative. This expresses one output as a function of several integral terms.
3. Group the terms on the right side in the form of a nested sequence of integrations.
4. Draw the simulation diagram by inspection. Use the fact that terms inside an integral sign constitute inputs to an integrator and terms already containing an integral are the outputs of other integrators.

To illustrate the above procedure, the simulation diagram for exponential population growth described by $\dot{N} = kN$ will be drawn. Step 1 is unnecessary. Step 2 gives

$$N(t) = k \int_0^t N(s)\, ds + N(0)$$

which is already in the form required by step 3. The required simulation diagram appears in Fig. 2.14.

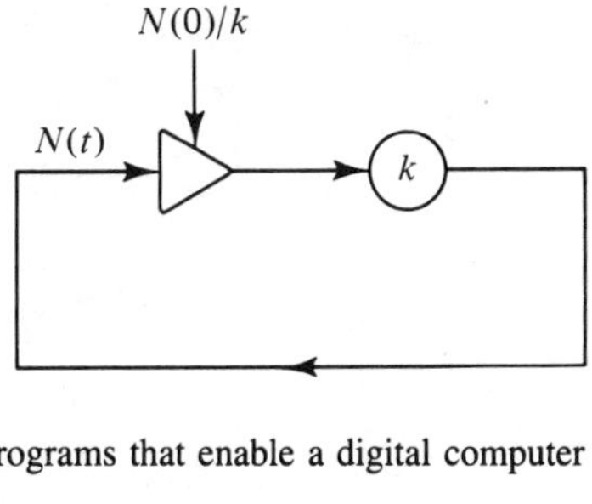

Figure 2.14 Simulation diagram for $\dot{N}(t) = kN(t)$.

* There are programs that enable a digital computer to be used as an analog computer.

As another illustration the simulation diagram for Newton's equation $m\ddot{y} = u$ will be drawn. Step 1 gives $\ddot{y} = u(t)/m$. According to step 2, two integrations are required, which gives the following expression,

$$y(t) = \frac{1}{m}\int_0^t \left[\int_0^{s_2} u(s_1)\, ds_1 + \dot{y}(0)\right] ds_2 + y(0)$$

in the form required by step 3. The required simulation diagram is shown in Fig. 2.15. Frequently, the integration constants are not shown in the simulation diagram.

The simulation diagram for $y(k) = ay(k-1)$ appears in Fig. 2.16.

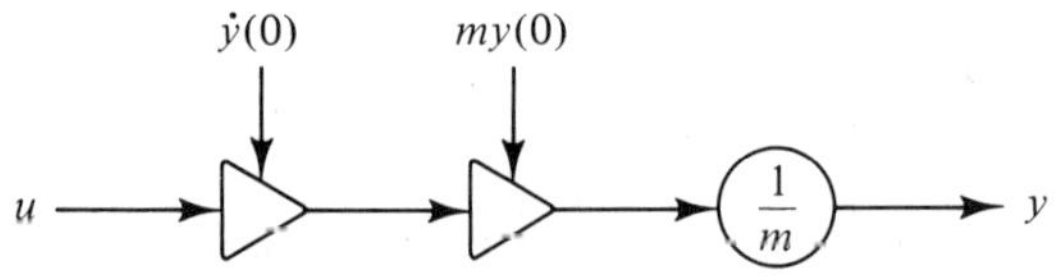

Figure 2.15 Simulation diagram for $m\ddot{y}(t) = u(t)$.

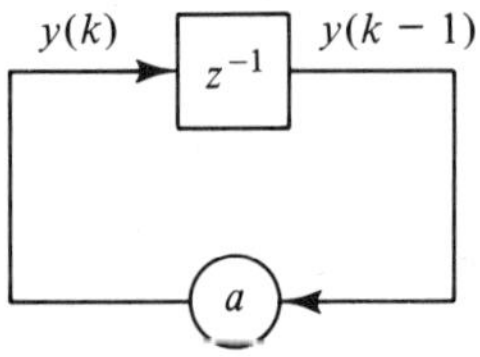

Figure 2.16 Simulation diagram for $y(k) = ay(k) = ay(k-1)$.

As a final example, consider the first-order discrete system,

$$y(k) = u(k) + 2u(k-1) - 3y(k-1) \tag{17}$$

First decompose the right side of (17) into two terms, $u(k) + 2u(k-1)$ and $-3y(k-1)$. Since $y(k)$ is just the sum of these two terms, it may be generated by a summation unit as shown in Fig. 2.17(a). The signal $-3y(k-1)$ may be generated by applying $y(k)$ to the delay and multiplier units as indicated in Fig. 2.17b. The schematic representation of the system (17) is completed by generating the signal $u(k) + 2u(k-1)$ as shown in Fig. 2.17(c).

The representation of system (17) given in Fig. 2.17 is not unique. It will be shown that the simulation diagram in Fig. 2.18 also represents system (17). To determine the relation between $y(k)$ and $u(k)$ for this simulation diagram, it is convenient to denote the signal generated by the leftmost summing junction by $e(k)$. Then the relationships governing the two summation units are formulated as

$$e(k) = u(k) - 3e(k-1) \tag{18}$$

$$y(k) = u(k) - e(k-1) \tag{19}$$

The elements $e(k)$ and $e(k-1)$ will be eliminated from these two identities in order to obtain the difference equation relating $y(k)$ and $u(k)$. Solving (19) for $e(k-1)$ yields

$$e(k-1) = u(k) - y(k)$$

which when substituted into (18) results in

$$e(k) = u(k) - 3u(k) + 3y(k) \tag{20}$$

It follows from (20) that

$$e(k-1) = -2u(k-1) + 3y(k-1)$$

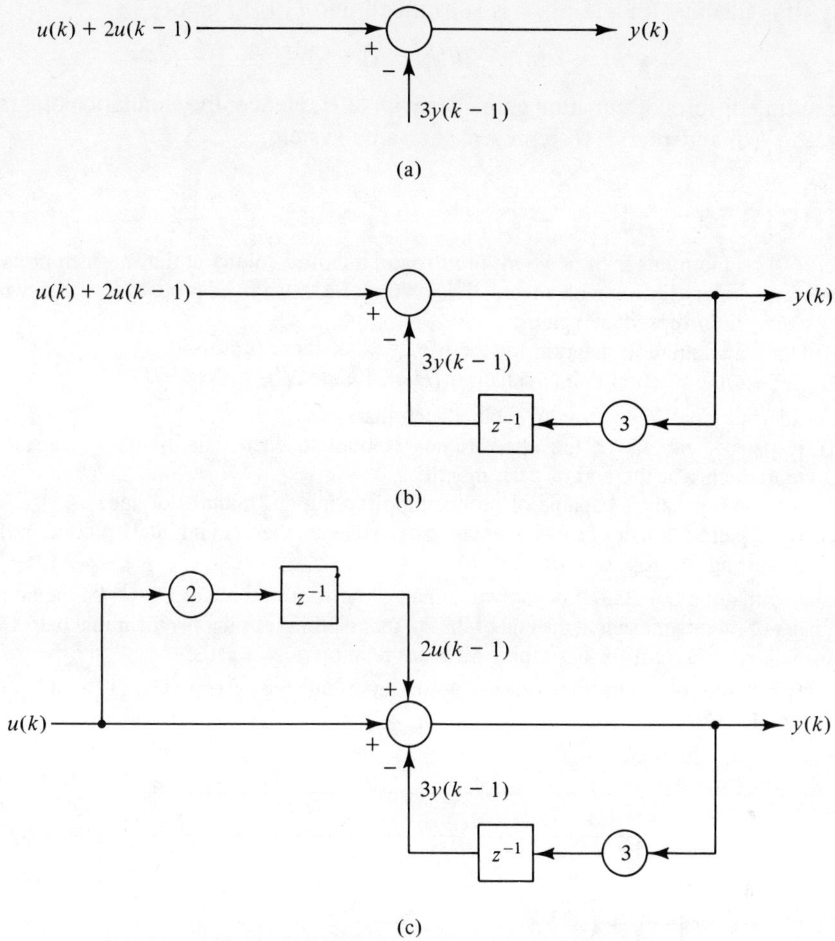

Figure 2.17 Simulation diagram for system (17).

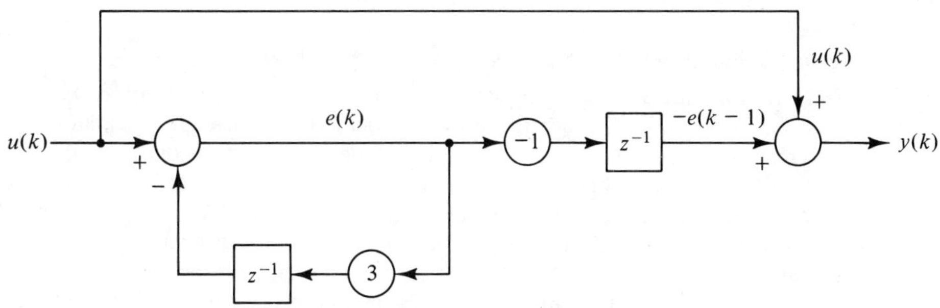

Figure 2.18 Simulation diagram for a first-order system.

Finally, this identity for $e(k-1)$ is substituted into (19) to give

$$y(k) = u(k) + 2u(k-1) - 3y(k-1)$$

The resulting difference equation coincides with (17). Hence, the simulation diagrams in Fig. 2.17(c) and Fig. 2.18 represent the same system.

EXERCISES

1. Let $N(t)$ be the number of newborn protozoa in a closed colony at time t. Each protozoa undergoes binary fission with a fixed division time T. Only a fixed proportion γ of newborn cells survive to reproduce again.
   (a) Draw a simulation diagram for the life cycle of these protozoa.
   (b) Deduce the relation that N satisfies. [*Hint:* Relate $N(t)$ to $N(t-T)$.]
2. Consider an ideal population of rabbits such that:
   (a) A pair of rabbits of the opposite sex is born to each pair of rabbits capable of reproducing at the end of each month;
   (b) A newborn pair is capable of producing offspring at 2 months of age;
   (c) Once paired, a pair of rabbits remain true to one another and infinitely produce rabbits according to properties (a) and (b).

   Assume that the population began with a pair of newborn rabbits. Let $y(k)$ be the number of pairs of rabbits present at the end of the kth month after introducing the initial pair. Draw a simulation diagram for describing y. What relation does y satisfy?
3. Verify that the following simulation diagram represents the system (11), (12), and (13) of Chapter 1.

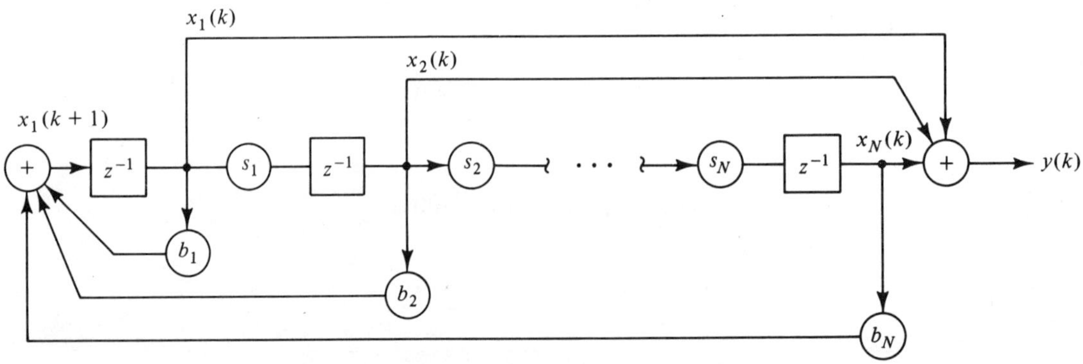

4. Draw a simulation diagram for $\ddot{y}(t) + a_1\dot{y}(t) + a_0 y(t) = b_0 u(t) + b_1 \dot{u}(t)$.
5. **(a)** Verify that the simulation diagram at the top of p. 55 represents the idealized life cycle of the female herring gull under the following assumptions. There is a maturation time of 36 months between fledglings and the production of the first fledged brood and a nesting interval of 12 months. On the average, the number of female survivors that return to breed again is γ_2 times the number of nesting females. To this flow of possible breeders there is added a flow of females that have just matured and an input of adults from an extraneous source (e.g., migration from another nesting colony). On the average 0.1 of this total flow are newborn. The proportion of newly matured gulls that survive to breed is γ_1. The output is the observed flow of returning breeders.

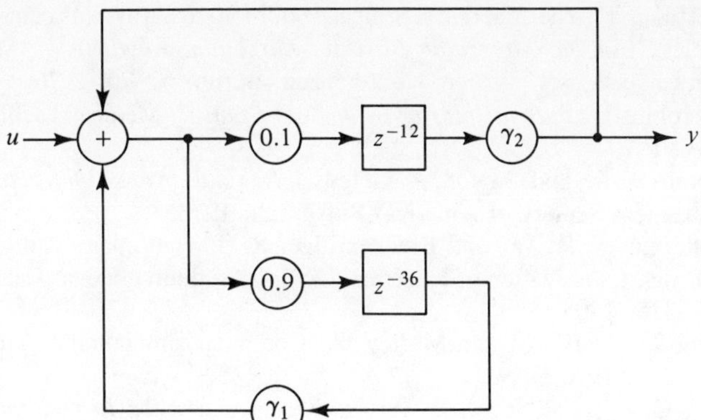

(b) Write down a system of equations representing the above simulation diagram.

REFERENCES

1. Blessner, W. B. *A Systems Approach to Biomedicine*. McGraw-Hill, New York, 1969.
2. Boiteux, A., and Hess, B. Oscillation in glycolysis. Cellular inspiration and communication. In: *Faraday Symposium No. 9, Physical Chemistry of Oscillatory Phenomena,* 202–214, 1974.
3. Brown, J. H. U., and Gann, D. *Engineering Principles in Physiology,* Vol. 1–3. Academic Press, New York, 1973.
4. Brown, J. H. U., Jacobs, J. E., and Stark, L. (eds.). *Biomedical Engineering*. F. A. Davis, Philadelphia, 1971.
5. Clynes, M., and Milsum, J. H. (eds.). *Biomedical Engineering Systems*. McGraw-Hill, New York, 1970.
6. Davie, E. W., and Kirby, E. P. Molecular mechanisms in blood coagulation. *Curr. Top. Cell. Reg. 7,* 51–86, 1973.
7. Dixon, M., Massey, V., and Webb, E. C. Unpublished observation cited in [9], p. 80.
8. Dixon, M., and Thurlow, W. Studies on xanthine oxidase II. The dynamics of the oxidase system. *Biochem. J. 18,* 976–988, 1924.
9. Dixon, M., and Webb, E. C. *Enzymes*. Longman, London, 1964.
10. Esnouf, M. P., and MacFarlane, R. G. Enzymology and the blood clotting mechanism. *Advan. Enzymol. 30,* 255–315, 1968.
11. Goldbeter, A., and Segel, L. A. Unified mechanism for delay and oscillation of cyclic AMP in *Dictyostelium discoideum*. *Proc. Natl. Acad. Sci. 74,* 1543–1547, 1977.
12. Goodwin, B. C. Oscillatory behavior in enzymatic control processes. In: *Advances in Enzyme Regulation,* G. Weber (ed.), vol. 3, 425–438, 1965.
13. Grodins, F. S. *Control Theory and Biological Systems*. Columbia University Press, New York, 1963.
14. Guyton, A. C. *Textbook of Medical Physiology*. Saunders, Philadelphia, 1971.
15. Guyton, A. C., and Crowell, J. W. Cardiac deterioration in shock: I. Its progressive nature. In: *Shock,* S. G. Hershey (ed.). Little, Brown, Boston, 1–12, 1964.
16. Guyton, A. C., Taylor, A. E., and Granger, H. J. *Circulatory Physiology. I. Cardiac Output and Its Regulation*. Saunders, Philadelphia, 1973. *II. Dynamics and Control of the Body Fluids,* Saunders, Philadelphia, 1975.
17. Haldane, J. B. S. *Enzymes*. Longmans, London, 1930.

18. Hoffman, K. The adaptive significance of biological rhythms corresponding to geophysical cycles. In: *The Molecular Basis of Circadian Rhythms*. J. W. Hastings and H. G. Schweiger (eds.). Dahlem Konferenzen, Berlin, 63–76, 1976.
19. Horrobin, D. F. *Principles of Biological Control*. Medical Technical Publishing, Aylesbury, 1970.
20. Iberall, A. S., and Guyton, A. C. (eds.). *Regulation and Control in Physiological Systems*. Instrument Society of America, Pittsburgh, 1973.
21. Kellermeyer, R. W., and Breckenridge, R. T. The inflammatory process in acute gouty arthritis, I. Activation of Hageman factor by sodium urate crystals. *J. Lab Clin. Med. 65,* 307–315, 1965.
22. King-Smith, E. A., and Morley, A. Computer simulation of granulopoiesis, *Blood 36,* 254–262, 1970.
23. Lin, C. C., and Segel, L. A. *Mathematics Applied to Deterministic Problems in the Natural Sciences*. Macmillan, New York, 1974.
24. Mandelstam, J. Coordination: Induction, repression and feedback inhibition. In: *Biochemistry of Bacterial Growth,* J. Mandelstam and K. McQuillen (eds.). Blackwell Scientific Publications, Oxford, 1968.
25. Massey, V. Studies on fumerase. 2. The effects of inorganic anions on fumerase activity. *Biochem. J. 53,* 67–71, 1953; and 4. The effects of inhibitors on fumerase activity. *Biochem. J. 55,* 172–177, 1953.
26. Mees, A. I., and Rapp, P. E. Periodic metabolic systems. Oscillations in multiple-loop negative feedback biochemical control networks. *J. Math. Biology 5,* 99–114, 1978.
27. Milhorn, H. T. *Applications of Control Theory to Physiological Systems*. Saunders, Philadelphia, 1966.
28. Milsum, J. H. *Biological Control System Analysis*. McGraw-Hill, New York, 1966.
29. Monod, J., Changeux, J. P., and Jacob, F. Allosteric proteins and cellular control systems. *J. Mol. Biol. 6,* 306–329, 1963.
30. Müller-Eberhard, H. J. Complement. *Ann. Rev. Biochem. 44,* 697–724, 1975.
31. Murray, D. R. P. Inhibition of esterases by excess substrate. *Biochem. J. 24,* 1890–1896, 1930.
32. Rapp, P. E. Biological applications of control theory. In *Continuum Models in Molecular and Cellular Biology,* I. A. Segel (ed.). Cambridge University Press, New York, 1984.
33. Savageau, M. A. *Biochemical System Analysis: A Study of Function and Design in Molecular Biology*. Benjamin/Addison Wesley Advanced Book Program, New York, 1976.
34. Seegmiller, J. E., Laster, L., and Howell, R. R. Biochemistry of uric acid and its relation to gout. *New Engl. J. Med. 268,* 712–716, 764–773, 821–827, 1963.
35. Stolwijk, J., and Hardy, J. Temperature regulation in man—A theoretical study. *Pflugers Archiv. 291,* 129–162, 1966.
36. Talbot, S. A., and Gessner, U. *Systems Physiology*. Wiley Interscience, New York, 1973.
37. Toates, F. *Control Theory in Biology and Experimental Psychology*. Hutchinson Education, London, 1975.
38. Umbarger, J. E. Evidence for a negative-feedback mechanism in the biosynthesis of isoleucine. *Science 123,* 848, 1956.
39. Wang, J. H. C., Teo, T. S., and Wang, T. H. Hysteretic substrate activation of bovine c-AMP phosphodiesterase. *Biochem. Biophys. Res. Commun. 46,* 1306–1311, 1972.
40. Watson, J. D. *Molecular Biology of the Gene,* 2nd ed. W. A. Benjamin, Menlo Park, CA, 1976.

3

Ordinary Difference Equations

3.1 INTRODUCTION

The introductory chapters contained examples of biological phenomena that were described by ordinary recursive or difference equations. Such equations involve a single dependent variable evaluated at different values of the single independent variable, usually time. In this chapter ordinary difference equations are defined formally. It is shown that, under appropriate conditions, certain types of difference equations always have a unique solution. Methods for finding the solutions of linear difference equations are presented. Finally, a special form of a difference equation occurring in population studies is analyzed graphically.

The use of difference equations instead of differential equations to study complex systems is gaining in popularity. The reason is that difference equations can be solved directly via a digital computer, while an approximating algorithm must be used or developed before differential equations can be solved numerically.

3.2 DEFINITIONS CONCERNING DIFFERENCE EQUATIONS

The following definitions and facts are required in order to understand difference equations.

Definition 1. Let $y(k)$ be a real-valued function* defined for $k = 0, 1, 2, 3, \ldots$. The variable k will be called the *index;* $y(k)$ are called *unknowns*. The *nth-order difference operator* Δ^n is defined recursively by

$$\Delta^0 y(k) = y(k)$$

$$\Delta^1 y(k) = \Delta y(k) = y(k+1) - y(k)$$

$$\Delta^n y(k) = \Delta[\Delta^{n-1} y(k)] \qquad (n = 1, 2, 3, \ldots)$$

The proof of the following lemma by mathematical induction is omitted.

*In many books the notation y_k is used instead of $y(k)$.

Lemma 1

$$\Delta y(k) = y(k+1) - y(k)$$
$$\Delta^2 y(k) = y(k+2) - 2y(k+1) + y(k)$$

and, in general,

$$\Delta^n y(k) = y(k+n) - ny(k+n-1) + \frac{n(n-1)}{1\cdot 2}y(k+n-2) + \cdots$$
$$+ (-1)^r \frac{n(n-1)(n-2)\cdots(n-r+1)}{1\cdot 2\cdot 3\cdots r}y(k+n-r) + \cdots + (-1)^n y(k)$$

Definition 2. The operator E is defined by

$$Ey(k) = y(k+1)$$

The proofs of the next two lemmas follow directly from Definition 2.

Lemma 2

$$E^n y(k) = y(k+n)$$

Lemma 3

(a) $\Delta = E - 1$
(b) $E = 1 + \Delta$

Corollary

$$y(k+n) = (1+\Delta)^n y(k) = (1+\Delta)^{2n} y(k-n)$$

where

$$(1+\Delta)^p = 1 + p\Delta + \frac{p(p-1)}{1\cdot 2}\Delta^2 + \cdots$$
$$+ \frac{p(p-1)(p-2)\cdots(p-r+1)}{1\cdot 2\cdots r}\Delta^r + \cdots + \Delta^p$$

Definition 3. A *difference* (or *recurrence*) *equation* is an equation relating the values of $y(k+n), y(k+n-1), \ldots, y(k), y(k-1), \ldots, y(k-m)$, where m and n are fixed positive integers. In other words,

$$f[y(k+n), y(k+n-1), \ldots, y(k), y(k-1), \ldots, y(k-m), k] = 0$$

where f is a real-valued function of $m + n + 2$ variables. The *order* of the difference equation is the difference between the highest and lowest indices that appear in the equation.

The reason for the name "difference equation" should be apparent from the corollary to Lemma 3. The unknowns whose indices are higher than the lowest index can be replaced by difference operators, operating on the unknowns with the least index. For example,

$$y(k+2) + 3y(k+1) - y(k) = 0$$

can be written in the form

$$[(1 + \Delta)^2 + 3(1 + \Delta) - 1]y(k) = 0 \quad \text{or} \quad (\Delta^2 + 5\Delta + 3)y(k) = 0$$

In many practical applications, events of interest occur at $t = 0, T, 2T, 3T, \ldots$. However, a simple change of the dependent variable will allow the standard set of time points $k = 0, 1, 2, 3, \ldots$ to be used.

Definition 4. A difference equation is *linear* if it has the form

$$a_n(k)y(k + n) + a_{n-1}(k)y(k + n - 1) + \cdots + a_1(k)y(k + 1) + a_0(k)y(k) = g(k) \tag{1}$$

for some given functions $a_i(k)$, $i = 0, 1, 2, \ldots, n$ and $g(k)$; otherwise the equation is *nonlinear*. The $a_i(k)$'s are called the *coefficients* of the equation. The function $g(k)$ is called the *forcing term,* the *driving term,* or simply the *right-hand side.*

Definition 5. A linear difference equation is said to have *constant coefficients,* be *time-invariant,* or be *stationary* if the coefficients a_i do not depend on k.
A similar definition holds for nonlinear equations.

Definition 6. The general linear difference Eq. (1) is said to be *homogeneous* if, and only if, $g(k) = 0$ for all k for which the equation is defined; otherwise it is *nonhomogeneous.*

Example 1

1. $2y(k + 3) - y(k + 2) + y(k - 2) = 0$ is a fifth-order linear difference equation with constant coefficients.
2. $4\Delta^3 y(k) - 2\Delta y(k) + 8y(k) = 0$ is a third-order linear difference equation with constant coefficients.
3. $(k + 1)y(k + 1) - ky(k) = 0$ is a first-order difference equation with non-constant coefficients. It is not time-invariant and is nonstationary.
4. $3y(k + 2) - 4y(k + 1) + y(k) = 8$ is a nonhomogeneous linear difference equation with constant coefficients. The equations appearing in 1–3 are homogeneous.
5. $[y(k + 2)]^2 + y(k + 1) - 2\sqrt{y(k - 1)y(k)} = 0$ is a nonlinear difference equation of order 3.

Definition 7. A *solution* of a difference equation is a function $y(k)$ that reduces the equation to an identity.

Example 2

1. Consider the difference equation

$$y(k + 1) = 3y(k) \qquad (k = 0, 1, 2, \ldots)$$

It has the solution $y(k) = 3^k$. To check this, note that $y(k + 1) = 3^{k+1}$ and so the given equation is reduced to the identity $3^{k+1} = 33^k$. Similarly, it can be shown that $y(k) = A3^k$ is a solution for any constant A.

2. The linear difference equation

$$(k + 1)y(k + 1) - ky(k) = 2 \qquad (k = 1, 2, 3, \ldots)$$

has a solution

$$y(k) = 2 + \frac{1}{k} = \frac{2k+1}{k}$$

since this function reduces the given difference equation to the identity

$$(k+1)\left(\frac{2k+3}{k+1}\right) - k\left(\frac{2k+1}{k}\right) = 2$$

It is easily seen that $y(k) = 2 + A/k$ is a solution for any constant A.

3. The nonlinear difference equation

$$[2y(k+1)]^2 + [y(k)]^2 = -1$$

has no solution $y(k)$, a real-valued function, since the right-hand side is non-negative while the left-hand side is negative.

An alternative definition of a solution to a difference is a sequence of numbers. The elements of this sequence must satisfy certain relations imposed by the difference equation. For example,

$$y(k+1) = 3y(k) \qquad (k = 0, 1, 2, \ldots)$$

has the solution

$$1, 3, 9, 27, 81, \ldots$$

Here $y(0) = 1$ and the relation between the elements of the sequence are

$$\begin{aligned} y(1) &= 3y(0) \\ y(2) &= 3y(1) \\ y(3) &= 3y(2) \\ &\vdots \end{aligned}$$

These relations are just the infinite set of equations equivalent to the given difference equation. This solution can be written $y(k) = 3^k$. In general, a solution of a difference equation is not easily expressible as a simple function of k. However, it can always be written as a sequence. Therefore, regarding a solution as a sequence is often convenient. Of course, the two viewpoints of a solution as a sequence and as function of k are equivalent.

The difference equations appearing in part 3 of Example 2 show that a difference equation need not have a solution. Parts 1 and 2 of Example 2 illustrate that if a solution exists, it need not be unique. The uniqueness and existence of solutions will be investigated in the next section.

EXERCISES

1. Complete the following table using the indicated abbreviations.

L: linear   S: stationary   H: homogeneous
NL: nonlinear   NS: nonstationary   NH: nonhomogeneous

Difference equation	L or NL	S or NS	H or NH	Order
$y(k+2) + 4y(k-1) = k^{-1}$				
$y(k+2) + [(y(k+1)]^{-1} = ky(k)$				
$y(k+4) + y(k+2) + 3y(k-1) = k^3$				
$y(k+4) + [y(k+2)]^2 = 0$				
$y(k-1) + \cos y(k) = 5$				
$y(k-2) + k^3 y(k-3) = 0$				

2. Show that the following functions are solutions of the given difference equations (c is a constant).
(a) $y(k) = c(-2)^k; \quad y(k) + 2y(k-1) = 0$
(b) $y(k) = k(k+2); \quad y(k+1) - y(k-1) = 3(k+1)$
(c) $y(k) = (k+c)^{-2}; \quad y(k+1) = y(k)\{1 + [y(k)]^{1/2}\}^{-2}$

3. Show that the following sequences are solutions of the given difference equations and find the next two elements in the sequence.
(a) $y(k) - 6y(k-1) + 11y(k-2) - 6y(k-3) = 0$

$$2, 5, 13, 35, 97, \ldots$$

(b) $y(k+1) - y(k-1) = 4(k+1)$

$$2, 3, 14, 19, 34, \ldots$$

(c) $y(k-2) + y(k+1) - 2y(k) = 0$

$$2, -1, 5, -8, 9, \ldots$$

4. Show that the solution of the difference equation

$$y(k) - y(k-1) = b(k)$$

is

$$y(k) = y(0) + \sum_{i=1}^{k} b(i)$$

5. (a) Show that the solution of the difference equation

$$y(k+1) = b(k)y(k)$$

is

$$\begin{aligned} y(k) &= b(k-1)b(k-2)\cdots b(0)y(0) \\ &= b(k-1)b(k-2)\cdots b(1)y(1) \end{aligned}$$

(b) Deduce that the solution of the difference equation

$$a(k+1)y(k+1) = a(k)y(k)$$

is

$$y(k) = \frac{a(0)y(0)}{a(k)}$$

6. Show that the difference equation

$$y(k) - ay(k-1) = b(k)$$

has the solution

$$y(k) = y(0)a^k + \sum_{i=1}^{k} a^{k-i}b(i)$$

3.3 EXISTENCE AND UNIQUENESS OF SOLUTIONS OF DIFFERENCE EQUATIONS

Theorem 1 (Existence and Uniqueness). Let a difference equation of the form

$$y(k+n) = f[y(k+n-1), y(k+n-2), \ldots, y(k), k] \tag{2}$$

where f is an arbitrary real-valued function of n variables, be defined over a finite or infinite sequence of consecutive integral values of k ($k = k_0, k_0 + 1, k_0 + 2, \ldots$). The equation has one and only one solution corresponding to each arbitrary specification of the so-called *n initial values* $y(k_0), y(k_0 + 1), \ldots, y(k_0 + n - 1)$.

Proof. If the specified values $y(k_0), y(k_0 + 1), \ldots, y(k_0 + n - 1)$ are substituted in the right-hand side of (2), $y(k_0 + n)$ is uniquely determined. Then the difference equation (2) with $k = k_0 + 1$ can be solved for $y(k_0 + n - 1)$ since $y(k_0 + n)$ is known. The procedure can be repeated to find $y(k_0 + n + 2)$, and so forth for all consecutive values of k.

Corollary. The linear difference equation

$$y(k+n) + a_{n-1}(k)y(k+n-1) + \cdots + a_0(k)y(k) = g(k)$$

has a unique solution given the n initial values:

$$y(k_0), y(k_0+1), \ldots, y(k_0+n-1)$$

Proof. Apply Theorem 1 with

$$f(x_1, x_2, \ldots, x_n) = -[a_{n-1}(k)x_1 + a_{n-2}(k)x_2 + \cdots + a_0(k)x_n] + g$$

Example 3 (Discrete Cell Growth by Binary Fission)

The discrete version of exponential growth, considered in Example 3 of Chapter 1, will now be formulated. A newborn cell divides after a certain maturation period, T, called the *doubling time*. For simplicity, it will be assumed that T is a constant that is the same for each cell. The size of the population, $Y(kT) = y(k)$, is measured at times $k = 0, 1, 2, \ldots$. It is also assumed that only a fixed proportion, γ, of newborn cells survive to reproduce again.

Under the assumption that the population began with $y(0)$ newborn cells, the size of the population at time k, $y(k)$, will be found. Let $n(k)$ be the number of newborn cells at time k. Since each parent cell produces two daughter cells in the time interval T,

$$n(k) = 2y(k - 1)$$

However, the population at time k is

$$y(k) = \gamma n(k)$$

since only a fixed proportion γ of the newborn cells survive. It follows that the difference equation for the population size is

$$y(k) = ay(k - 1) \qquad (a = 2\gamma)$$

for $k = 1, 2, 3, \ldots$. Setting $k = 1, 2, 3, \ldots$ in this last difference equation yields

$$y(1) = ay(0)$$
$$y(2) = ay(1) = a^2y(0)$$
$$y(3) = ay(2) = a^3y(0)$$

Hence, its solution is

$$y(k) = a^k y(0)$$

If $a > 1$, $a = 1$, $a < 1$ the population increases, does not change, decreases, respectively.

It should be noted that no restrictions are placed on the real-valued function f: It can be nonlinear. The method of solution of the difference equation (2) given in the proof is numerical and independent of the form of f. Moreover, it is particularly suited for implementation on a digital computer. In some situations an analytic form of the solution may be guessed by induction. In general, however, it is difficult to find an analytic solution of nonlinear equations. For example, consider a population of size $y(k)$ described by a nonlinear difference equation, the *logistic difference equation,*

$$y(k) - y(k - 1) = ay(k - 1) - b[y(k - 1)]^2 \qquad (3)$$

where a and b are positive constants. This type of growth is the discrete analog of the logistic growth law discussed in Example 2 of Chapter 2 and can be derived in a similar manner. Setting $k = 1, 2$ in (3) yields

$$\begin{aligned} y(1) &= (a + 1)y(0) - by(0)^2 \\ y(2) &= (a + 1)y(1) - by(1)^2 \\ &= (a + 1)^2y(0) - (a + 1)by(0)^2 - b[(a + 1)y(0) - by(0)^2]^2 \end{aligned}$$

Continuing in this manner, the values of $y(3)$, $y(4)$, . . . can be evaluated in terms of $y(0)$. However, an analytic formula is not obvious. There are, however, methods of finding analytic solutions of linear equations with constant coefficients. Some of these methods will be presented in Section 3.4.

The following example is another application of Theorem 1 involving two initial conditions.

Example 4

Consider the second-order difference equation

$$y(k+2) = 3y(k) \qquad (k = 0, 1, 2, \ldots)$$

Setting $k = 0, 1, 2, 3, 4, 5$ in the difference equation gives

$$y(2) = 3y(0)$$
$$y(3) = 3y(1)$$
$$y(4) = 3y(2) = 3^2 y(0)$$
$$y(5) = 3y(3) = 3^2 y(1)$$
$$y(6) = 3y(4) = 3^3 y(0)$$
$$y(7) = 3y(5) = 3^3 y(1)$$

It can be seen that successive values of $y(k)$ are uniquely specified once $y(0)$ and $y(1)$ are fixed. The astute reader may guess that the analytic form of the solution is

$$y(k) = 3^{[k/2]}\left(\left(\frac{y(0) + y(1)}{2}\right) + (-1)^k\left(\frac{y(0) - y(1)}{2}\right)\right)$$

where $[x]$ denotes the largest interest in x.

The logistic growth model, described by (3), predicts an equilibrium size for the population. By definition, a *stationary* or *equilibrium size* of a population means that the size of the population does not change with time. Algebraically this means that

$$y(k) = y(k-1)$$

Hence, from (3),

$$0 = ay(k-1) - b[y(k-1)]^2 = by(k-1)\left[\frac{a}{b} - y(k-1)\right]$$

This last equation has two solutions. One solution is $y(k-1) = 0$, which corresponds to the population becoming extinct. The other solution occurs when the population approaches a finite nonzero size,

$$y(k-1) = y_e = \frac{a}{b}$$

The equilibrium values may not be reached in a finite time. Further insight into the behavior of the solution of (3) will be obtained in Section 3.6 by a graphical method. An elementary investigation of the behavior of the solution near equilibrium, as in Exercise 11, can be found in [3]. Such problems will also be discussed in Volume II, Chapter 4.

EXERCISES

1. Solve the following difference equations with the given initial conditions by finding the first five elements of the solution sequence:

(a) $y(k) + 4y(-1) = 0$; $y(0) = -1$
(b) $y(k) - 2y(k-1) = k^2$; $y(0) = 2$
(c) $y(k+1) + \sin y(k) = k(k-1)$; $y(0) = \pi/2$
(d) $y(k+2) + 4y(k+1) + y(k) - 2y(k-1) = 0$; $y(0) = 1$, $y(1) = 0$, $y(2) = -1$.

2. A forest ranger wants to reforest a park. He writes a letter to two of his friends asking each of them to contribute the price of a tree and to write a similar letter to two of their friends. The recipients of their letters belong to generation 1, and so on.
(a) Assume that everyone contacted donates a tree. Set up a difference equation for $t(k)$, the total number of trees donated by generation k.
(b) Find the total number of trees contributed by 10 generations, that is, $t(1) + t(2) + \cdots + t(10)$.

3. **(a)** Express $y(2)$, $y(3)$, and $y(4)$, where

$$y(k+2) - 3y(k+1) + 2y(k) = 0 \qquad (k = 0, 1, 2, \ldots)$$

in terms of $y(0)$ and $y(1)$.
(b) Verify that

$$y(k) = (2 - 2^k)y(0) + (2^k - 1)y(1)$$

is a solution of the difference equation.

4. **(a)** Let r be the growth rate of an animal population between year 0 and year 1. If the growth rate doubles each year, what is the growth rate between year k and year $k + 1$?
(b) Write down a difference equation describing the size of the population in year $k + 1$ in terms of the population in year k.
(c) If the size of the population at year 0 is 1000 and the size at year 1 is 1100, what is the size in year k?

5. **(a)** An animal population increases by 25% per year. Its size in the year $k = 0$ is 1000. If 400 of the animals are trapped each year, find the size of the population after k years.
(b) After how many years is the population size reduced to below 350?

6. **(a)** A certain population of birds is decreasing as a result of competition for limited resources. The decrease in population size between the $(k-1)$th year and the kth year is equal to one-third of the size at the $(k-2)$th year. Find a difference equation for the size of the population.
(b) If the original population sizes were 15,000 in year 0 and 14,500 in year 1, what is the size of the population in the fifth year?

7. **(a)** An insect population is increasing in such a way that its growth in any year is twice its growth in the previous year. What is the difference equation describing the population size?
(b) If the size of the population in years 0 and 1 is 1000 and 1100, respectively, find the size in year 4.

8. Assume that each pair of adult rabbits in a rabbit colony produces a pair (one male and one female) every month and that the rabbits start to produce young when they are 2 months old. Furthermore, once paired, a pair of rabbits remain true to one another and produce offspring indefinitely. Set up a difference equation for
(a) $c(k)$, the number of pairs of children after k months;
(b) $a(k)$, the number of pairs of adults after k months;
(c) $t(k)$, the total number of pairs of rabbits after k months.
(d) If the colony beings with one adult pair, find $c(6)$, $a(6)$, and $t(6)$. (The generated sequences are called Fibonacci sequences.)

9. Repeat Exercise 8 under the assumption that each month every pair of rabbits produces three pairs (three males and three females) of young rabbits.

10. A population has an initial size of 100 and increases by 30%.
    (a) What is the difference equation describing the population size?
    (b) Solve the difference equation in (a).
    (c) In how many years will the population size exceed 1000?

11. Consider the logistic difference equation (3). Show that if $y(k-1) > a/b$, $y(k-1) < a/b$, then the population size decreases, increases, respectively, in $[k-1, k]$ (c.f. Fig. 2.3).

12. The growth of a population of blue whales is described by the logistic difference equation

$$y(k) - y(k-1) = [0.4 - 0.0002y(k-1)]y(k-1)$$

    (a) What is the equilibrium population size?
    (b) Find $y(10)$ if $y(0) = 1000$.
    (c) Find $y(10)$ if $y(0) = 3000$.

3.4 LINEAR DIFFERENCE EQUATIONS

Recall that a difference equation is linear if it has the form

$$a_n(k)y(k+n) + a_{n-1}(k)y(k+n-1) + \cdots + a_1(k)y(k+1) + a_0(k)y(k) = g(k) \tag{4}$$

Linearity makes it possible to express the general solution of (4) in terms of certain solutions of the corresponding homogeneous equation,

$$a_n(k)y(k+n) + a_{n-1}(k)y(k+n-1) + \cdots + a_1(k)y(k+1) + a_0(k)y(k) = 0 \tag{5}$$

and a particular solution of the nonhomogeneous equation (4). In general, only a numerical solution can be obtained. However, if the coefficients are constants, an analytic solution can be found. Linear equations with constant coefficients will be studied in the next section.

Theorem 2. Assume that $a_n(k) \neq 0$ for any k $(k = k_0, k_0 + 1, \ldots)$. Let $y(k)$ be a solution of the linear difference equation (4). Let $y_p(k)$ be a particular solution of (4) and $y_h(x)$ a solution of the homogeneous equation (5) satisfying the initial conditions

$$y_h(k) = y(k) - y_p(k) \qquad (k = k_0, k_0 + 1, \ldots, k_0 + n - 1) \tag{6}$$

Then

$$y(k) = y_h(k) + y_p(k) \qquad (k = k_0, k_0 + 1, \ldots)$$

Proof. According to the hypothesis, $y(k)$ and $y_h(k) + y_p(k)$ satisfy the same initial conditions. Since

$$\begin{aligned} a_n(k)[y_h(k) + y_p(k)] + \cdots + a_0(k)[y_h(k) + y_p(k)] \\ &= [a_n(k)y_h(k) + \cdots + a_0(k)y_h(k)] + [a_n(k)y_p(k) + \cdots + a_0(k)y_p(k)] \\ &= 0 + g(k) = g(k) \end{aligned}$$

$y_h(k) + y_p(k)$ is a solution of (4). According to Theorem 1, the difference equation (4) has a unique solution. Hence, $y(k) = y_h(k) + y_p(k)$.

Another justification of the adjective *linear* appears in the course of the proof of Theorem 2. Let

$$L(y) = a_n(k)y(k + n) + \cdots + a_0(k)y(k)$$

For example, if

$$L(y) = k^2y(k + 2) + y(k + 1) + ky(k)$$

then for $y(k) = k^2$,

$$L(k^2) = k^2(k + 2)^2 + (k + 1)^2 + k^3$$

It was shown that

$$L[y_1(k) + y_2(k)] = L[y_1(k)] + L[y_2(k)] \tag{7}$$

The reader should verify that

$$L[cy(k)] = cL[y(k)] \tag{8}$$

where c is an arbitrary constant. It follows from (7) and (8) that if c_1 and c_2 are constants, then

$$L[c_1y_1(k) + c_2y_2(k)] = c_1L[y_1(k)] + c_2L[y_2(k)] \tag{9}$$

In other words, the operator L acts linearly on linear combinations of functions.

Theorem 2 leads to a numerical method of solving the nonhomogeneous equation (4). The procedure requires two steps:

1. Find a particular solution $y_p(k)$ of the nonhomogeneous equation. This can be done by arbitrarily specifying the initial values: $y_p(k_0), \ldots, y_p(k_0 + n - 1)$.
2. Solve the homogeneous equation (5) and choose y_h to satisfy the initial conditions given by (6), where $y(k_0), \ldots, y(k_0 + n - 1)$ are the given initial conditions.

Then the general solution is $y_h + y_p$.

In general, the numerical method of Theorem 1 should be used, since the solution of only one difference equation is involved. The procedure in Theorem 2 is easily applicable for linear difference equations with constant coefficients or if a particular solution is obvious or if certain solutions of the homogeneous equation (5) are known. The subsequent definitions and theorems in this section will be concerned with the solutions of the corresponding homogeneous equation.

Example 5

Consider the linear difference equation

$$ky(k) - (k - 1)y(k - 1) = 1 \qquad (k = 2, 3, 4, \ldots)$$

with the initial condition $y(1) = 3$. By inspection, one solution is $y_p(k) = 1$. The corresponding homogeneous equation is

$$ky(k) - (k-1)y(k-1) = 0$$

This equation is of the same form as the difference equation in Exercise 5 of Section 3.2 with $b(k-1) = (k-1)/k$. Hence, its solution is

$$y_h(k) = \left(\frac{k-1}{k}\right)\left(\frac{k-2}{k-1}\right) \cdots \left(\frac{1}{2}\right)y_h(1) = \frac{1}{k}y_h(1)$$

According to (6), $y_h(1) = 3 - 1$, and so $y_h(1) = 2$.

From Theorem 2, the general solution is

$$y(k) = y_p(k) + y_h(k) = 1 + \frac{2}{k}$$

Theorem 3. If $z_1(k), z_2(k), \ldots, z_m(k)$ are all solutions to the homogeneous equation (5), then any linear combination

$$z(k) = c_1z_1(k) + c_2z_2(k) + \cdots + c_mz_m(k)$$

where $c_1, c_2, \ldots, c_m$ are arbitrary constants, is also a solution of (5).

Proof. The result follows by repeated application of (9).

Theorem 3 shows that a large collection of solutions to the homogeneous equation can be derived from a few known solutions. This raises the question of whether it is possible to find a special finite number of solutions that can be used to generate all other solutions. This problem will be considered next.

The situation in which $a_n(k) \neq 0$ will be considered. Equation (5) can be divided by $a_n(k)$. Without loss of generality, it can be assumed that the homogeneous equation under investigation is

$$z(k+n) + b_{n-1}(k)z(k+n-1) + \cdots + b_0(k)z(k) = 0 \tag{10}$$

for $k = 0, 1, 2, \ldots$.

The following definition is required to describe the finite number of solutions that generate all solutions of (10).

Definition 8. The functions $z_1(k), z_2(k), \ldots, z_m(k)$, defined for $k = 0, 1, 2, \ldots$, are *linearly independent* if it is impossible to find a solution

$$c_1z_1(k) + c_2z_2(k) + \cdots + c_mz_m(k) = 0$$

valid for all $k = 0, 1, 2, \ldots$ except by setting $c_1 = c_2 = \cdots = c_m = 0$. If it is possible to find such a relation, the set of functions is said to be *linearly dependent*.

Example 6

(a) The functions $z_1(k) = 1$, $z_2(k) = 2^k$, $z_3(k) = 2^{k+1} - 3$ are linearly dependent because

$$3z_1(k) - 2z_2(k) + z_3(k) = 0$$

(b) The functions $z_1(k) = 1$, $z_2(k) = 2^k$, $z_3(k) = 3^k$ are linearly independent because

$$c_1 + c_22^k + c_33^k = 0$$

implies (by setting $k = 0, 1, 2$) that

$$c_1 + c_2 + c_3 = 0$$
$$c_1 + 2c_2 + 3c_3 = 0$$
$$c_1 + 2^2c_2 + 3^2c_3 = 0$$

This last set of equations has the solution $c_1 = c_2 = c_3 = 0$.

Definition 9. A *fundamental set* of solutions of the nth-order linear homogeneous equation (10) is any set of n linearly independent solutions.

Lemma 4. There exists a fundamental set of solutions of (10).

Proof. The fundamental set of solutions $f_1(k), f_2(k), \ldots, f_n(k)$ can be constructed as follows. Let $f_1(k)$ be the solution corresponding to the initial conditions

$$f_1(0) = 1, f_1(1) = 0, f_1(2) = 0, \ldots, f_1(n-1) = 0$$

In general, let $f_i(k)$ be the solution corresponding to the initial conditions that are all zero except $f_i(i-1)$, which equals 1. These solutions can be generated by solving the difference equation (10) n times, once for each of the special sets of initial conditions.

It remains to show that these solutions are linearly independent. Suppose that

$$c_1f_1(k) + c_2f_2(k) + \cdots + c_nf_n(k) = 0 \tag{11}$$

Setting $k = 0$ in (11) yields $c_1 = 0$, since $f_i(1) \neq 0$ for $i = 1$. Similarly, if $k = 1$, (11) reduces to $c_2 = 0$. In general, if $k = i - 1$, (11) becomes $c_i = 0$, since $f_i(k) = 1$ for $k = i - 1$ and 0 otherwise. Therefore, the only linear combination that is identically zero is one having all zero coefficients.

Theorem 4. Let $z_1(k), \ldots, z_n(k)$ be a fundamental set of solutions of the homogeneous equation (10). Then any solution $z(k)$ to (10) can be expressed as a linear combination

$$z(k) = c_1z_1(k) + c_2z_2(k) + \cdots + c_nz_n(k)$$

for some constants $c_1, c_2, \ldots, c_n$.

Proof. It will be shown in Volume II, Chapter 1 that any $n + 1$ solutions of (10) are linearly dependent. Hence, $f_i, z_1, \ldots, z_n$ are linearly dependent, and so

$$f_i(k) = d_{i1}z_1(k) + d_{i2}z_2(k) + \cdots + d_{in}z_n(k) \qquad (i = 1, 2, \ldots, n) \tag{12}$$

for the fundamental set of solutions $f_i(k)$ constructed in Lemma 4, where the d_{i1}, $d_{i2}, \ldots, d_{in}$ for $i = 1, 2, \ldots, n$ are constants that are not all zero.

Corresponding to the initial values of $z(k)$, let

$$a_i = z(i-1) \qquad (i = 1, 2, \ldots, n)$$

Now consider the special solution

$$y(k) = a_1f_1(k) + a_2f_2(k) + \cdots + a_nf_n(k)$$

This solution has the same n initial conditions as $z(k)$. Therefore, by Theorem 1,

$y(k) = z(k)$. Hence $z(k)$ is a linear combination of the $f_1(k), \ldots, f_n(k)$ and so a linear combination of $z_1(k), \ldots, z_n(k)$ from (12).

Example 7

Consider the linear homogeneous equation

$$z(k+2) - 2z(k+1) + z(k) = 0$$

By inspection it is clear that the two functions

$$z_1(k) = 1 \quad \text{and} \quad z_2(k) = k$$

are both solutions. These two solutions are linearly independent. Since the given equation is of order 2, the two solutions form a fundamental set. Theorem 4 shows that the general solution of the given equation is

$$z(k) = c + dk$$

where c and d are arbitrary constants.

Theorems 2 and 4 imply that a general solution y of the nonhomogeneous equation (4), given a particular solution $\overline{y}$ of the nonhomogeneous equation (4) and a fundamental set of solutions $y_1, y_2, \ldots, y_n$ of the homogeneous equation (5), is

$$y(k) = \overline{y}(k) + c_1 y_1(k) + c_2 y_2(k) + \cdots + c_n y_n(k)$$

where the constants $c_1, c_2, \ldots, c_n$ are chosen to satisfy the given initial conditions. The next example illustrates this general result.

Example 8

Find the solution of the linear nonhomogeneous equation

$$y(k+2) - 2y(k+1) + y(k) = 2$$

such that $y(0) = 1$ and $y(1) = 3$.

The corresponding linear equation was studied in Example 7. The reader can verify that $\overline{y}(k) = k^2$ is a particular solution of the given nonhomogeneous equation. Hence, the general solution is

$$y(k) = k^2 + c + dk$$

The constants c and d are determined from their initial conditions. Setting $k = 0, 1$ in the general formula for $y(k)$ gives

$$1 = c$$

$$3 = 1 + c + d$$

The solution of these equations is $c = d = 1$. Hence

$$y(k) = 1 + k + k^2$$

EXERCISES

1. Consider the difference equation

$$y(k)y(k+1) + y(k+2) = k^2 + k + 1$$

with the initial conditions $y(0) = 1$ and $y(1) = 2$.

(a) Find a particular solution sequence $y_p(k)$ for $k = 2, 3, 4$ with initial conditions $y_p(0) = y_p(1) = 3$.

(b) Find a homogeneous solution sequence $y_h(k)$ for $k = 2, 3, 4$ with initial conditions $y_h(0) = -2$ and $y_h(1) = -1$.

(c) What are the elements $y(k)$ of the general solution sequence for $k = 2, 3, 4$?

2. Verify directly that the difference equation

$$y(k + 2) - 4y(k) = 0$$

has the solutions

(a) $y(k) = 2^k$

(b) $y(k) = (-2)^k$

(c) $y(k) = a2^k + b(-2)^k$, where a and b are constants

3. Verify directly that the difference equation

$$y(k + 3) - 10y(k + 2) + 31y(k + 1) - 30y(k) = 0$$

has the solutions $y(k) = 2^k, 3^k, 5^k$ and $a2^k + b3^k + c5^k$, where a, b, and c are constants.

4. (a) Show that a particular solution of the difference equation

$$y(k + 1) - y(k - 1) = 4(k + 1)$$

is $y_p(k) = k(k + 2)$.

(b) Given that the general solution of the corresponding homogeneous equation is $y_h(k) = a + (-1)^k b$, where a and b are constants, find the solution $y(k)$ of the difference equation satisfying $y(0) = 2$ and $y(1) = 1$.

5. Which of the following sets of functions are linearly independent? Why?

(a) $z_1 = 2^k$, $z_2 = 3^k$

(b) $z_1 = k$, $z_2 = k(k - 1)$, $z_3 = k(k - 1)(k - 2)$

(c) $z_1 = 2$, $z_2 = 4^k$, $z_3 = 4^{k+1} - 10$

(d) $z_1 = 2^k$, $z_2 = 3^k$, $z_3 = 2^{k+1} + 5 \cdot 3^{k+2}$

6. (a) Verify that the functions $z_1 = 2^k$, $z_2 = k2^k$, and $z_3 = k^2 2^k$ are linearly independent.

(b) Show that the above functions are a fundamental set for

$$y(k + 3) - 6y(k + 2) + 12y(k + 1) - 8y(k) = 0$$

7. (a) Verify that the functions $z_1 = k$, $z_2 = k(k + 1)$, and $z_3 = k(k + 1)(k + 2)$ are linearly independent.

(b) Prove that the above functions are a fundamental set for the difference equation

$$y(k + 3) - 3y(k + 2) + 3y(k + 1) - y(k) = 0$$

8. Find the solution of the difference equation

$$y(k + 2) - 6y(k + 1) + 9y(k) = 5$$

satisfying $y(0) = 2$ and $y(1) = -2$ given that $y_p(k) = c$, a constant, and the fundamental set $z_1(k) = 3^k$; $z_2(k) = k3^k$.

3.5 LINEAR DIFFERENCE EQUATIONS WITH CONSTANT COEFFICIENTS

In this section it will be shown how to find a fundamental set of solutions of a linear homogeneous equation with constant coefficients. This provides a means of calculating the general solution of nonhomogeneous equations with constant coefficients as described at the end of the last section.

Theorem 5. To solve the linear homogeneous equation

$$a_n z(k+n) + a_{n-1}z(k+n-1) + \cdots + a_0 z(k) = 0 \tag{13}$$

where $a_0, \ldots, a_n$ are constants and $a_n \neq 0$:

Step 1. Replace $z(k+i)$ by r^i in (13), obtaining

$$a_n r^n + a_{n-1}r^{n-1} + \cdots + a_1 r + a_0 = 0 \tag{14}$$

the so-called *characteristic equation*.

Step 2. Factor the left side of (14) into distinct linear factors $r - \lambda_i$, obtaining

$$(r-\lambda_1)^{p_1}(r-\lambda_2)^{p_2}\cdots(r-\lambda_m)^{p_m} = 0 \qquad (p_1 + p_2 + \cdots + p_m = n) \tag{15}$$

The λ_i are called *characteristic roots*.

Step 3.(a) If $p_1 = p_2 = \cdots = p_m = 1$, then $m = n$ (there are no repeated roots) and the general solution of (13) is

$$z(k) = c_1\lambda_1^k + c_2\lambda_2^k + \cdots + c_n\lambda_n^k \tag{16}$$

where $c_1, \ldots, c_n$ are constants.

(b) If at least one $p_i > 1$, the general solution of (13) is (16) with the following modification: If $p_i > 1$, replace $c_i\lambda_i^k$ in (16) by

$$\lambda_i^k(b_0 + b_1 k^{(1)} + b_2 k^{(2)} + \cdots + b_{p_i-1}k^{(p_i-1)}) \tag{17}$$

where $b_0, b_1, \ldots, b_{p_i-1}$ are constants and $k^{(1)} = k$ while $k^{(s)} = k(k-1)\cdots(k-s+1)$.

Note. Suppose that all of the coefficients a_i of the difference equation are real numbers. Let α and β be real numbers and $i = \sqrt{-1}$. It is known from the theory of polynomial equations that if the characteristic equation has a complex root $r = \alpha + i\beta$, of multiplicity m, then the characteristic equation also has the complex conjugate $\alpha - i\beta$ as a root of the same multiplicity. Applying Theorem 5, the general solution contains summands of the form

$$(b_0 + b_1k^{(1)} + \cdots + b_{m-1}k^{(m-1)})(\alpha + i\beta)^k + (c_0 + c_1k^{(1)} + \cdots + c_{m-1}k^{(m-1)})(\alpha - i\beta)^k \tag{18}$$

It can be shown, by separating $(\alpha + i\beta)^k$ and $(\alpha - i\beta)^k$ into real and imaginary parts (see Exercise 11), that the above summands can be rewritten in the form

$$(d_0 + d_1k^{(1)} + \cdots + d_{m-1}k^{(m-1)})A^k \sin k\theta + (e_0 + e_1k^{(1)} + \cdots + e_{m-1}k^{(m-1)})A^k \cos k\theta \tag{19}$$

where all the d_i and e_i are real constants and

$$A = \sqrt{\alpha^2 + \beta^2} \tag{20}$$

$$\theta = \tan^{-1} \frac{\beta}{\alpha} \tag{21}$$

From this result and Theorem 5, it follows that the general solution of a linear, homogeneous difference equation with real and constant coefficients can always be expressed as a linear combination of real-valued functions.

To illustrate the above result, consider the difference equation

$$z(k + 4) - 4z(k + 3) + 8z(k + 2) - 8z(k + 1) + 4z(k) = 0$$

with its characteristic equation,

$$0 = r^4 - 4r^3 + 8r^2 - 8r + 4 = [r - (1 + i)]^2[r - (1 - i)]^2$$

Here, $\alpha + i\beta = 1 + i$, and so $A = 2^{1/2}$ and $\theta = \tan^{-1} 1 = \pi/4$. Thus, the general solution of the given difference equation is

$$(d_0 + d_1 k^{(1)})2^{k/2} \sin \frac{k\pi}{4} + (e_0 + e_1 k^{(1)})2^{k/2} \cos \frac{k\pi}{4}$$

There is an alternative form for (17) in which $k^{(s)}$ is replaced by k^s, the ordinary sth power of k. The reason for using the polynomial $k^{(s)}$ is that the coefficients b_i are easier to find than if k^s were used. This will be illustrated in the following examples.

Example 9

(a) Solve the difference equation

$$z(k + 3) - 13z(k + 1) - 12z(k) = 0 \tag{22}$$

The equation can be solved by applying the steps in Theorem 5.
Step 1. Replacing $z(k + i)$ by r^i in (18) yields $r^3 - 13r + 12 = 0$.
Step 2. Factoring the above equation gives

$$(r - 1)(r - 3)(r + 4) = 0 \tag{23}$$

Step 3. The exponent of each of the factors of (23) is 1. Solution (a) therefore applies. Hence, the general solution of (22) is

$$z(k) = c_1 1^k + c_2 3^k + c_3(-4)^k$$

(b) Solve the difference equation

$$z(k + 3) - 6z(k + 2) + 12z(k + 1) - 8z(k) = 0 \tag{24}$$

with the initial conditions $z(0) = 1$ and $z(1) = z(2) = 2$.

Once again Theorem 5 will be applied.

Step 1. Replacing $z(k + i)$ by r^i in (24) gives

$$r^3 - 6r^2 + 12r - 8 = 0$$

Step 2. The above equation can be written as

$$(r - 2)^3 = 0$$

Step 3. The exponent is 3 and hence by part (b) the general solution of (24) is

$$z(k) = 2^k[b_0 + b_1 k + b_2 k(k - 1)]$$

The constants can be determined from the initial conditions

$$1 = z(0) = b_0$$

$$2 = z(1) = 2b_0 + 2b_1$$

$$2 = z(2) = 4(b_0 + 2b_1 + 2b_2)$$

Since $b_0 = 1$, the second equation in the above system yields $b_1 = 0$ and the third equation gives $b_2 = -\frac{1}{4}$.

(c) The characteristic equation of the difference equation

$$z(k + 4) - 5z(k + 3) + 9z(k + 2) - 7z(k + 1) + 2z(k) = 0 \tag{25}$$

is

$$r^4 - 5r^3 + 9r^2 - 7r + 2 = (r - 1)^3(r - 2) = 0$$

By using Theorem 5, the general solution of (25) can be written as

$$z(k) = c2^k + b_0 + b_1k + b_2k(k - 1)$$

From the theory developed in the last section, the general solution of the nonhomogeneous linear equation

$$a_n y(n + k) + a_{n-1} y(n + k - 1) + \cdots + a_1 y(k + 1) + a_0 y(k) = f(k) \tag{26}$$

with constant coefficients $a_0, a_1, \ldots, a_n$ is the sum of the general solution of the related homogeneous equation (13) and a particular solution of (26). The difficulty in applying this result is finding a particular solution. Sometimes this may be done by inspection. Unfortunately, there is no general method for finding a particular solution. The following methods can be used if $f(k)$ has a certain special form.

Algorithm for finding a particular solution to (26) if $f(k) = c\mu^k$. If $f(k) = c\mu^k$, where c and μ are nonzero constants, then a particular solution $y_p(k)$ of (26) can be found as follows:

Step 1. Replace $y(n + i)$ by x^i on the left side of (22), forming the *characteristic polynomial*

$$h(x) = a_nx^n + a_{n-1}x^{n-1} + \cdots + a_1x + a_0$$

Step 2. Replace μ by x on the right side of (26), forming the polynomial

$$g(x) = cx^k$$

Then

(a) if $\mu \neq \lambda_i (i = 1, 2, \ldots, m)$, a characteristic root, a particular solution is

$$y_p(\mu) = \frac{g(\mu)}{h(\mu)}$$

(b) if $\mu = \lambda_i$, where $x - \lambda_i$ is a factor that is repeated s times in (15), then

$$y_p = \frac{g^{(s)}(\mu)}{h^{(s)}(\mu)}$$

where $g^{(s)}$ is the sth derivative of g.

Example 10

Find a particular solution of

(a) $y(k+3) - 8y(k+2) + 20y(k+1) - 16y(k) = 5$
(b) $y(k+3) - 8y(k+2) + 20y(k+1) - 16y(k) = 5 \cdot 4^k$
(c) $y(k+3) - 8y(k+2) + 20y(k+1) - 16y(k) = 5 \cdot 2^k$

Each of the problems can be solved by using the preceding algorithm. All of the difference equations have the same characteristic polynomial,

$$h(x) = x^3 - 8x^2 + 20x - 16 = (x-2)^2(x-4)$$

(a) Here $f(k) = 5 = 5 \cdot 1^k$ and so $c = 5$ and $\mu = 1$. Therefore $g(x) = 5$. Since 1 is not a characteristic root, part (a) of the algorithm gives

$$y_p(k) = \frac{g(1)}{h(1)} = \frac{5}{1 - 8 + 20 - 16} = -\frac{5}{3}$$

(b) Noting that $f(k) = 5 \cdot 4^k$ and $\mu = 4$ is a characteristic root, part (b) of the algorithm must be applied. Since $g(x) = 5 \cdot x^k$, it follows that

$$y_p(k) = \frac{g'(4)}{h'(4)} = \frac{5k4^{k-1}}{4} = 5k(4)^{k-2}$$

(c) Here $f(k) = 5 \cdot 2^k$, $\mu = 2$, and $g(x) = 5x^k$. Since $\mu = 2$ is a repeated characteristic root ($s = 2$), part (a) of the algorithm gives

$$y_p(k) = \frac{g''(2)}{h''(2)} = \frac{5k(k-1)2^{k-2}}{6(2) - 16} = -5k(k-1)2^{k-4}$$

A method for finding a particular solution of (26) when $f(k)$ is a polynomial. Suppose that f is a polynomial of degree $s \geq 1$. The situation in which f is a constant has already been treated in the previous algorithm. A particular solution of (26) can be found as follows:

(a) If $z(k) = k$ is not a solution of the corresponding homogeneous (13), then a particular solution of (26) is

$$y_p(k) = \gamma_0 + \gamma_1 k + \cdots + \gamma_s k^s$$

(b) If $z(k) = k$ is a solution of (13), then keep trying successive powers k^2, k^3, $k^4, \ldots$, until $z(k) = k^e$ ($e > 1$) is not a solution of (13). Then a particular solution of (26) is

$$y_p(k) = \gamma_1 k^e + \gamma_2 k^{e+1} + \cdots + \gamma_{s-1} k^{e+s}$$

The unknowns, γ_i, are determined by substituting y_p into the left side of (26). The resulting expression is a polynomial in k whose coefficients are linear equations involving the unknown γ_i. These coefficients must be equal to the known numerical coefficients of equal powers of k in $f(k)$. This leads to a system of equations in γ_i that can then be solved and the γ_i determined.

Example 11

Find a particular solution for

(a) $y(k + 1) - 2y(k) = 3k^2 + 2k - 1$
(b) $y(k + 3) - 2y(k + 2) + y(k + 1) = 6k + 8$

(a) Since $z(k) = k$ is not a solution of the corresponding homogeneous equation, a particular solution is of the form

$$y_p(k) = \gamma_0 + \gamma_1 k + \gamma_2 k^2$$

Substituting $y_p(k)$ into the left side of the given difference equation gives

$$\gamma_0 + \gamma_1(k + 1) + \gamma_2(k + 1)^2 - 2(\gamma_0 + \gamma_1 k + \gamma_2 k^2)$$
$$= (-\gamma_0 + \gamma_1 + \gamma_2) + (-\gamma_1 + 2\gamma_2)k - \gamma_2 k^2$$

Equating the coefficients of the resulting polynomial to the coefficients of like powers of k appearing on the right side of the given difference equations leads to the following system of equations:

$$-\gamma_0 + \gamma_1 + \gamma_2 = 3$$
$$-\gamma_1 + 2\gamma_2 = 2$$
$$-\gamma_2 = -1$$

The solution of the above system of equations is $\gamma_2 = 1$, $\gamma_1 = 0$, and $\gamma_0 = -2$. Hence

$$y_p(k) = -2 + k^2$$

(b) Since $z(k) = k$ satisfies the corresponding homogeneous equation but $z(k) = k^2$ does not, a particular solution of the given equation is of the form

$$y_p(k) = \gamma_1 k^2 + \gamma_2 k^3$$

Substituting y_p into the left side of the given difference equation gives

$$\gamma_1[(k + 3)^2 - 2(k + 2)^2 + (k + 1)^2] + \gamma_2[(k + 3)^3 - 2(k + 2)^3 + (k + 1)^3]$$
$$= (2\gamma_1 + 12\gamma_2) + 6\gamma_2 k$$

Equating the resulting coefficients of powers of k to the coefficients of like powers appearing on the right side of the given difference equation gives the following system of equations:

$$6\gamma_2 = 6$$
$$2\gamma_1 + 12\gamma_2 = 8$$

The solution of these equations is $\gamma_2 = 1$ and $\gamma_1 = -2$.

Example 12 (Study of a Population of Beavers)

(a) A beaver population has an initial size of 100 and increases by 25% per year. If 10 of the beaver are trapped each year, find an expression for the size of the population after k years.

(b) Repeat (a) for the situation in which the number of trapped animals increases by 5 per year (15 are trapped in the second year, 20 in the third year, etc.).

(a) Let $y(k)$ be the size of the population at the beginning of year k. According to the assumptions in (a),

$$\begin{pmatrix}\text{change in population}\\ \text{size in year } k\end{pmatrix} = \begin{pmatrix}\text{population size}\\ \text{increase}\end{pmatrix} - \begin{pmatrix}\text{number of}\\ \text{trapped beaver}\end{pmatrix}$$

or symbolically

$$y(k) - y(k-1) = 0.25y(k-1) - 10$$

which reduces to

$$y(k) - \tfrac{5}{4}y(k-1) = -10$$

Since the characteristic equation is $r - \frac{5}{4} = 0$, the general solution is

$$y(k) = c(\tfrac{5}{4})^k + \left(\frac{-10}{1-\frac{5}{4}}\right) = c(\tfrac{5}{4})^k + 40$$

by the algorithm for finding a particular solution for $f(k) = -10 \cdot 1^k$. Using the given initial conditions,

$$100 = y(0) = c + 40$$

it follows that $c = 60$. Hence, the size of the population after k years is

$$y(k) = 60(\tfrac{5}{4})^k + 40$$

(b) The forcing term $f(k) = -10 - 5(k-1) = -k - 5$ is a polynomial, and so, by the algorithm, a particular solution is of the form

$$y_p(k) = a + bk$$

since $y(k) = k$ does not satisfy the homogeneous equation. The constants a and b are determined from the fact that $y_p(k)$ must satisfy the nonhomogeneous equation, namely,

$$a + bk - \tfrac{5}{4}[a + b(k-1)] = -5k - 5$$

Hence, by equating powers of k, the following equations are obtained:

$$-a + 5b = -20$$

$$b = 20$$

Thus $a = 120$. The general solution is

$$y(k) = c(\tfrac{5}{4})^k + 120 + 20k$$

Using the given initial conditions,

$$100 = y(0) = c + 120$$

it follows that $c = -20$. The general solution is

$$y(k) = -20(\tfrac{5}{4})^k + 120 + 20k$$

A particular solution to the nonhomogeneous linear difference equation (26) can be found for other functions f if a fundamental set of solutions of the linear homogeneous difference equation (13) is known. The technique, called the *method of variation of parameters,* replaces the constants appearing in a linear combination of the fundamental set of solutions by unknown functions of k. These unknown functions are determined by substituting the linear combination in (26) and deriving a first-order difference equation for each unknown function of k. The method will be illustrated for a second-order linear difference equation but can easily be generalized to higher-order equations.

Consider the second-order linear difference equation,

$$y(k+2) + p(k)y(k+1) + q(k)y(k) = r(k) \tag{27}$$

Let $y_1(k)$ and $y_2(k)$ be two independent solutions of the homogeneous equation corresponding to (27). Assume that

$$y(k) = c_1(k)y_1(k) + c_2(k)y_2(k) \tag{28}$$

is a particular solution of (27). Then

$$\begin{aligned} y(k+1) &= c_1(k+1)y_1(k+1) + c_2(k+1)y_2(k+1) \\ &= c_1(k)y_1(k+1) + c_2(k)y_2(k+1) + [c_1(k+1) - c_1(k)]y_1(k+1) \\ &\qquad + [c_2(k+1) - c_2(k)]y_2(k+1) \end{aligned}$$

Since a second-order difference equation for c_1 and c_2 is not desired, set

$$\Delta c_1(k)y_1(k+1) + \Delta c_2(k)y_2(k+1) = 0 \tag{29}$$

Hence

$$y(k+1) = c_1(k)y_1(k+1) + c_2(k)y_2(k+1) \tag{30}$$

and so

$$y(k+2) = c_1(k)y_1(k+2) + c_2(k)y_2(k+2) + \Delta c_1(k)y_1(k+2) + \Delta c_2(k)y_2(k+2) \tag{31}$$

Substituting the values of $y(k)$, $y(k+1)$, and $y(k+2)$ from (28), (30), and (31), respectively, into (27) yields

$$\Delta c_1(k)y_1(k+2) + \Delta c_2(k)y_2(k+2) = r(k) \tag{32}$$

since y_1 and y_2 are solutions of the homogeneous equation corresponding to (27) with $r(k) = 0$. Solving Eqs. (29) and (32) for $\Delta c_1(k)$ and $\Delta c_2(k)$ gives

$$\Delta c_1(k) = \frac{-r(k)y_2(k+1)}{y_1(k+1)y_2(k+2) - y_1(k+2)y_2(k+1)} \tag{33}$$

$$\Delta c_2(k) = \frac{r(k)y_1(k+1)}{y_1(k+1)y_2(k+2) - y_1(k+2)y_2(k+1)} \tag{34}$$

These last two difference equations can be solved for $c_1(k)$ and $c_2(k)$, since the right-hand sides of these equations are known functions.

Example 13

Find a particular solution of $y(k+2) - 3y(k+1) + 2y(k) = 3^k$.

An independent set of solutions of the corresponding homogeneous equation with 3^k replaced by 0 is $y_1 = 1$ and $y_2 = 2^k$. According to the method of variation of parameters, a particular solution is of the form

$$y = c_1(k) + c_2(k)2^k$$

According to (33) and (34), the unknown coefficients satisfy

$$\Delta c_1(k) = \frac{-3^k 2^{k+1}}{2^{k+2} - 2^{k+1}} = -3^k$$

$$\Delta c_2(k) = 2^{-1}(\tfrac{3}{2})^k$$

Solving these equations gives

$$c_1(k) = -(1 + 3 + 3^2 + \cdots + 3^{k-1}) = 2^{-1}(1 - 3^k)$$

$$c_2(k) = 2^{-1}[1 + \tfrac{3}{2} + \cdots + (\tfrac{3}{2})^{k-1}] = (\tfrac{3}{2})^k - 1$$

Hence

$$y = 2^{-1}(1 - 3^k) + 2^k\left[\left(\frac{3}{2}\right)^k - 1\right] = \frac{1}{2} + \frac{3^k}{2} - 2^k$$

Since $\frac{1}{2}$ and 2^k are solutions of the homogeneous equation, a particular solution is $3^k/2$.

EXERCISES

1. Find the general solution of the following difference equations:
   (a) $y(k + 2) - 2y(k + 1) + y(k) = 0$
   (b) $y(k + 2) - y(k + 1) - 6y(k) = 0$
   (c) $y(k + 3) - 7y(k + 2) + 35(k + 1) - 50y(k) = 0$
   (*Hint:* A characteristic root is 2.)
   (d) $y(k + 3) - 15y(k + 2) + 75y(k + 1) + 125y(k) = 0$

2. Find the general solution of the following difference equations:
   (a) $y(k + 2) + y(k + 1) - 2y(k) = 3 \cdot 5^k$
   (b) $y(k + 2) + y(k + 1) - 2y(k) = 5(-2)^k$
   (c) $y(k + 3) - 15y(k + 2) + 75y(k + 1) + 125y(k) = -2 \cdot 3^k$
   (d) $y(k + 3) - 15y(k + 2) + 75y(k + 1) + 125y(k) = 4 \cdot 5^k$

3. Solve the following nonhomogeneous difference equations:
   (a) $y(k + 2) - 4y(k) = 3k^2 + 2k - 2$
   (b) $y(k + 2) - 2y(k + 1) + y(k) = 2k + 3$

4. Solve Exercise 2(a) using the method of variation of parameters and compare your answer with the algorithmic solution.

5. Find the general solution of the difference equation

$$y(k + 2) - 2y(k + 1) + 6y(k) = 3 \sin k$$

by the method of variation of parameters.

6. The number of new cases occurring during the kth week of an epidemic is equal to four times the number of cases $y(k - 2)$ that existed at the end of the $(k - 2)$th week of an epidemic.
   (a) What is the difference equation satisfied by $y(k)$?
   (b) Solve the difference equation with the initial conditions $y(0) = 1$ and $y(1) = 4$.
   (c) Solve (a) and (b) under the additional assumption that during the kth week half of the cases that existed at the end of the previous week are cured.

7. The conduction of an electrical impulse along a myelinated nerve fiber is shown in the figure below.

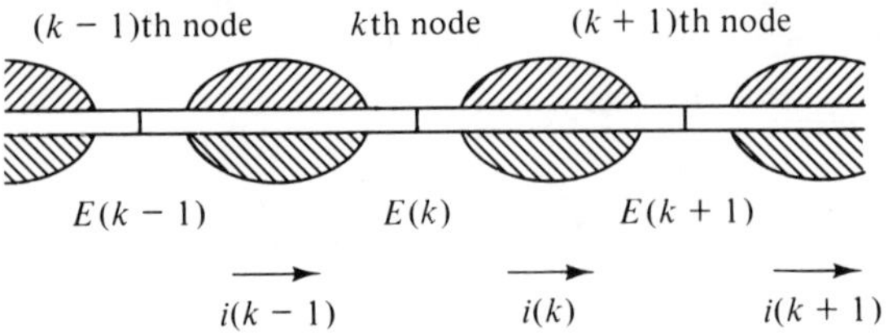

$E(k)$: electrical potential at node k
$i(k)$: current between node k and node $k + 1$
R: resistance of fiber between two successive nodes
r: leakage resistance of a node

(a) Prove that $i(k) = [E(k) - E(k + 1)]R$. (Recall Ohm's law, $E = IR$.)
(b) Show that the leakage current at the $(k + 1)$th node is $i(k + 1) - i(k) = E(k)/r$.
(c) Show that

$$i(k + 1) - \left(2 + \frac{R}{r}\right)i(k) + i(k - 1) = 0$$

(d) Solve the above difference equation for $i(k)$.

8. A fasting individual loses 2% of his weight at the beginning of that day.
(a) If he weighed 100 kg initially, find a difference equation for his weight after k days of fasting.
(b) How many days does it take him to reduce to 75 kg?

9. A herd of animals increases by 20% each year by normal birth and death processes and receives 15 animals per year through immigration. Initially the herd size is 300. Calculate the size after k years.

10. The CO_2 content of blood entering the lungs can be measured as follows. The subject breathes repeatedly in and out of a closed plastic bag. After each breath the partial pressure of CO_2 in the plastic bag is measured. Let $p(k)$ denote the partial pressure after k breaths. A graph of $p(k)$ versus $p(k - 1)$ is a straight line whose equation is

$$p(k) = ap(k - 1) + b$$

where $a < 1$ and b are two constants that can be determined from the graph.
(a) Solve the above difference equation.
(b) Show that

$$\lim_{k\to\infty} p(k) = \frac{b}{1 - a}$$

is the partial pressure of CO_2 in the venous blood entering the lungs.

11. **(a)** Show that $a(\alpha + i\beta)^k + b(\alpha - i\beta)^k$ can be written in the form

$$A^k(c \sin k\theta + d \cos k\theta)$$

where $A = \sqrt{\alpha^2 + \beta^2}$ and $\theta = \tan^{-1} \beta/\alpha$, and c and d are real numbers. {*Hint:* Use De Moivre's theorem [5], $(\cos \theta + i \sin \theta)^k = \cos k\theta + i \sin k\theta$, and let a and b be complex conjugates of each other.}
(b) Justify the Note following Theorem 5.

3.6 GRAPHICAL ANALYSIS OF ONE-LOOP DISCRETE SYSTEMS

The analysis described in this section is applicable to networks of the form shown in Fig. 3.1.

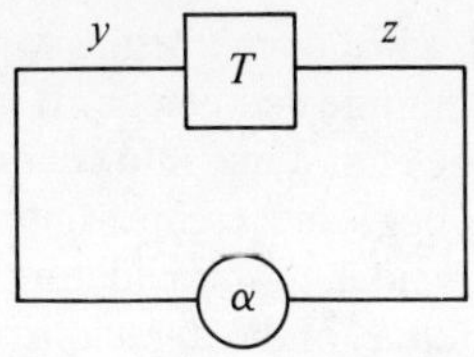

Figure 3.1 Simulation diagram for a one-loop discrete system.

More complicated single-loop networks with a series of delays,

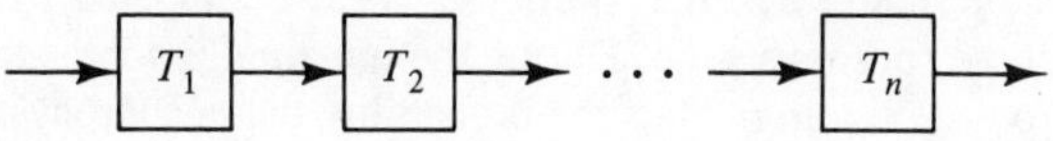

or a series of scalars,

can be reduced to the form shown in Fig. 3.1, because the given series of delays is equivalent to

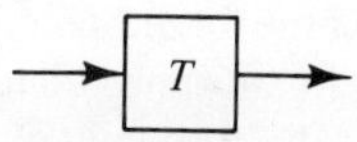

where $T = T_1 + T_2 + \cdots + T_n$, and the series of scalars is equivalent to

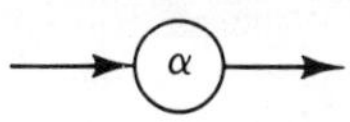

where $\alpha = \alpha_1 \alpha_2 \cdots \alpha_n$.

The difference equation satisfied by y can be derived from Fig. 3.1 by noting that

$$z(t) = y(t - T) \qquad \alpha z(t) = y(t)$$

Hence

$$y(t) = \alpha y(t - T) \tag{35}$$

The analysis that follows is still applicable if α is a function of z. More generally, it can be applied to relations of the form

$$y(t) = F[y(t - T)] \tag{36}$$

for a given function F.

Relations of the form (35) often occur when the life cycle of an animal is simplified. This reduction to a single loop is illustrated by the following example on the life cycle of blood flukes (schistosomes).

Example 14 (Life Cycle of Schistosomes)

Schistosomiasis (bilharziasis) is a common parasitic disease of humans [2]. The disease is acquired by working or wading in waters containing infective cercariae. These larvae escape from their snail intermediate host and penetrate the unbroken skin of a mammalian host. Once inside the new host, the parasites mature and migrate to various tissue locations in the body.

There are three species of schistosomes that infect people. They have similar life cycles. An idealized life cycle will be presented below in order to show how the life cycle can be simulated.

The life cycle of a schistosome begins when an egg containing fully developed miracidia are excreted by an infected person. It is assumed that a fixed proportion α_1 of the eggs are excreted. If the eggs come into contact with water, the miracidium becomes very active, ruptures the egg, and escapes into the water. Only a fraction α_2 of the excreted eggs become miracidia. A miracidium swims about in search of an appropriate snail host. A certain fraction α_3 of the miracidia penetrate the intermediate host, develop into sporocysts, and migrate to the visceral mass of the mollusk. The sporocysts mature for 1 month in the snails. During November, December, January, February, and March, the maturation rate of the sporocysts is slow compared with the death rate of the snails, and all the sporocysts die. During the remaining months there is a fixed probability k of a sporocyst dividing. Thus the probability that a sporocyst survives to undergo division is

$$\gamma_1(\tau) = \begin{cases} k & 12n \le \tau \le 6 + 12n \\ 0 & 7 + 12n \le \tau \le 11 + 12n \end{cases} \tag{37}$$

where n and τ are nonnegative integers and $\tau = 0$ corresponds to April.

The final larval stage, cercariae, arises from sporocysts after the 1-month maturation. A fraction β_1 of the forked-tailed cercariae escape from the snail and swim freely in the water, searching for an appropriate host. A fraction α_4 penetrate the unbroken skin of a mammalian host. The fork tails of the cercariae are discarded.

With the aid of their various enzyme secretions and their capacity alternatively to contract and elongate, these parasites migrate through the body. They enter the small lymphatic vessels and eventually reach the systemic circulation. Subsequently the young flukes (schistosomules) are carried to the portal system, where they mature in the sinusoids of the host's liver. The total maturation period takes 2 months, and only a fixed proportion γ_2 of the schistosomules survive to become adults. The adults return to the circulatory system, specifically to the superior mesenteric veins. Here the adults mate. It is assumed that only one-half of the new adults mate. The remaining new adults are of the same sex as any old leftover adults and so must wait 1 month for a subsequent group of new adults in order to find a mate. A mated pair must also wait 1 month before remating occurs. The probability, γ, that an adult survives from one month to the next is constant with respect to time and does not vary from individual to individual. Widows and widowers do not remate.

The egg-bearing females enter the smaller branches of the superior mesenteric veins to lay eggs. It is assumed that the egg production rate, β, is the same for all females. The deposited eggs can become dislodged and carried by the blood to intestinal blood vessels. The ova penetrate the intestinal wall, causing traumatic injury, then enter the intestinal lumen and leave their host via the feces.

The simulation diagram of the idealized life cycle of schistosomes is shown in Fig. 3.2. Time is measured in periods of 1 month. The reader should trace through the network, beginning with the flow of eggs, to verify that the simulation diagram actually represents the idealized life cycle.

The model represented in Fig. 3.2 can be studied numerically with the aid of a computer. An analytic investigation is difficult. In order to obtain a model that is amenable to analysis, the average behavior over an interval of 1 year is studied. Figure 3.2 simplifies to Fig. 3.3. The fluctuations in the parameter γ_1, defined in (37), are eliminated by replacing γ_1 by its average value over its annual cycle:

$$\gamma_1 = \frac{7k}{12} \tag{38}$$

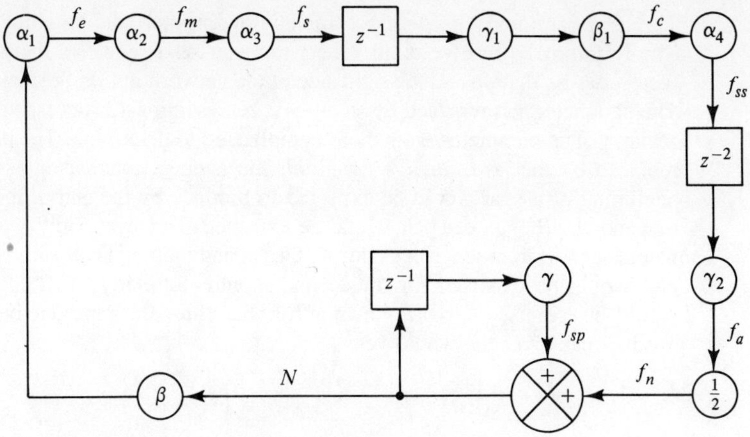

Notation

α_1 = proportion of eggs excreted by infected mammal
f_e = flow of eggs from mated adult pairs
α_2 = fraction of excreted eggs that become miracidia
f_m = flow of newly hatched miracidia into fresh water system
α_3 = fraction of miracidia that invade snails
f_s = flow of sporocysts in snail
z^{-1} = delay of 1 month to represent maturation of sporocysts
γ_1 = proportion of sporocysts that survive to become cercariae
β_1 = proportion of cercariae that escape from a snail
f_c = flow of cercarial larvae out of snail
α_4 = proportion of cercariae that penetrate mammals
f_{ss} = flow of schistosomules in primary host
z^{-2} = delay of 2 months to represent maturation of schistosomules
γ_2 = proportion of schistosomules that survive to become adults
f_a = proportion of worms in adult phase
$\frac{1}{2}$ = fraction of new adults that mate
f_n = proportion of newly mated pairs that pass into reproductive pool
z^{-1} = delay of 1 month before a mated pair can remate
γ = proportion of adults that survive for 1 month
f_{sp} = proportion of surviving pairs from the pool during previous month
N = number of mated pairs
β = egg production rate for females

Figure 3.2 Simulation diagram of life cycle of schistosomes.

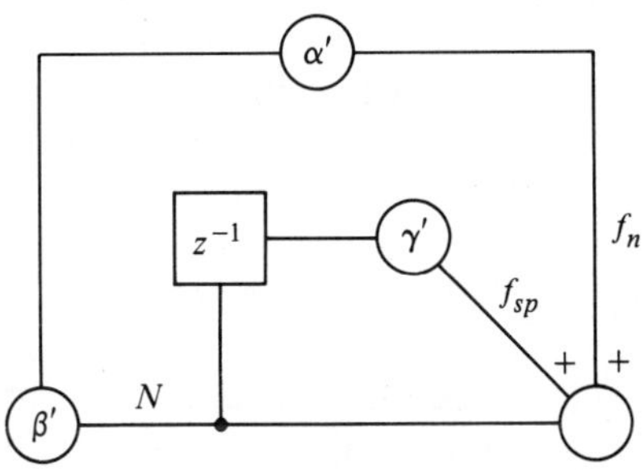

Figure 3.3 Simplified version of the model of Fig. 3.2.

This is a commonly used device for eliminating periodic variations in parameters by simply failing to resolve them. The parameter α' is equal to the product of the scale factors across the top and the right side of the network in Fig. 3.2, with γ_1 given by (38). The parameter γ is replaced by $\gamma' = \gamma^{12}$, representing 12 successive passes through the scalar γ. The parameter β' is more complicated to determine. The parameter β must be replaced by the *reproductive potential,* the average number of eggs that pairs at the beginning of a year would be expected to produce by the end of the same year. In the first month, β eggs per pair would be expected. However, only γ of the original pairs would survive to produce eggs during the second month. Thus, only $\gamma\beta$ eggs per original pair would be expected for the second month. Similarly, $\gamma^n\beta$ eggs per original pair would be expected for the nth month. Therefore, the reproductive potential of each original pair over the whole year is

$$\beta' = \sum_{n=0}^{11} \gamma^n \beta = \left(\frac{1-\gamma^{12}}{1-\gamma}\right)\beta \tag{39}$$

The difference equation satisfied by N, the number of mated pairs, for the network in Fig. 3.3 can be found as follows:

$$N(t) = f_{sp}(t) + f_n(t) \tag{40}$$

$$\alpha' N(t) = f_n(t) \tag{41}$$

$$\gamma' N(t-1) = f_{sp}(t) \tag{42}$$

Substituting the values of f_n and f_{sp} from (41) and (42) into (40) yields, after some simplification,

$$N(t) = \left(\frac{\gamma'}{1-\alpha'\beta}\right) N(t-1) \tag{43}$$

Equation (43) is of the form (36).

The behavior of the solution $y(t)$ of the general equation (36) will be studied graphically. The graphical method is useful when the difference equation satisfied by y is difficult to solve or the solution is so complicated that its behavior is difficult to determine. To apply the graphical method, set $y = y(t)$ and $x = y(t-T)$ and plot the graph of $y = F(x)$ using (36) either analytically or by computer. An example of such a graph along with the line $y = x$ is shown in Fig. 3.4.

Next choose an initial value of t, say, $t = T$, and mark $x = y(0)$ on the x axis. The corresponding ordinate on the graph is $y = y(T)$ and is obtained by finding the intersection of the vertical line $x = y(0)$ with the graph as shown in Fig. 3.4. The point $x = y(T)$ is found by intersecting the horizontal line $y = y(T)$ with the line $y = x$ as shown in Fig. 3.4. By repeatedly drawing vertical lines from the intersection on $y = x$ to the graph and horizontal lines from the intersection on the graph to $y = x$, the values $y(T)$, $y(2T)$, $y(3T)$, . . . can be obtained. When the graph of $y(t-T)$ versus $y(t)$ lies above the line $y = x$, as in Fig. 3.4, $y(t)$ increases as t increases. It is easy to verify that when the graph of (36) lies below the line $y = x$, $y(t)$ decreases as t increases.

Although difference equation (36) is simple in form, its solution can exhibit many varied patterns of behavior if F is nonlinear. Various possibilities and examples are presented in [9] and the references cited in that paper. Here only three possible

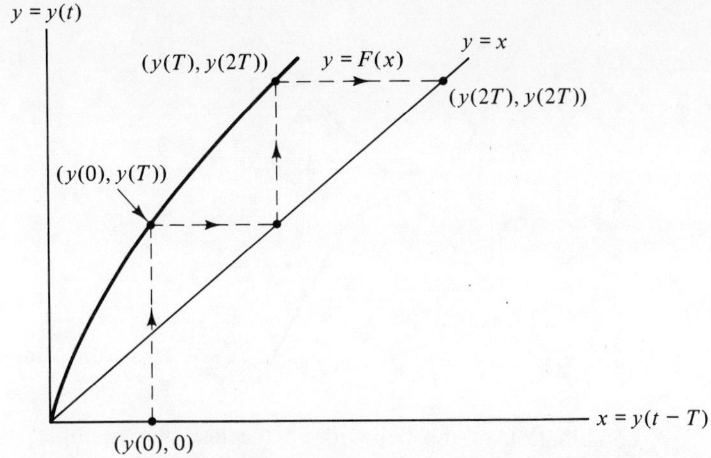

Figure 3.4 Graphical analysis of the dynamics of a discrete one-loop system.

modes of behavior will be illustrated graphically. The behavior of the solution will be studied in the neighborhood of an equilibrium point, that is, points determined by values of y that satisfy $y = F(y)$. Graphically, the equilibrium points correspond to the points of intersection of the line $y = x$ and the graph of $y = F(x)$.

A *stable limit cycle* is displayed in Fig. 3.5. Suppose that the initial value $y(0)$ corresponds to a point on the graph of $y = F(x)$ at either corner of the square. The graphical method of Fig. 3.4 shows that subsequent values of y will oscillate back and forth between the two values corresponding to the diagonally opposite corners of the square. If the initial $y(0)$ lies outside or inside the square, subsequent values of $y(t)$ will tend toward the corners of the square is illustrated in Fig. 3.5. Since the dynamics of the model tend to carry $y(t)$ toward the limit cycle (the boundary of the square), the limit cycle is said to be *stable*.

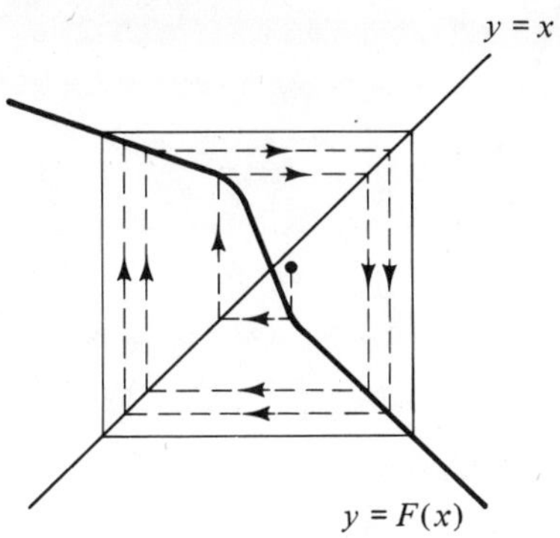

Figure 3.5 Stable limit cycle of $y(t) = F[y(t - T)]$.

Figure 3.6 shows an *unstable limit cycle*. Once again, if an initial point corresponds to a point on the graph at either corner of the square, subsequent values of $y(t)$ will oscillate between these values. However, for any other initial values, subsequent values of $y(t)$ will move away from the boundary of the square. For example, if an initial point lies within the square, it will tend to the intersection of $y = x$ and $y = F(x)$ as shown in Fig. 3.6.

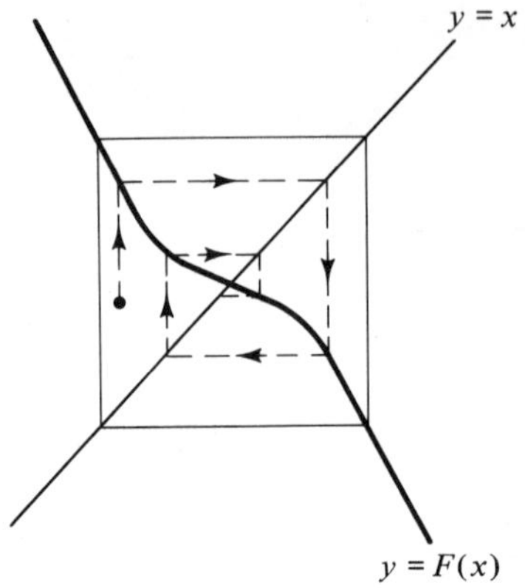

Figure 3.6 Unstable limit cycle of $y(t) = F[y(t - T)]$.

A combination of the behavior in the last two examples is also possible. A necessary, but not sufficient, condition for this to occur is that part of the graph of $y = F(x)$ between the corners of the square lies above (or below) the square as illustrated in Fig. 3.7. For some initial points both inside and outside the square, subsequent values of $y(t)$ will tend toward the limit cycle, while for other initial values subsequent values of $y(t)$ will move away from the limit cycle. There are also an infinite number of points $y(0)$ that give totally aperiodic but bounded trajectories. The sequential values of $y(t)$ do not form a repeated pattern and are indistinguishable from the values generated by a random function. Such a situation has been termed "chaotic" by Li and Yorke [8].

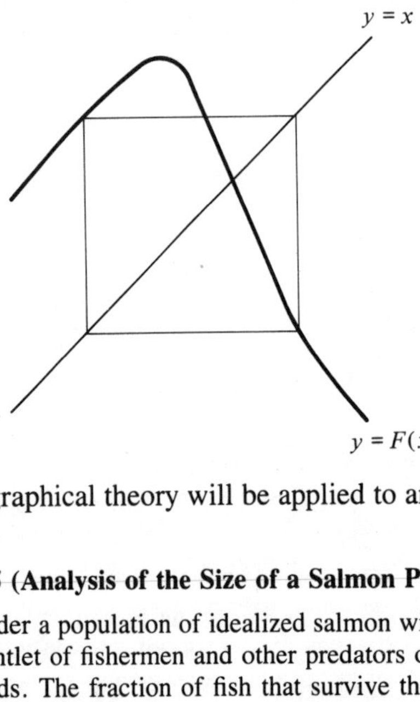

Figure 3.7 Chaotic dynamics of $y(t) = F[y(t - T)]$.

The graphical theory will be applied to analyze the behavior of an ideal salmon population.

Example 15 (Analysis of the Size of a Salmon Population)

Consider a population of idealized salmon with the following life cycle. The adults run a gauntlet of fishermen and other predators on their way from the sea to the spawning grounds. The fraction of fish that survive the journey to become spawners is σ. Each adult female that survives to reproduce produces α female eggs. Having spawned, the adult fish dies. Two years are required for a newly produced female egg to become a sexually mature adult. The fraction, γ, of eggs that survive to become adults is dependent on the number of eggs produced, decreasing exponentially with increasing number of eggs (N_e):

$$\gamma = \gamma_0 \exp(-N_e)$$

The simulation diagram for the life cycle appears in Fig. 3.8, where $N(t)$ is the number of female spawners at time t. From the description of the life cycle,

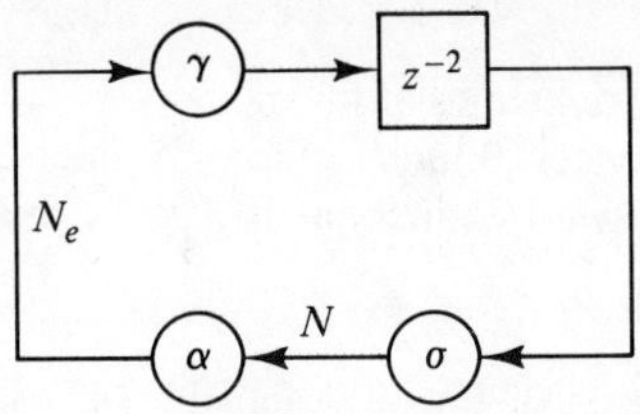

Figure 3.8 Idealized life cycle of salmon.

$$N_e(t) = \alpha N(t)$$

$$N(t) = \sigma\gamma_0 \exp[-\beta N_e(t-2)]N_e(t-2)$$

Hence, the equation for N is

$$N(t) = \sigma\alpha\gamma_0 \exp[-\beta N(t-2)]N(t-2) \tag{44}$$

Letting $M = \alpha\beta N$ and $A = \alpha\sigma\gamma_0$, Eq. (44) simplifies to

$$M(t) = AM(t-2)\exp[-M(t-2)] \tag{45}$$

To study the behavior of the solution of (45), its graph, $y = Axe^{-x}$, where $y = M(t)$ and $x = M(t-2)$, is plotted in Fig. 3.9 for several values of A.

Since

$$y' = A(1-x)e^{-x} \tag{46}$$

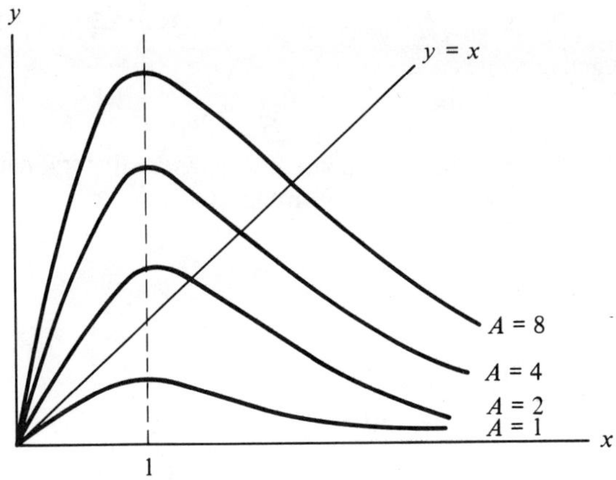

Figure 3.9 Graph for analyzing the ideal salmon population.

it follows that the local maximum occurs at $x = 1$. The graph intersects the reference line $y = x$ when $x = Axe^{-x}$. This equation has the solutions $x = 0$ and $x = \ln A$. The second solution, $x = \ln A > 0$ when $A > 0$. From (46), the slopes of the graphs when $x = 0$ and $x = \ln A$ are A and $1 - \ln A$, respectively.

The behavior of the solutions of (45) will be studied using the points of intersection.

(a) $0 < A \leq 1$

There is only one intersection at the origin. The graph at the origin is approximated by the tangent line at the origin whose slope is A. Applying the graphical technique illustrated in Fig. 3.4 to the tangent line yields

$$y(kT) = A^k y(0) \qquad (T = 2) \tag{47}$$

which describes the dynamics of the model very close to the origin. Therefore, for all values of A between 0 and 1, the only intersection is at the origin and the population tends toward extinction. If $A = 1$, the population also tends toward extinction.

(b) $A > 1$

An examination of Fig. 3.9 and Eq. (47) shows that the solution of (45) moves away from the origin toward the second intersection with $y = x$ at $x = \ln A$. The curve at the point $x = \ln A$ can be approximated by the tangent line at this point. The slope of the tangent line is $1 - \ln A$, and its equation is $y = \ln A + (1 - \ln A)(x - \ln A)$. Applying the graphical technique illustrated in Fig. 3.4 to the tangent line yields

$$y(kT) = \ln A + (1 - \ln A)^k[y(0) - \ln A] \tag{48}$$

The behavior of the solution in a small neighborhood of the second intersection depends on the value of $1 - \ln A$. The following possibilities can occur.

(i) $$0 < \ln A < 1 \qquad (1 < A < 2.71828)$$

The solution monotonically approaches the second intersection, since $(1 - \ln A)^k$ approaches zero as k increases.

(ii) $$1 < \ln A < 2 \qquad (2.71828 < A < 7.389056)$$

The solution approaches $\ln A$ in an oscillatory manner, since $(1 - \ln A)^k$ oscillates between positive and negative values.

(iii) $$\ln A > 2 \qquad (A > 7.389056)$$

In this situation, $1 - \ln A < -1$ and the solution grows in an oscillatory manner. The possibility of a stable limit cycle exists as shown in Fig. 3.10.

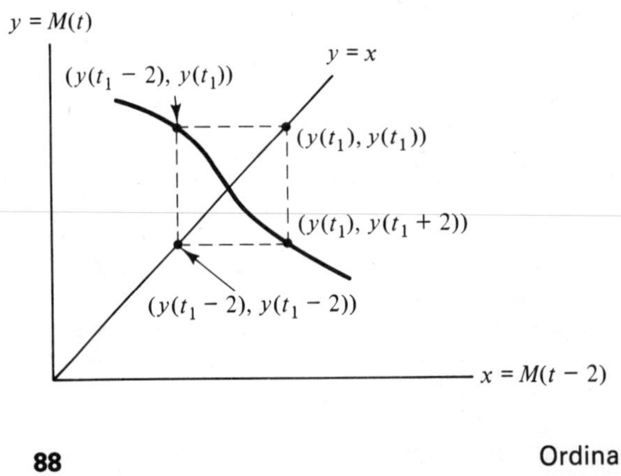

Figure 3.10 A stable limit cycle.

The points of such a limit cycle could be located by the graphical methods already described. However, an analytic approach can also be taken. Assume that the lower corner of the square has coordinates $(y(t_1 - 2), y(t_1 - 2))$. Then the coordinates of the remaining corners of the square are uniquely determined as described by Fig. 3.4. It follows from Fig. 3.10 that

$$y(t_1 - 2) = y(t_1 + 2) \quad \text{or} \quad y(t_1) = y(t_1 - 4)$$

From (45),

$$M(t_1 - 2) = AM(t_1 - 4)\exp[-M(t_1 - 4)] \tag{49}$$

Replacing $M(t_1 - 2)$ in (45) with its value from (49) gives

$$M(t_1) = A^2 M(t_1 - 4)\exp[-M(t_1 - 4)]\exp[-AM(t_1 - 4)]\exp[-M(t_1 - 4)] \tag{50}$$

Letting $y(t_1) = y(t_1 - 4) = x_c$, Eq. (50) becomes

$$1 = A^2 \exp(-x_c)\exp[-Ax_c \exp(-x_c)] \tag{51}$$

When a cycle exists, (51) should exhibit three solutions, one at the second intersection ($x_c = \ln A$) and two equidistant from the second intersection. If a square is constructed with the latter two points at opposite corners, then the cycle will be stable as long as no part of the model's graph lies above or below the square—in other words, if the peak of the graph lies on or to the left of the left-hand side of the square (i.e., if the smaller of the two values of x_c corresponding to the cycle is greater than or equal to 1).

Investigation of the roots of (51) is facilitated by taking natural logarithms of both sides of the equation, which after some rearrangement gives

$$\frac{2\ln A - x_c}{A} = x_c e^{-x_c} \tag{52}$$

A root x_c of (52) is the x coordinate of the point of intersection of the straight line

$$y = \frac{2\ln A - x}{A} \tag{53}$$

and the curve

$$y = xe^{-x} \tag{54}$$

The peak of (54) occurs at $x = 1$, and the ordinate at this point is e^{-1} (c.f. Fig. 3.9). It is evident from the graph of (54) that the straight line (55) will intersect it at $x < 1$ if and only if the ordinate of (53) is less than e^{-1} at $x = 1$; in other words,

$$\frac{2\ln A - 1}{A} < e^{-1} \tag{55}$$

The last inequality can be solved by plotting $(2\ln A - 1)/A$ versus A as shown in Fig. 3.11. Since only values of $A > 7.389$ are being considered, the second intersection ($A \approx 9.5$) is of interest. It can be shown that $A \approx 9.549$ at this intersection by using the Newton-Raphson method [6] for approximating the root of

$$\frac{2\ln(A - 1)}{A} - e^{-1} = 0$$

It follows that if $A > 9.549$, then the solution of (51) lying farthest to the left will be

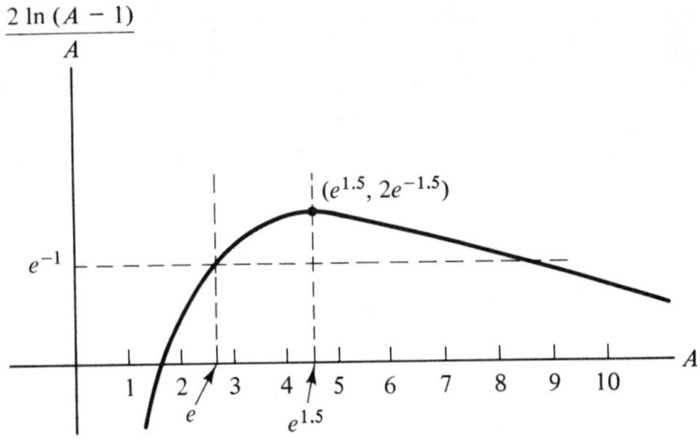

Figure 3.11 The graph of $2\ln(A - 1)/A$.

less than 1 (that is, $x_c < 1$). However, for $7.389 < A < 9.549$, $x_c > 1$. Hence, $7.389 < A < 9.549$ produces a stable limit cycle, while $A > 9.549$ leads to chaotic behavior.

This section has been adapted from [7]. Further details and other examples of population studies can be found in [7] and its comprehensive references.

EXERCISES

1. Use graphical methods to deduce the following facts about the logistic equation

$$y_k - y_{k-1} = (a - by_{k-1})y_{k-1}$$

where $0 < a < 1$ and $b > 0$ [3]:

(a) If $y_0 < a/b$, then the y_k form a decreasing sequence that approaches the limiting value a/b as k gets larger.

(b) If $a/b < y_0 < 1/b$, then the y_k form a decreasing sequence that approaches the limiting value a/b as k gets larger.

(c) If $y_0 = 1/b$, then $y_k = a/b$ for all $k \geq 1$.

(d) If $1/b < y_0 < (a + 1)b$, then $y_1 < a/b$ and for $k \geq 1$ the y_k form an increasing sequence that again approaches the limiting value a/b as k increases.

(e) If $y_0 \geq (a + 1)/b$, then y_1 is zero or negative; so this situation leads to the immediate extinction of the population.

2. Prove the following results concerning the logistic equation described in Exercise 1, where $b > 0$ and $1 < a < 2$, by drawing a figure similar to Fig. 3.4.

(a) If $1/b < y_0 < 2/b$, then y_n gets closer and closer to a/b as n increases, but oscillates from one side of this limiting value to the other as n goes from one value to the next.

(b) If $y_0 < 1/b$, then the y_n at first increase steadily until they exceed $1/b$ and thereafter they follow the same type of oscillatory behavior as in (a), approaching the limit a/b as $n \to \infty$.

(c) If $2/b < y_0 < (a + 1)b$, then the sequence immediately falls below the equilibrium value (that is, $y_1 < a/b$) and thereafter the y_n follow the same pattern as in (a) or (b), depending on whether $y_1 > 1/b$ or $y_1 < 1/b$.

(d) If $y_0 \geq (a + 1)/b$, then $y_1 \leq 0$ and the population immediately becomes extinct.

3. (a) Plot the curve $y = xe^{a(1-x)}$.

(b) Apply the graphical technique to study the behavior of the solutions of the difference equation

$$y(t+1) = y(t)\exp[a(1 - y(t))]$$

for $y(0) = 0.25$ and $a = 0.5$.

(c) Repeat (b) for $a = 2$ and $y(0) = 2$.

3.7 CONCLUSION

This chapter concentrated mainly on the direct solution of difference equations or the quantitative theory of difference equations. The reader interested in pursuing this topic should consult [1, 4, 10]. In many situations an analytic solution cannot be found. Sometimes only certain properties of the solution are of interest. For example, does the solution have a limit cycle? Such questions may be difficult to answer using only numerical methods. Some analytic techniques for studying certain properties of the solution without actually solving the difference equation will be presented in Volume II, Chapter 4. This theory is known as the qualitative theory of difference equations.

REFERENCES

1. Brand, L. *Differential and Difference Equations*. Wiley, New York, 1966.
2. Brown, J. W. *Basic Clinical Parasitology*. Appleton-Century-Croft, New York, 1969.
3. Eisen, M. *Graphical Analysis of Some Difference Equations in Mathematical Biology*, Unit 553. UMAP, Newton, MA, 1981.
4. Goldberg, S. *Introduction to Difference Equations*. Wiley, New York, 1958.
5. Kaplan, W. *Introduction to Analytic Functions*. Addison-Wesley, Reading, MA, 1966.
6. Kellison, S. G. *Fundamentals of Numerical Analysis*. Richard D. Irwin, Homewood, IL, 1975.
7. Lewis, E. R. *Network Models in Population Biology*. Springer-Verlag, Berlin, 1977.
8. Li, T.-Y., and Yorke, J. A. Period three implies chaos. *Am. Math. Monthly 83*, 985–992, 1975.
9. May, R. M. Simple mathematical models with very complicated dynamics. *Nature 261*, 459–467, 1976.
10. Miller, K. S. *Linear Difference Equations*. W. A. Benjamin, New York, 1968.

4

Ordinary Differential Equations

4.1 INTRODUCTION

Many biological processes can be described by ordinary differential equations. Such equations involve a single dynamic variable, which is a function of time only, and the derivatives of the variable. Some examples appeared in Chapters 1 and 2. In this chapter the different types of ordinary differential equations are defined. Conditions under which certain types of differential equations have a unique solution are stated. Methods for finding the solutions of linear differential equations are presented. The theory is illustrated by examples from biology. It will be seen that the theory of differential equations is analogous to that of difference equations as presented in Chapter 3.

4.2 DEFINITIONS CONCERNING DIFFERENTIAL EQUATIONS

In this chapter a real-valued function y defined on an interval $t_0 \leq t \leq t_1$ will be considered. The variable t is usually interpreted as time. If a derivative of a certain order is written, then the function is assumed to possess a derivative of that order on the given interval. The dot notation will be used to denote the derivative of a function, for example, $\dot{y} = dy/dt$, $\ddot{y} = d^2y/dt^2$.

Definition 1. An *ordinary differential equation* is an equation connecting a real-valued function and some of its derivatives for all values of t lying in a given interval. In other words, a relation of the form

$$f\left(\frac{d^n y}{dt^n}, \frac{d^{n-1} y}{dt^{n-1}}, \ldots, \frac{dy}{dt}, y, t\right) = 0 \qquad (t_0 \leq t \leq t_1)$$

where f is a real-valued function of $n + 2$ variables, is valid.

The *order* of a differential equation is the order of the highest derivative that appears in the equation.

Definition 2. A differential equation is *linear* if it has the form

$$a_n(t)\frac{d^n y}{dt^n} + a_{n-1}(t)\frac{d^{n-1}y}{dt^{n-1}} + \cdots + a_1(t)\frac{dy}{dt} + a_0(t)y = g(t) \tag{1}$$

for some functions $a_i(t)$, $i = 0, \ldots, n$ and $g(t)$ defined on the given interval. The a_i are referred to as *coefficients,* and $g(t)$ is called the *forcing term,* the *driving term,* or simply the *right-hand side*.

Differential equations not of the form (1) are called *nonlinear*.

Definition 3. A linear differential equation is said to have *constant coefficients,* be *time-invariant,* or be *stationary* if the coefficients a_i do not depend on the time t.

Definition 4. The linear differential equation (1) is *homogeneous* if, and only if, $g(t) = 0$ for $t_0 \leq t \leq t_1$ and *nonhomogeneous* otherwise.

The following example illustrates the above definitions.

Example 1

(a) $3(d^5y/dt^5) - 3(d^2y/dt^2) + (dy/dt) - 10y = 0$ is a fifth-order linear differential equation with constant coefficients.

(b) $(t + 1)^2(d^3y/dt^3) + (t^2 - 6t + 1)(d^2y/dt^2) - (t + 3)y = 0$ is a third-order linear differential equation with nonconstant coefficients. It is nonstationary.

(c) $2(d^2y/dt^2) + 6(dy/dt) - 8y = 8 + \sin t$ is a second-order nonhomogeneous linear differential equation with constant coefficients.

(d) The equations appearing in (a)–(b) are homogeneous.

(e) The equations appearing in (a)–(c) are linear. The following differential equation is an example of a nonlinear equation of order 3:

$$2\left(\frac{d^3y}{dt^3}\right)\left(\frac{dy}{dt}\right) + \frac{d^2y}{dt^2} - \sqrt{y\,\frac{d^2y}{dt^2}} = 0$$

Definition 5. An (*explicit*) *solution* of a differential equation, defined on $t_0 \leq t \leq t_1$, is a function that possesses, on the given interval, all derivatives appearing in the differential equation and that when substituted into the differential equation reduces it to an identity. A function defined implicitly on $t_0 \leq t \leq t_1$ is an (*implicit*) *solution* of a differential equation if at least one root of its defining equation is an explicit solution of the given differential equation on $t_0 \leq t \leq t_1$.

Example 2

(a) The differential equation

$$\frac{dy}{dt} = ay$$

has the solution $y(t) = Ce^{at}$, where C is an arbitrary constant, since

$$\frac{dy}{dt} = aCe^{at} = ay$$

(b) The second-order differential equation

$$\frac{d^2y}{dt^2} = 0 \qquad (-\infty < t < \infty)$$

has a solution of the form $y(t) = A + Bt$, where A and B are arbitrary constants.

(c) The solution of $\ddot{y} - 4\dot{y} + 3y = 0$ is $y = 7e^{3t}$, since

$$\ddot{y} - 4\dot{y} + 3y = 7 \cdot 9e^{3t} - 4 \cdot 7 \cdot 3e^{3t} + 3 \cdot 7e^{3t} = 0$$

(d) The function y defined implicitly by

$$t^2y^3 + y^2 - 6 \sin t + 3 = 0$$

is an implicit solution of the differential equation

$$(3t^2y^2 + 2y)\dot{y} - 6 \cos t + 2ty^3 = 0$$

This can be verified by differentiating both sides of the defining equation with respect to t, which gives

$$2ty^3 + 3t^2y^2\dot{y} + 2y\dot{y} - 6 \cos t = 0$$

A simple type of differential equation is

$$\frac{dy}{dt} = P(y)Q(t)$$

Equations of this form are said to be *separable,* since they can be solved by "separating" the variables,

$$\frac{dy}{P(y)} = Q(t)\,dt$$

and simply integrating. This defines the implicit solution

$$\int \frac{dy}{P(y)} = \int Q(t)\,dt + C$$

The next two examples illustrate the method of separation of variables.

Example 3 (Radioactive Dating of Fossils)

The radioactive decay of certain elements (for example, the isotope ^{14}C can be used as a means of dating certain archaeological objects (c.f. Exercises 12, 14, and 15 in Chapter 1). The carbon dating method is based on the following facts. The proportion of ^{12}C to the radioactive isotope ^{14}C occurring in the earth's atmosphere remains at a constant level.* The amount of ^{14}C that decays is exactly balanced by the formation of new ^{14}C resulting from bombardment of the earth's atmosphere by cosmic rays. Living plants and animals absorb carbon from the atmosphere and eventually reach an equilibrium so that the proportion of ^{14}C is very close to that in the atmosphere. However, when a plant or animal dies, the absorption of new ^{14}C ceases. Thus, the proportion of

*The testing of nuclear weapons has significantly increased the amount of ^{14}C in the atmosphere since the mid-1950s. The simple method used in Example 3 is no longer valid. There are also other corrections that must be made, for example, for pollution by other isotopes. A critical discussion of carbon dating appears in [16] and [17].

^{14}C in dead organic matter decreases according to the law of radioactive decay. By measuring the ratio of ^{14}C to ^{12}C in dead organic matter, it is possible to calculate, as shown below, how long ago the plant or animal died.

The basis of this dating calculation is the law of radioactive decay, which can be derived as follows. Let y be the amount of radioactive material remaining at time t. The number of nuclei that decay in a small time interval, and so the decay rate, is proportional to the number of radioactive nuclei that remain undecayed. The mathematical form of this last assertion is

$$\frac{dy}{dt} = -ky \tag{2}$$

since the mass is proportional to the number of nuclei. The constant k is called the *decay constant* and depends on the radioactive element.

Equation (2) can be solved by the method of separation of variables,

$$\int \frac{dy}{y} = -k\,dt + \ln C$$

which yields $\ln y = -kt + \ln C$, or

$$y = Ce^{-kt} \tag{3}$$

where $C = y(0)$ is the amount of material present initially.

To illustrate the dating method, assume that the amount of radioactive carbon in a certain plant fossil is measured to be 58% of the natural amount in the atmosphere. Let $t = 0$ be the instant of time that the plant died and let C be the amount of ^{14}C that was then present. Then (3) with $k = 1.24 \times 10^{-4}$ gives the amount of ^{14}C present t years later. If t is the time of measurement, then $y = 0.58C$, since the amount of ^{14}C is measured to be 58% of the amount that occurs naturally in the atmosphere. Hence

$$Ce^{-kt} = 0.58C$$

By taking natural logarithms of both sides of this last equation,

$$kt = -\ln(0.58) = 0.545$$

Finally, solving for t gives

$$t = \frac{0.545}{k} = \frac{0.545}{1.24 \times 10^{-4}} = 4400$$

Therefore the plant fossil is 4400 years old.

Example 4 (Logistic Law of Growth)

The method of separation of variables will be applied to solve the logistic differential equation

$$\frac{dy}{dt} = ay - by^2 \tag{4}$$

where $a > 0$ and $b > 0$. This equation was discussed in Example 2 of Chapter 2. Separating the variables in (4) gives

$$\frac{dy}{y - (b/a)y^2} = a\,dt \tag{5}$$

For convenience, set

$$y_e = \frac{a}{b}$$

Then the fraction on the left side of (5) is

$$\frac{1}{y - by^2/a} = \frac{1}{y - y^2/y_e} = \frac{y_e}{y(y_e - y)} = \frac{1}{y} + \frac{1}{y_e - y}$$

Substitute this result in (5) and then integrate both sides. On the right, let time vary from $t = 0$ to an arbitrary time t. On the left, y must vary from its initial value y_0 to its value at time t:

$$\int_{y_0}^{y} \left(\frac{1}{y} + \frac{1}{y_e - y}\right) dy = \int_0^t a\, dt \tag{6}$$

Carrying out the indicated integrations in (6) gives

$$\ln \frac{y}{y_0} - \ln \frac{y - y_e}{y_0 - y_e} = at$$

which can be rewritten as

$$\frac{y}{y - y_e} = \frac{y_0}{y_0 - y_e} e^{at}$$

Solving for y yields the *logistic law* of growth,

$$y = \frac{y_0 y_e}{y_0 + (y_e - y_0)e^{-at}} = y_e - (y_e - y_0)e^{-at} + \cdots \tag{7}$$

If $y_0 < y_e$, then y never exceeds y_e and the graph of (7) has a characteristic S-shaped appearance as shown in Fig. 2.3. If $y_0 > y_e$, then y is never less than y_e and decays to the equilibrium value y_e. Equation (7) indicates that the approach is approximately exponential.

A bacterial colony grown in a nutrient medium usually grows according to the logistic law [10]. The growth of the infusorian* *Parmecium caudatum* also follows this law [4]. However, if bacteria are transferred to a different nutrient medium, there may be an initial phase of growth, called *lag phase,* during which time the growth of the population is retarded. The lag phase is believed to be due to the formation of new enzyme systems required for growth using the new nutrients.

A population very close to its stationary value, y_e, is said to be in its *stationary phase* of growth. After a long time the population usually decays to zero from its equilibrium value y_e. New growth and death conditions have arisen, such as depletion of a necessary nutrient or the accumulation of a new toxic metabolic product, which are not accounted for by Eq. (7). Other models of limited growth appear in Exercises 10 and 11. A generalized growth law is described in [3].

Another type of differential equation for which a general solution can be found

*Infusoria are a class of protozoa that have cilia. Protozoa are single-celled animals (eukaryotes), the lowest division of the animal kingdom. They are more complex and generally larger than bacteria. Higher-order animal populations, such as fruit flies *(Drosophilia),* also grow according to the logistic law [9].

is the following first-order, linear, nonstationary, nonhomogeneous equation,

$$\frac{dy}{dt} = a(t)y + b(t) \tag{8}$$

where $a(t)$ and $b(t)$ are real functions of time that are integrable. If both sides of (8) are multiplied by $\exp[-\int a(s)\,ds]$, a so-called *integrating factor,* then (8) can be written as

$$\frac{d\{\exp[-\int a(s)\,ds]\,y\}}{dt} = b(t)\exp[-\int a(s)\,ds] \tag{9}$$

since the left side of (9) reduces to

$$[\dot{y} - a(t)y]\exp[-\int a(s)\,ds]$$

Multiplying both sides of (9) by dt and integrating gives

$$y\exp[-\int a(s)\,ds] = b(t)\exp[-\int a(s)\,ds] + c \tag{10}$$

where c is the constant of integration. Solving (10) for y yields

$$y = \exp[\int a(s)\,ds]\int b(t)\exp[-\int a(s)\,ds]\,dt + c\exp[\int a(s)\,ds] \tag{11}$$

The use of (11), the general solution of (8), will be illustrated in the following example on water pollution in the Great Lakes.

Example 5 (Water Pollution in Lakes)

One of the problems created by industrialization is water pollution. A polluted river will cleanse itself fairly rapidly once the pollution is stopped, because the steady flow of water carries the pollutant away. Cleansing a lake of pollutants that presently can be removed only by the natural process of flow is a more difficult task, since a considerable amount of water is involved. For example, it would take about 189 years to drain Lake Superior if its outflow were unchanged and its inflow stopped. A model will be formulated to study how long it takes to clean up a lake when the major clean-up mechanism is the natural process of gradually replacing the water in the lake.

The model is adapted from Rainey [15] and is based on the following assumptions:

1. The average inflow and outflow rates are approximately equal. The common flow rate will be denoted by r.
2. The pollutants are uniformly distributed in the water. Their concentration in the inflow and lake will be denoted by C_i and C_1, respectively.
3. The pollutants are removed only by the natural process of freely flowing out of the lake dissolved in the water.

By the above assumptions, during the time interval Δt,

change in total pollution = total inflow of pollutants − total outflow of pollutants

The mathematical form of the above relation is

$$\Delta(VC_1) = (C_i - C_1)r\,\Delta t + o(\Delta t) \tag{12}$$

where V is the volume of the lake and $o(\Delta t)$ denotes a function of Δt such that $o(\Delta t)/\Delta t$ tends to zero as Δt tends to zero. The term $o(\Delta t)$ is required since both C_i and C_1 are, in general, functions of t; hence $C_i r\,\Delta t$ and $C_1 r\,\Delta t$ are just approximations to the total flow of pollutants in a time interval Δt that approaches zero with Δt^2. Dividing both sides of (12) by Δt and letting Δt tend to zero gives

$$\dot{C}_1 = \frac{(C_i - C_1)r}{V}$$

This differential equation is of the form (8) with $a = -r/V$ and $b = rC_i/V$. Hence, by (11), its solution is

$$C_1(t) = e^{-t/T}\left[C_1(0) + T^{-1}\int_0^t C_i(x)e^{x/T}\,dx\right] \tag{13}$$

where $T = V/r$ is the number of years required to drain the lake if outflow is unchanged and inflow stops. Rainey's values of T for some of the Great Lakes are given in Table 4.1.

TABLE 4.1 NUMBER OF YEARS REQUIRED TO DRAIN THE LAKES IF INFLOW STOPS

Lake	T, years
Superior	189
Michigan	30.8
Erie	2.6
Ontario	7.8

If all pollution inflow ceases ($C_i = 0$), the fastest possible clean-up will occur. In this situation (13) reduces to

$$t = T\ln\left(\frac{C_1(0)}{C_1(t)}\right) \tag{14}$$

Using (14), Rainey calculated how long it would take to reduce pollution to a given percentage of its present level. For example, it would take 2 years to reduce the pollution in Lake Erie to 50% of its present level. The above results are not applicable to Lake Ontario, since approximately 84% of its inflow comes from Lake Erie, a source of pollution that cannot be controlled directly. The modifications required for Lake Ontario are outlined in Exercise 12.

The above model is a simplification of reality even for pollutants that are removed solely by flow. For example, if a lake has only one inlet and one outlet, the flow will move from source to the outlet in a pipeline fashion. Hence the concentration of pollutants will not be uniformly distributed.

Research is required to formulate models for pollutants that are not removed solely by flow but also by sedimentation, decay, or decomposing into harmless compounds. Such models would be more complex than the example above, as can be readily seen from the following brief description of the behavior of a few such pollutants.

The insecticide DDT and several other chlorinated hydrocarbons take a long time to decompose into relatively benign compounds. An organism retains most of the DDT it ingests, since DDT is readily soluble in body fat. Consequently, this insecticide can become concentrated in higher predators with harmful effects. Slow decomposition into less harmful compounds may be the main factor in removing DDT from a lake. However, consumption of fish by birds and humans may be significant.

Mercury also tends to remain in an organism and so moves up the food chain. Mercury may be even harder to remove from a lake than DDT, since it is an element and so does not decay. It disappears slowly due to sedimentation, outflow, and the removal of fish by birds and humans.

Large amounts of phosphorus are present in detergents, in many fertilizers, and in human wastes. Excessive quantities of this element can cause algae blooms. Eventually the algae die and settle to the bottom, removing much of the phosphorus in this manner. Unfortunately, some of the removal is only temporary, since the phosphorus may be released by the process of decay.

Not all differential equations have a solution. The differential equation

$$\left(\frac{dy}{dt}\right)^2 + y^2 = -1$$

has no solution $y(t)$ that is real-valued, since the right-hand side is nonnegative while the left-hand side is negative. Part (b) of Example 2 shows that a differential can have many solutions. Hence, just as for linear difference equations, the existence and uniqueness of the solutions of a differential equation must be investigated.

EXERCISES

1. The growth of microorganisms is described by the differential equation

$$\dot{y} = ky$$

where k is a constant and y is the population size at time t. Suppose that organisms divide on the average every 15 min and the initial population size is 10^3. Find the population size at time t.

2. The growth rate at time t (measured in hours) of certain bacteria is $500(1 + 2t)^{-1}$. Initially the population size is 200. What will be the population size after 5 hr?

3. The rate of change of population of squirrels is $y = 2^{t/4}$, where t is measured in half-years. Find the size of the population in 18 months if the initial population size is 50.

4. The number of people y infected at time t (measured in weeks) in a town of 600 people because of an influenza epidemic satisfies the differential equation

$$\frac{dy}{dt} = 0.003y(600 - y)$$

If initially four people are infected, find how long it is before two-thirds of the population are infected.

5. The rate of the biochemical reaction

$$A + B \rightarrow C + D$$

is given by

$$-\frac{da}{dt} = ka(t)b(t)$$

where k is a constant, $a(t)$ is the number of gram-molecules of A, and $b(t)$ is the number of gram-molecules of B present at time t. Solve this equation by eliminating $b(t)$. [*Hint:* $a(0) - a(t) = b(0) - b(t)$, since each of these quantities is the number of gram-molecules of A or B that have been used up in the reaction up to time t.]

6. The rate at which an enzyme reaction takes place is described by the Michaelis-Menten equation,

$$\dot{y} = -\frac{My}{k + y}$$

where y is the amount of the substrate present at the time that is transformed by the enzyme, and M and k are constants. Solve this equation.

7. A population has an initial size of 1 million and a specific growth rate of 0.04 when time is measured in years. The population is losing members through emigration at the rate of $50{,}000(1 + \sin \pi t/5)$ at time t. Find the size of the population at time t.

8. Determine the pressure, P, in the aorta (see Example 1 of Chapter 1) as a function of t by following the outline given below. Assume that the rate at which the heart pumps blood into the aorta is

$$R(t) = \begin{cases} A \sin \omega t & 0 < t < \pi/\omega \quad \text{during systole} \\ 0 & \text{during diastole} \end{cases}$$

(a) Let V be the volume of the aorta. Verify that

$$\frac{dV}{dt} = R(t) - cP$$

(*Hint:* From fluid mechanics, the rate at which blood flows out of the aorta is proportional to the pressure driving the blood. The constant of proportionality is c.)

(b) Under the assumption that

$$V = V_0 + kP$$

where V_0 is the volume of the aorta when $P = 0$ and k is a constant, show that

$$k\dot{P} + cP = R(t)$$

(c) Show that during the systolic phase,

$$P(t) = P_0 e^{-ct/k} + A(c^2 + \omega^2 k^2)^{-1}(ke^{-ct/k} + c \sin \omega t - k \cos \omega t)$$

[*Hint:* Solve the differential equation in (b), noting that it is of the form (8).]

(d) Show that during the diastolic phase,

$$P(t) = P_1 e^{-ct/k}$$

where P_1 is the value of P at the end of the systolic phase.

9. Some nonlinear equations can be reduced to first-order linear equations.

(a) The equation can be solved for y. The solved form is differentiated with respect to t to give a first-order linear equation. Apply this method to solve

$$3p^5 - py + 1 = 0 \qquad (p = \dot{y})$$

[*Hint:* Solve in parametric form $t = t(p)$ and $y = y(p)$.]

(b) The equation can be solved for t. Differentiate the solved form with respect to y and rewrite dt/dy in the form $1/p$ to get a linear equation. Solve

$$p^2 - tp + 1 = 0$$

(c) The equation can be solved for $p = \dot{y}$. Use this method to solve

$$p^2 + pt + py + ty = 0$$

10. Some tumors grow according to the Gompertz law of limited growth [8], described by the differential equation

$$\frac{dy}{dt} = ky \ln\left(\frac{y_e}{y}\right)$$

(a) What is the sign of dy/dt if $y > y_e$?
(b) What is the sign of dy/dt if $y < y_e$?
(c) What conclusions can you draw from (a) and (b)?
(d) Solve the differential equation by substituting $u = \ln(y/y_e)$.

11. The limited growth model of von Bertalanffy is described by

$$\frac{dy}{dt} = 3ky^{2/3}(y_e^{1/3} - y^{1/3})$$

Solve the differential equation by substituting $u = y^{1/3}$.

12. (a) Let the subscripts e, o, i refer to Erie, Ontario, non-Erie inflow to Ontario, respectively. Prove that the differential equation for the concentration of pollutants in Lake Ontario is

$$\dot{C}_o = V_o^{-1}(C_e r_e + C_i(r_o - r_e) - C_o r_o)$$

(b) If about 5/6ths of the inflow to Ontario is the outflow from Erie, prove that

$$C_o(t) = e^{-t/T}\left(C_o(0) + \frac{1}{6T}\int_o^t (5C_e(x) + C_i(x))e^{x/T}\,dx\right)$$

(c) Assume that all the inflow of pollution to Erie and Ontario ceases except for the uncontrollable flow from Erie to Ontario. Find the 50% clean-up times for Ontario in terms of $C_e(0)/C_o(0)$.

4.3 EXISTENCE AND UNIQUENESS OF SOLUTIONS OF DIFFERENTIAL EQUATIONS

It is usually necessary, just as for difference equations, to specify a set of auxiliary conditions, called *initial conditions,* in order to completely specify a unique solution to a differential equation. In Example 2(a), C must be fixed in order to determine a unique solution. One way to do this is to specify the initial value $y(0)$. This determines C by $C = y(0)$.

Higher-order differential equations require additional initial conditions. These additional conditions frequently are specified by assigning initial conditions to the derivatives of the functions as well as assigning its initial value. In Example 2(b), the constants A and B can be determined by specifying $y(0)$ and $\dot{y}(0)$. In fact, this gives $A = y(0)$ and $B = \dot{y}(0)$.

The reason the initial values uniquely determine the solution of a difference equation is obvious. However, it is not obvious how the initial values uniquely determine the solution of a differential equation. The analog of Theorem 1 of Chapter 3 for differential equations is a Taylor's series expansion of the unknown function. The initial values determine the values of the function and its derivatives, which appear as coefficients in the Taylor's series (see Exercise 7). The relation between the initial values and the solution can also be seen by interpreting the differential equation geometrically.

The first-order differential equation

$$\dot{y} = f(t, y) \qquad (t_0 \le t \le T)$$

has a simple geometric interpretation. The solution $y = \phi(t)$ is a curve called the *solution curve*. The slope of the tangent to the curve $y = \phi(t)$ at the point (t, y) is given by $f(t, y)$. Accordingly, even though the solutions are unknown, tangents can be drawn through each point at which f is defined. By drawing many such short tangent lines, called *line elements,* a good idea of the behavior of the solutions is obtained. A line element diagram is shown in Fig. 4.1. Such an assignment is called a *direction field*.

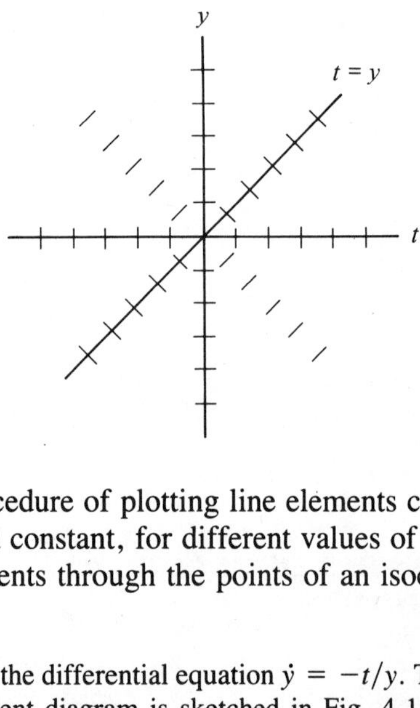

Figure 4.1 Line element diagram of $\dot{y} = -t/y$.

The procedure of plotting line elements can be shortened by plotting the curve $f(t, y) = m$, a constant, for different values of m. These curves are called *isoclines*. The line elements through the points of an isocline all have slope m.

Example 6

Consider the differential equation $\dot{y} = -t/y$. The isoclines are the lines $-t/y = m$. The line element diagram is sketched in Fig. 4.1. Note that on the line $t = y$ all the line elements have slope $m = -1$, and on $t = 0$ they have slope $m = 0$.

Suppose that only the solution through the point (t_0, y_0) is required. In other words, the initial condition $y(t_0) = y_0$ is specified. Then a line element diagram is not necessary. Simply draw a line through (t_0, y_0) with slope $m_0 = f(t_0, y_0)$. A nearby point (t_1, y_1) is chosen on the line with $t_1 > t_0$. Analytically, the point (t_1, y_1) is the intersection of the line $y - y_0 = m_0(t - t_0)$ and $t = t_1$. Next evaluate $m_1 = f(t_1, y_1)$ and draw a line segment from (t_1, y_1) to a nearby point on this line segment (t_2, y_2). Repeating this procedure, a polygonal line is obtained as shown in Fig. 4.2. The points $t_0, t_1, \ldots$ are usually chosen such that $t_1 - t_0 = t_2 - t_1 = \ldots$. It is clear that once (t_0, y_0) is specified, the polygonal line is uniquely determined. Moreover, the shorter the individual line segments, the better the polygonal line approximates the solution to the

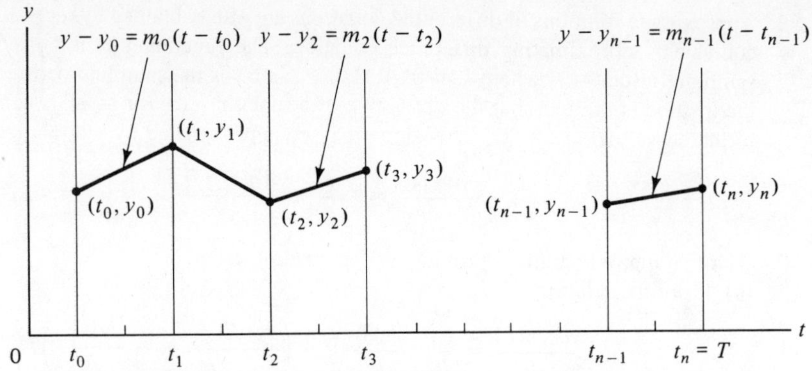

Figure 4.2 Graphical approximation to the solution of $\dot{y} = f(t, y)$.

differential equation. It can be shown, by passage to the limit in the approximation, that a unique solution through (t_0, y_0) is obtained if $f(t, y)$ is continuous and has continuous first partial derivatives with respect to y inside some rectangle containing (t_0, y_0) [2].

A similar result is valid for higher-order differential equations. It will be seen that such an equation can be reduced to a system of first-order equations that contain more dependent variables than the original equation. The same procedure described for a single equation can be carried out for a system of equations, but it is difficult to visualize the procedure graphically in higher dimensions. The following existence theorem for a linear differential equation will be adequate for our needs.

Theorem 1. Consider a linear differential equation of the form

$$\frac{d^n y}{dt^n} + a_{n-1}(t)\frac{d^{n-1}y}{dt^{n-1}} + \cdots + a_0(t)y = f(t) \tag{15}$$

with coefficients $a_i(t)$ $(i = 0, 1, \ldots, n - 1)$ and right-hand side $f(t)$ that are continuous on the interval $0 \le t \le T$. The differential equation (15) has one and only one solution on the given interval corresponding to each set of initial conditions $y(0)$, $dy(0)/dt, \ldots, d^{n-1}y(0)/dt^{n-1}$.

EXERCISES

1. Find a graphical approximation to

$$\dot{y} = 1 + yt^{-1} \qquad [y(1) = 1]$$

as illustrated in Fig. 4.2 according to the following steps.

(a) Find the equation of the line segment through $(1, 1)$ with slope $\dot{y}(1) = 2$.
(b) Verify that its intersection with the line $t = 1.1$ is $y = 1.2$.
(c) Find the equation of the line segment through $(1.1, 1.2)$ with slope $\dot{y}(1.1)$.
(d) Find its intersection with the line $t = 1.2$ and find the next line segment through this point of intersection with slope $\dot{y}(1.2)$.
(e) Compare the values of the polygonal line obtained in (a)–(d) at $t = 1.1, 1.2, 1.3$ with the corresponding values of the exact solution $y = (t + 1/t)/2$.

2. Approximate solutions of differential equations are often obtained by replacing them by an equivalent approximating difference equation. The function $y(t)$ is considered only at sampling instants kT, where $k = 0, 1, 2, \ldots$ and T is the sampling period. The basis of the method is the fact that the derivative of a funciton $y(t)$ at $t = kT$ is simply the slope of the curve $y(t)$ at $t = kT$. This slope may be approximated by the relationship

$$\frac{dy(kT)}{dt} \approx \frac{y(kT + T) - y(kT)}{T}$$

with the approximation improving as T is made smaller.

(a) From the relation

$$\frac{d^2y(kT)}{dt} = \frac{d}{dt}\left(\frac{dy(kT)}{dt}\right) \approx \left(\frac{dy(kT + T)}{dt} - \frac{dy(kT)}{dt}\right)T^{-1}$$

show that

$$\frac{d^2y(kT)}{dt^2} \approx \frac{y(kT + 2T) - 2y(kT + T) + y(kT)}{T^2}$$

(b) What is an approximation for $d^3y(kT)/dt^3$?

3. Consider the differential equation for $t \geq 0$,

$$\frac{dy(kT)}{dt} + y(t) = 1$$

with $y(0) = 0$. The above relation evaluated at $t = kT$ is

$$\frac{dy(kT)}{dt} + y(kT) = u(kT) \qquad (k = 1, 2, 3, \ldots)$$

where $u(kT) = 1$ for $k = 0, 1, 2, 3, \ldots$

(a) Using Exercise 2, show that the equivalent approximating difference equation is

$$y_a[(k + 1)T] + (T - 1)y_a(kT) = Tu(kT)$$

where the subscript a is used to emphasize the fact that $y_a(kT)$ is an approximation of $y(kT)$ and $y_a(0) = 0$.

(b) Show that the solution of the difference equation is

$$y_a(k) = 1 - (1 - T)^k \qquad (k = 0, 1, 2, 3, \ldots)$$

where the explicit appearance of T in $y_a(kT)$ has been dropped for convenience.

(c) Let $T \to 0$ and $k \to \infty$ so that $kT \to t$. Prove that the approximate solution approaches the exact solution $y(t) = 1 - e^{-t}$ [*Hint:* $\lim(1 - T)^k = \lim(1 - \mathrm{T})^{(1/T)/(kT)} = e^{-t}$).]

4. **(a)** Show that the approximating difference equation to

$$\frac{d^2y}{dt^2} - 3\frac{dy}{dt} + 2y = 0$$

with initial conditions $y(0) = 0$ and $\dot{y}(0) = 1$ is

$$y_a(k + 2) - (2 + 3T)y_a(k + 1) + (1 + 3T + 2T^2)y_a(k) = 0$$

with initial conditions $y_a(0) = 0$ and $y_a(1) = T$. {*Hint:* $\dot{y}(0) = 1$ is approximated by $[y_a(1) - y_a(0)]T^{-1} = 1$.}

(b) Solve the approximating difference equation and show that its solution approaches the solution of the differential equation as $T \to 0$ and $k \to \infty$ so that $kT \to t$.

5. A child's head grows more slowly than her body. Allometry studies this different pattern of growth. Let $x(t)$ and $y(t)$ be the sizes of two different parts in the same individual at time t. Empirical data support the *allometric law*, which states that the specific growth rates are proportional; that is,

$$\frac{1}{y}\frac{dy}{dt} = k \cdot \frac{1}{x}\frac{dx}{dt}$$

is a good approximation. The constant $k > 0$ depends only on the nature of the two parts being considered. Eliminate dt and solve the resulting differential equation.

6. Plot the family of trajectories for the following differential equations:
   (a) $\dot{y} = 3$
   (b) $\dot{y} = t$
   (c) $\dot{y} = 1/y$
   (d) $\dot{y} = t/y$
   (e) $\dot{y} = y/t$

7. The following procedure is the analog for differential equations of the method of finding a solution used in the existence theorem (Theorem 1 of Chapter 3) for difference equations. Find $y^{(k)}(0)$ for $k = 0, 1, 2, 3, \ldots$ from the initial conditions and the given differential equation. The solution $y(t)$ is then given as the Taylor's series,

$$y(t) = y(0) + \dot{y}(0)t + \frac{\ddot{y}(0)t^2}{2!} + \cdots + \frac{y^{(k)}(0)t^k}{k!} + \cdots$$

Apply the above method to find the series solution of $\dot{y} = y$ if $y(0) = 1$. [*Hint:* $y(0) = \dot{y}(0) = \ddot{y}(0) = 1$.]

8. Using the method described in Exercise 6, find the first four terms of the series expansion of the solution of $\dot{y} = ty$ with initial conditions $y(0) = \dot{y}(0) = 1$.

9. A rapid and accurate sketch of the family of solutions of a differential equation (solution curves) can be obtained by using other curves besides the isoclines—namely, the locus of inflections and the locus of maxima and minima. Apply this technique to $\dot{y} = -t(y - 1)$ according to the following steps.
   (a) Show that the solution curves are concave down above the line $y = 1$ and concave up below the line of $|t| < 1$; if $|t| > 1$, the situation is reversed.
   (b) Prove that the lines $t = \pm 1$ are the loci of the points of inflection.
   (c) Show that the maxima or minima lie on the line $t = 0$.
   (d) Using the results of (a)–(c) and the fact that the characteristics near $y = 1$ are flatter than those farther away, sketch the family of solutions of the given differential equation.

10. Apply the method of Exercise 9 to sketch the solution curves of $\dot{y} = y + t^2$ [*Hint:* The curve of maxima and minima is the parabola $y = -t^2$. The curve of inflections is the parabola $y - 2 = -(t + 2)^2$. Sketch both of these curves and consider solution curves that cut both, one, and neither of these curves.]

4.4 LINEAR DIFFERENTIAL EQUATIONS

Theorems analogous to those appearing in Section 3.4 can be formulated for linear differential equations.

Theorem 2. Let y be a solution of the linear differential equation (15) and let y_p be a particular solution of (15). Let y_h be a solution of the homogeneous equation $f(t) = 0$ on $0 \le t \le T$ corresponding to (15) satisfying the initial conditions

$$y_h(0) = y(0) - y_p(0),\ \dot{y}_h(0) = \dot{y}(0) - \dot{y}_p(0),\ \ldots,\ \frac{d^{n-1}y_h(0)}{dt^{n-1}}$$

$$= \frac{d^{n-1}y(0)}{dt^{n-1}} - \frac{d^{n-1}y_p(0)}{dt^{n-1}}$$

Then

$$y = y_p + y_h \qquad (0 \le t \le T)$$

The proofs in this section will be omitted, since they are analogous to the proofs of the corresponding theorems for difference equations in Section 3.4.

Example 7 (Growth of a Field Mouse Population)

A population of field mice has a specific growth rate of 0.5 and suffers a loss of 10 mice/month due to predators and emigration. The size of the population, y, satisfies the differential equation

$$\dot{y} = 0.5y - 10$$

where the time, t, is in months. Find the size of the population at time t if initially there were 100 mice.

It can verified that a particular solution of the given equation is $y_p = 20$ and that a solution of the homogeneous equation is $y_h(t) = ce^{0.5t}$, where c is a constant. According to Theorem 2, $y_h(0) = 100 - 20 = 80$. Hence $c = 95$ and the general solution is

$$y = y_h + y_p = 80e^{0.5t} + 20$$

Theorem 3. If $z_1(t), \ldots, z_m(t)$ are all solutions of a linear homogeneous differential equation, then any linear combination of these m solutions,

$$z(t) = c_1 z_1(t) + \cdots + c_m z_m(t)$$

where $c_1, \ldots, c_m$ are arbitrary constants, is also a solution.

Just as for linear difference equations, the following definition is required to describe the finite number of solutions that generate all solutions of the homogeneous equation corresponding to (15).

Definition 6. The functions $z_1(t), \ldots, z_m(t)$ defined on $t_0 \le t \le t_1$ are *linearly independent* if it is impossible to find a relation

$$c_1 z_1(t) + \cdots + c_m z_m(t) = 0$$

valid on $t_0 \le t \le t_1$ except by setting $c_1 = c_2 = \ldots = c_m = 0$. If it is possible to find such a relation, the set of functions is said to be *linearly dependent*.

Definition 7. A *fundamental set* of solutions of an nth-order linear homogeneous equation is any set of n linearly independent solutions.

Lemma 1. There exists a fundamental set of solutions.

Proof. The set of fundamental solutions $f_1(t), \ldots, f_n(t)$ are defined by letting $f_i(t)$ be the solution corresponding to the initial conditions,

$$f_i(0) = 0, \ldots, f_i^{(i-2)}(0) = 0, f_i^{(i-1)}(0) = 1, f_i^{(i)}(0) = 0, \ldots, f_i^{(n-1)}(0) = 0 \tag{16}$$

(The zeroth derivative is defined to be the function itself.)

To show that these solutions are linearly independent, suppose that

$$c_1 f_1(t) + c_2 f_2(t) + \cdots + c_n f_n(t) = 0 \tag{17}$$

Setting $t = 0$ in (17) yields $c_1 = 0$, since $f_i(0) = 0$ for $i \neq 1$ while $f_1(0) = 1$. Differentiating (17) with respect to t and setting $t = 0$ gives

$$c_1 \dot{f}_1(0) + c_2 \dot{f}_2(0) + \cdots + c_n \dot{f}_n(0) = 0$$

Hence $c_2 = 0$, since $\dot{f}(0) = 1$ while $\dot{f}_1(0) = 0$ for $i \neq 2$. In general, differentiating (17) $i - 1$ times with respect to t and setting $t = 0$ yields $c_i = 0$ using (16).

Example 8

Two independent solutions of the differential equation

$$\ddot{y} = 3\dot{y} + 2y = 0$$

are e^t and e^{2t}. To prove this assertion, assume that these solutions are dependent; then

$$c_1 e^t + c_2 e^{2t} = 0 \tag{18}$$

for some constants c_1 and c_2 that are not both zero. Differentiating (18) yields

$$c_1 e^t + 2c_2 e^{2t} = 0 \tag{19}$$

Subtracting (18) from (19) gives $c_2 e^{2t} = 0$, and so $c_2 = 0$. This implies that $c_1 = 0$ from (18), which is a contradiction. Hence, e^t and e^{2t} are independent.

Theorem 4. Let $z_1(t), \ldots, z_n(t)$ be a fundamental set of solutions of an nth-order linear homogeneous equation. Then any solution $z(t)$ of the given linear homogeneous equation can be expressed as a linear combination

$$z(t) = c_1 z_1(t) + c_2 z_2(t) + \cdots + c_n z_n(t)$$

for some constants $c_1, c_2, \ldots, c_n$.

Theorem 4 follows immediately from Example 22 and Theorem 4 in Volume II, Chapter 1.

Combining Theorems 2 and 4 yields the following theorem.

Theorem 5. Let $y_p(t)$ be a particular solution of the nonhomogeneous equation (15) and let $z_1(t), \ldots, z_n(t)$ be n independent solutions of the corresponding homogeneous equation. Then every solution $y(t)$ of (15) is of the form

$$y(t) = y_p(t) + c_1 z_1(t) + \cdots + c_n z_n(t)$$

where the constants $c_1, \ldots, c_n$ are determined from the initial conditions.

EXERCISES

1. **(a)** Verify that $y_h = c_1e^{-t} + c_2e^{3t}$, where c_1 and c_2 are constants, is a solution of the homogeneous differential equation obtained from $\ddot{y} - 2\dot{y} - 3y = 4$.
   (b) Verify that a particular solution of the differential equation in (a) is $y_p(t) = -\frac{4}{3}$.
   (c) Find the general solution of the given differential that satisfies $y(0) = 0$ and $\dot{y}(0) = 2$.
2. Show that the functions e^{-t} and e^{3t} are independent.
3. Show that the functions t^m and t^n are independent if $m \neq n$.
4. Find a set of linearly independent solutions of the differential equation $t^2\ddot{y} + 5t\dot{y} + 3y = 0$. [*Hint:* Try $y(t) = t^k$.]
5. Show that the functions $6, e^t, e^{2t} + e^t, 3e^{2t} + 4$ are linearly dependent.
6. **(a)** Verify that $y_1(t) = t^{-1}$ and $y_2(t) = t^{1/2}$ for $0 < t < \infty$ are solutions of the differential equation
$$2t^2\ddot{y} + 3t\dot{y} - 3y = 0$$
   (b) Do y_1 and y_2 form a fundamental set of solutions?
   (c) Solve the differential equation
$$2t^2\ddot{y} + 3t\dot{y} - y = 3$$
   with the initial conditions $y(0) = 1$ and $\dot{y}(0) = 2$.

4.5 LINEAR DIFFERENTIAL EQUATIONS WITH CONSTANT COEFFICIENTS

In this section it will be shown how to find a fundamental set of solutions of a linear homogeneous equation with constant coefficients. By Theorem 4 this enables one to find the general solution, y_h, of the given homogeneous equation. It will also be shown how the fundamental set of solutions can be used to find a particular solution, y_p, of the corresponding nonhomogeneous equation (15). The general solution of the nonhomogeneous equation is then $y = y_h + y_p$ by Theorem 5.

Theorem 6. To find the general solution of

$$a_n\frac{d^nz}{dt} + a_{n-1}\frac{d^{n-1}z}{dt} + \cdots + a_1\frac{dz}{dt} + a_0z = 0 \tag{20}$$

where $a_0, \ldots, a_n$ are constants and $a_n \neq 0$:

Step 1. Replace z by $e^{\lambda t}$ in (20). This has the effect of replacing d^iz/dt^i by $\lambda^ie^{\lambda t}$. Hence (20) becomes

$$a_n\lambda^n + a_{n-1}\lambda^{n-1} + \cdots + a_1\lambda + a_0 = 0 \tag{21}$$

the so-called *characteristic equation*.

Step 2. Factor the left-hand side of (21) into the distinct factors $\lambda - \lambda_i$ obtaining

$$(\lambda - \lambda_1)^{p_1}(\lambda - \lambda_2)^{p_2}\cdots(\lambda - \lambda_m)^{p_m} = 0 \qquad (p_1 + p_2 + \cdots + p_m = n) \tag{22}$$

The λ_i are called *characteristic roots*.

Step 3.

(a) If $p_1 = p_2 = \ldots = p_m = 1$, then $m = n$ and the general solution of (20) is

$$z(t) = c_1 e^{\lambda_1 t} + c_2 e^{\lambda_2 t} + \cdots + c_n e^{\lambda_n t} \tag{23}$$

(b) If at least one $p_i > 1$, the general solution of (20) is (23) with the following modification: If $p_i > 1$, replace $c_i e^{\lambda_i t}$ by

$$e^{\lambda_i t}(b_0 + b_1 t + b_2 t^2 + \cdots + b_{p_i - 1} t^{p_i - 1}) \tag{24}$$

where $b_0, b_1, \ldots, b_{p_i-1}$ are constants.

Example 9

(a) Solve the differential equation

$$\frac{d^2 z}{dt^2} - 3\frac{dz}{dt} + 2z = 0 \tag{25}$$

The equation can be solved by the steps in Theorem 6.

Step 1. Replace $d^i z/dt^i$ by $\lambda^i (i = 0, 1, 2)$ in (25), yielding

$$\lambda^2 - 3\lambda + 2 = 0$$

Step 2. Factoring the above equation gives

$$(\lambda - 1)(\lambda - 2) = 0$$

Step 3. Since the exponent of each factor in the above equation is one, part of step 1 applies. Thus, the general solution of (25) is

$$z(t) = c_1 e^t + c_2 e^{2t}$$

(b) Solve the differential equation

$$\frac{d^4 z}{dt^4} + 3\frac{d^3 z}{dt} - 7\frac{d^2 z}{dt} - 15\frac{dz}{dt} + 18z = 0$$

Once again Theorem 6 will be applied. Replacing $d^i z/dt^i$ by λ^i in the given differential equation yields

$$\lambda^4 + 3\lambda^3 - 7\lambda^2 - 15\lambda + 18 = (\lambda - 1)(\lambda - 2)(\lambda + 3)^2$$

Since -3 is a repeated characteristic root ($p_3 = 2$), Eq. (24) must be used. Hence the general solution is

$$z(t) = c_1 e^t + c_2 e^{2t} + e^{-3t}(b_0 + b_1 t)$$

Example 10

Find the solution of the linear differential equation

$$\frac{dy}{dt} = 6y + 2$$

satisfying the initial condition $y(0) = \frac{2}{3}$.

The general solution of the homogeneous equation $dy/dt = 6y$ is $y_h = ce^{6t}$, where

c is a constant. A particular solution, y_p, of the nonhomogeneous equation can easily be guessed: $y_p = -\frac{2}{6}$. Hence the general solution of the nonhomogeneous equation is $y = ce^{6t} - \frac{1}{3}$. The constant c is determined from the initial condition

$$\tfrac{2}{3} = y(0) = c - \tfrac{1}{3}$$

Hence $c = 1$.

Example 11

According to Theorem 6, the solution of the differential equation

$$\ddot{y} + 9y = 0$$

is

$$y = c_1 e^{3it} + c_2 e^{-3it} \tag{26}$$

where $i = \sqrt{-1}$. In order that the solution be real, c_1 and c_2 must be complex conjugates. Hence if

$$c_1 = a + bi \tag{27}$$

then

$$c_2 = a - bi \tag{28}$$

Recalling De Moivre's Theorem,

$$e^{i\omega t} = \cos \omega t + i \sin \omega t$$

Eq. (26) can be written as

$$y = (c_1 + c_2) \cos 3t + i(c_1 - c_2) \sin 3t \tag{29}$$

Substituting c_1 and c_2 from (27) and (28) into (29) gives

$$y = 2a \cos 3t - 2b \sin 3t \tag{30}$$

A similar method to that used in Example 11 can be used to prove the following theorem.

Theorem 7. Suppose that the characteristic equation (21) corresponding to the linear differential equation with constant coefficients has a simple complex root $\lambda = \sigma + i\omega$, where σ and ω are real. Then the characteristic equation must also have the conjugate root $\bar{\lambda} = \sigma - i\omega$, and the solution corresponding to

$$c_1 e^{\lambda t} + c_2 e^{\bar{\lambda} t}$$

can be written as

$$e^{\sigma t}(A \cos \omega t + B \sin \omega t)$$

where A and B are real constants.

Note. Suppose that all of the coefficients a_i of the differential equation (21) are real numbers. If the characteristic equation has a complex root $\lambda = \sigma + i\omega$ of multiplicity m, then the characteristic equation also has the complex conjugate root $\bar{\lambda} = \sigma - i\omega$ of the same multiplicity. Applying Theorem 6, the general solution has summands of the form

$$(b_0 + b_1 t + \cdots + b_{m-1}t^{m-1})e^{\lambda t} + (c_0 + c_1 t + \cdots + c_{m-1}t^{m-1})e^{\bar{\lambda} t}$$

It can be shown, by using De Moivre's theorem, that the above summands can be written in the form

$$e^{\sigma t}[(d_0 + d_1 t + \cdots + d_{m-1}t^{m-1}) \sin \omega t] + e^{\sigma t}[(e_0 + e_1 t + \cdots + e_{m-1}t^{m-1}) \cos \omega t]$$

where all the d_i and e_i are real constants. From this result and Theorem 6, it follows that the general solution of a linear, homogeneous differential equation with real and constant coefficients can always be expressed as a linear combination of real-valued functions.

To illustrate the above result, consider the differential equation

$$y^{(4)} - 4\dddot{y} + 8\ddot{y} - 8\dot{y} + 4y = 0$$

with its characteristic equation

$$0 = \lambda^4 - 4\lambda^3 + 8\lambda^2 - 8\lambda + 4 = [\lambda - (1 + i)]^2[\lambda - (1 - i)]^2$$

The general solution of the given differential equation is

$$(d_0 + d_1 t)e^t \sin t + (e_0 + e_1 t)e^t \cos t$$

In order to analyze the vibratory nature of the solution in Theorem 7, the following lemma will be required.

Lemma 2. Any function of the form

$$y(t) = A \cos \omega t + B \sin \omega t$$

can be written in the simpler form

$$y(t) = R \cos(\omega t - \delta)$$

where $R = \sqrt{A^2 + B^2}$ and $\delta = \tan^{-1}(B/A)$.

Notice that $y(t)$ always lies between $-R$ and R and that the motion is periodic—it repeats itself over every time interval of length $2\pi/\omega$. This type of motion is called *simple harmonic motion*. R is called the *amplitude* of the motion, δ the *phase angle* of the motion, $T = 2\pi/\omega$ the *natural period* of the motion, and ω the *natural frequency* of the motion.

In general, a particular solution of the nonhomogeneous equation

$$a_n \frac{d^n y}{dt^n} + a_{n-1}\frac{d^{n-1}y}{dt^{n-1}} + \cdots + a_0 y = f(t) \tag{31}$$

cannot be guessed. However, if a fundamental set of solutions of the corresponding homogeneous equation (20) is known, these solutions can be used to find a particular solution of (31). The technique is called the *method of variation of parameters,* since the constants c_i appearing in the linear combination of fundamental solutions are assumed to be functions of t. These functions can be found by substituting the solution in (31) and deriving a first-order equation for each c_i. The method will be illustrated for a second-order equation. The generalization to higher-order equations is evident.

Consider the second-order linear nonhomogeneous equation

$$\ddot{y} + p(t)\dot{y} + q(t)y = r(t) \tag{32}$$

Suppose that y_1 and y_2 form a fundamental set for the homogeneous equation corresponding to (32). Assume that a particular solution of (27) is of the form

$$y = c_1 y_1 + c_2 y_2$$

where c_1 and c_2 are functions of t that will be determined by substituting y in (32). Now

$$\dot{y} = c_1 \dot{y}_1 + c_2 \dot{y}_2 + \dot{c}_1 y_1 + \dot{c}_2 y_2$$

Since a first-order equation is desired, set

$$\dot{c}_1 y_1 + \dot{c}_2 y_2 = 0 \tag{33}$$

Hence, $\dot{y} = c_1 \dot{y}_1 + c_2 \dot{y}_2$ and so

$$\ddot{y} = c_1 \ddot{y}_1 + c_2 \ddot{y}_2 + \dot{c}_1 y_1 + \dot{c}_2 y_2$$

Finally, substituting y, $\dot{y}$, and $\ddot{y}$ in (32) gives

$$c_1 \ddot{y}_1 + c_2 \ddot{y}_2 + \dot{c}_1 \dot{y}_1 + \dot{c}_2 \dot{y}_2 + p(t)(c_1 \dot{y}_1 + c_2 \dot{y}_2) + q(t)(c_1 y_1 + c_2 y_2) = r(t)$$

which can be rearranged as follows:

$$c_1[\ddot{y}_1 + p(t)\dot{y}_1 + q(t)y_1] + c_2[\ddot{y}_2 + p(t)\dot{y}_2 + q(t)y_2] + \dot{c}_1 \dot{y}_1 + \dot{c}_2 \dot{y}_2 = r(t) \tag{34}$$

Since both y_1 and y_2 are solutions of the homogeneous equation, (34) reduces to

$$\dot{c}_1 \dot{y}_1 + \dot{c}_2 \dot{y}_2 = r(t) \tag{35}$$

The equations (33) and (35) can be solved for $\dot{c}_1$ and $\dot{c}_2$, resulting in

$$\dot{c}_1 = \frac{-ry_2}{y_1 \dot{y}_2 - \dot{y}_1 y_2} \tag{36}$$

$$\dot{c}_2 = \frac{ry_1}{y_1 \dot{y}_2 - \dot{y}_1 y_2} \tag{37}$$

The functions appearing on the right side in (36) and (37) are all known functions of t. Hence c_1 and c_2 can be determined by a single integration. Thus a particular solution has been found.

Example 12

Find the general solution of

$$\ddot{y} - y = e^t$$

A fundamental set of solutions of $\ddot{y} - y = 0$ is $y_1 = e^t$ and $y_2 = e^{-t}$. To find a particular solution of the nonhomogeneous equation, let $y_p = c_1 e^t + c_2 e^{-t}$, where c_1 and c_2 are functions of t. Equations (36) and (37) with $y_1 = e^t$, $y_2 = e^{-t}$, and $r = e^t$ reduce to

$$\dot{c}_1 = \frac{-1}{-2} = \frac{1}{2} \quad \text{or} \quad c_1(t) = \frac{t}{2}$$

$$\dot{c}_2 = \frac{e^{2t}}{-2} \quad \text{or} \quad c_2(t) = \frac{-e^{2t}}{4}$$

Hence, a particular solution of the nonhomogeneous equation is

$$y_p(t) = \frac{te^t}{2} - \frac{e^{2t}e^{-t}}{4}$$

Thus, the general solution of the nonhomogeneous equation is

$$y = c_1e^t + c_2e^{-t} + e^t\frac{(t - \frac{1}{2})}{2}$$

where c_1 and c_2 are constants. Since e^t is a solution of the homogeneous equation, the general solution can be written in the simpler form

$$y = b_1e^t + b_2e^{-t} + \frac{te^t}{2}$$

where b_1 and b_2 are constants.

Example 13 (Free-Swinging Human Limb)

Experimental evidence shows that the speed of relaxed walking is adjusted to that requiring minimum respiration. The average minimum energy expenditure velocity is 2.25 or 3.5 miles per hour (depending on whether or not the resting metabolism is deducted from the total energy expenditure rate). The rough model of a free-swinging leg will be formulated. Then the theoretical velocity of walking will be calculated and compared with the experimental velocity.

The following assumptions will be made. In walking, the rear leg is raised without shortening and allowed to swing forward freely. Walking is considered to be a passive controlled process (the neuromuscular feedback loop of the antagonistic pairs of effector muscles is ignored). The leg is assumed to be a thin, stiff, uniform cylinder. (See Fig. 4.3.) The large inertial mass of the foot compensates partially for the tapering nature of

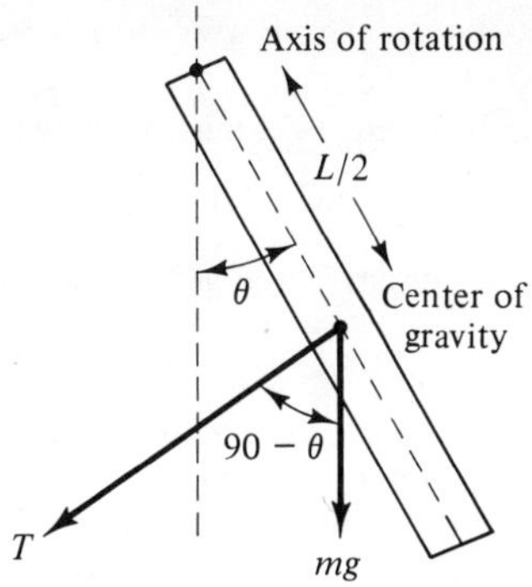

Figure 4.3 Freely swinging leg.

actual legs. The average leg length $L \approx 90$ cm. The diameter, D, of the leg is approximately 12 cm. The leg is assumed to be of uniform density, $\rho \approx 1.1$ g/cm^3. Hence the mass of the leg is

$$m = \rho\pi\left(\frac{D}{2}\right)^2 L = 11.2 \text{ kg}$$

The equation of motion of the leg will be derived using the analog of Newton's equation for angular motion. Recalling that the analog of a force is a torque, T, Newton's second law, $F = m\ddot{x}$, becomes

$$J\ddot{\theta} = T \tag{38}$$

The moment of inertia of a thin cylindrical leg swinging about the hip joint is

$$J = \frac{mL^2}{3} \tag{39}$$

The mass of the leg is concentrated at the center of gravity of the leg. The torque, T, perpendicular to the leg is obtained by resolving the force due to the weight of the leg, mg, which acts vertically downward, into a component perpendicular to the leg. Recalling that torque equals force times the distance that the force acts from the center of rotation gives

$$T = -mg\cos(90 - \theta)\left(\frac{L}{2}\right) = -\frac{(mgL\sin\theta)}{2} \tag{40}$$

Substituting the values of J and T from (39) and (40), respectively, into (38) gives

$$\ddot{\theta} + \frac{3g\sin\theta}{2L} = 0 \tag{41}$$

For a small angle θ, $\sin\theta \approx \theta$. For the swinging 90-cm leg with the assumption that a half-pace is 46 cm (18 in), the maximum value of θ is 0.51 radians, whereas $\sin\theta = 0.49$. Thus the maximum linearization error in replacing $\sin\theta$ by θ is 4.1%. Using the approximation $\sin\theta \approx \theta$ in (41) gives

$$\ddot{\theta} + \omega^2\theta = 0 \tag{42}$$

where the undamped natural frequency

$$\omega = \left(\frac{3g}{2L}\right)^{1/2} \tag{43}$$

Equation (42) describes the motion of the leg, which intuitively must swing back and forth. To verify this fact, Eq. (42) must be solved. The characteristic equation is

$$\lambda^2 + \omega^2 = (\lambda + \omega i)(\lambda - \omega i)$$

where $i = \sqrt{-1}$. From Theorem 7, the general solution of (42) is

$$\theta = c_1\cos\omega t + c_2\sin\omega t \tag{44}$$

where c_1 and c_2 are real constants. Lemma 6 shows that the motion is periodic. Hence the leg executes simple harmonic motion.

The walking speed can be calculated from the theoretical model as follows. From (43), $\omega \approx 4$ radians/sec. Hence, the frequency $f = \omega/2\pi = 0.64$ double-paces/sec. Assuming a 3-ft pace, the walking speed is 3.8 ft/sec or 2.6 mph. The rough model gives a good approximation to the experimental result!

Although energy dissipation effects are small in relaxed walking, they nevertheless exist. Such effects will be considered in the next example.

Example 14 (Damped Free-Swinging Leg)

Energy dissipation effects are included by adding the torque $-b\dot{\theta}$, where $b > 0$, to the right side of (38). A negative sign is required, since viscous resistance of the tissues opposes the motion. Thus (42) becomes

$$\ddot{\theta} + b\dot{\theta} + \omega^2\theta = 0 \tag{45}$$

The roots of the characteristic equation $\lambda^2 + b\lambda + \omega^2 = 0$ are

$$\lambda_1 = \frac{-b + \sqrt{b^2 - 4\omega^2}}{2} \qquad \lambda_2 = \frac{-b - \sqrt{b^2 - 4\omega^2}}{2}$$

Hence there are three possibilities, depending on whether $b^2 - 4\omega^2$ is positive, negative, or zero.

1. $b^2 - 4\omega^2 > 0$ (overdamped motion)

   In this situation both λ_1 and λ_2 are negative, and every solution of (45) is of the form

   $$\theta(t) = c_1 e^{\lambda_1 t} + c_2 e^{\lambda_2 t}$$

   Thus, after a long time, the leg practically ceases moving ($t \to \infty$ implies $\theta \to 0$).

2. $b^2 - 4\omega^2 = 0$ (critically damped motion)

   Since $\lambda_1 = \lambda_2 = -b/2$, every solution of (45) is of the form

   $$\theta(t) = (c_1 + c_2 t)e^{-bt/2}$$

   Once again, after a sufficiently long time the leg practically ceases moving.

3. $b^2 - 4\omega^2 < 0$ (underdamped motion)

   The characteristic equation has complex conjugate roots. Hence, by Theorem 7, the solution of (45) is

   $$\theta(t) = e^{-bt/2}(c_1 \cos \mu t + c_2 \sin \mu t)$$

   where $\mu = \sqrt{4\omega^2 - b^2}/2$

   The solution represents a damped vibration. To see this, rewrite the solution, by applying Lemma 2, in the form

   $$\theta(t) = Re^{-bt/2} \cos(\mu t - \delta)$$

   The angle θ oscillates between the curves $\theta = \pm Re^{-bt/2}$ and thus represents a cosine curve with decreasing amplitude, as shown in Fig. 4.4.

In relaxed walking the energy dissipated in small, and so b is small and case 3 holds. Although it may take some time for the motion to cease, nevertheless the motion will eventually cease. Hence, in steady walking, the appropriate leg muscles must pump in enough energy to compensate for this viscous dissipation as well as providing the various positionally required forces which the analysis has ignored. This is especially true for nonrelaxed walking. For example, the army marches at 120 paces/min or 4 mph. Extra muscle power must be expended to drive the system at a higher rate than its natural frequency. The effect of adding a suitable time-varying torque function, T_m, representing muscle torque, will be considered in Example 15.

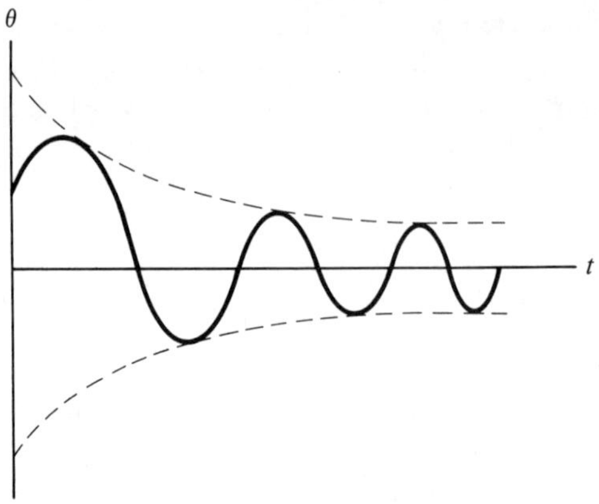

Figure 4.4 Underdamped motion described by $\theta = e^{-bt/2} \cos (\mu t - \delta)$.

The motion of many mechanical systems is described by an equation of the form (30), where θ is interpreted as a displacement distant. The term $-b\dot{\theta}$ is interpreted as resistance to motion. A mechanical device for producing such resistance is called a "dashpot" by engineers. The force term $-\omega^2\theta$ is often provided by a spring. The reason such spring-mass-dashpot systems are so useful in mechanical systems is that they can be used to damp out any undesirable disturbances. For example, the shock transmitted to an automobile by a pothole in the road is dissipated by the shock absorbers in the car. The momentum from the recoil of a large gun barrel is dissipated by a spring-mass-damper system attached to the gun.

Example 15 (Damped Forced Vibrations of a Leg)

Consider a time-varying muscle torque, T_m, of the form $T_m(t) = T_0 \cos \psi t$. The differential equation of the motion of the leg is now

$$\ddot{\theta} + b\dot{\theta} + \omega^2\theta = T_0 \cos \psi t \tag{46}$$

since the driving torque must be added to the right-hand side of (44).

The general solution of (46) is of the form

$$\theta = \theta_h + \theta_p$$

where θ_h is a solution of the corresponding homogeneous differential equation (45) and θ_p is a particular solution of (46). This substitution gives, after grouping the terms containing *cos* ψt and sin ψt,

$$(\omega^2 A + b\psi B - \psi^2 A) \cos \psi t + (\omega^2 B - b\psi A - \psi^2 B) \sin \psi t = T_0 \cos \psi t$$

Equating coefficients of cos ψt and sin ψt on both sides of the last equation yields

$$(\omega^2 - \psi^2)A + b\psi B = T_0$$

$$-b\psi A + (\omega^2 - \psi^2)B = 0$$

The solution of the above system of equations is

$$A = \frac{T_0(\omega^2 - \psi^2)}{(\omega^2 - \psi^2)^2 + b^2\psi^2}$$

$$B = \frac{T_0 b\psi}{(\omega^2 - \psi^2)^2 + b^2\psi^2}$$

and so

$$\begin{aligned}\theta_p &= \frac{T_0}{(\omega^2 - \psi^2)^2 + b^2\psi^2}[(\omega^2 - \psi^2)\cos\psi t + b\psi\sin\psi t] \\ &= \left(\frac{T_0}{(\omega^2 - \psi^2)^2 + b^2\psi^2}\right)[(\omega^2 - \psi^2)^2 + b^2\psi^2]^{1/2}(\cos\delta\cos\psi t + \sin\delta\sin\psi t) \\ &= \frac{T_0}{[(\omega^2 - \psi^2) + b^2\psi^2]^{1/2}}\cos(\psi t - \delta) \qquad (47)\end{aligned}$$

where

$$\tan\delta = \frac{b\psi}{\omega^2 - \psi^2}$$

In Example 14 it was shown that every solution θ_h of the corresponding homogeneous equation approaches zero as t approaches infinity. Thus, for large t, the equation $\theta(t) = \theta_p(t)$ describes the motion of the leg quite accurately, regardless of its initial position and velocity. For this reason θ_p is called the *steady-state* part of the solution θ, while θ_h is called the *transient* part of the solution. Note that the ultimate frequency of vibration of the leg is the same as the frequency of vibration of the forcing term.

Example 16 (Forced Free Vibrations)

Assume that there is no damping ($b = 0$), but there is still a forcing term of the form $T_0 \cos \psi t$. In this situation, the solution of the differential equation (46) with $b = 0$ using the particular solution (47) with $b = 0$ has the form

$$\theta(t) = c_1 \cos\omega t + c_2 \sin\omega t + \frac{T_0}{\omega^2 - \psi^2}\cos\psi t$$

This solution is the sum of two periodic functions of different periods. An interesting situation occurs when $\psi = \omega$, that is, when the frequency of the external force equals the natural frequency of the system. This situation is called *resonance,* and the differential equation of motion is

$$\ddot{\theta} + \omega^2\theta = T_0\cos\omega t \qquad (48)$$

A particular solution of (48) will be found as the real part of a complex-valued solution of the equation

$$\ddot{\theta} + \omega^2\theta = T_0 e^{i\omega t} \qquad (49)$$

The method of variation of parameters will be used. Two independent solutions of the homogeneous equation corresponding to (49) are $\theta_1 = e^{i\omega t}$ and $\theta_2 = e^{-i\omega t}$. Hence a particular solution of (48) is assumed to be of the form $c_1\theta_1 + c_2\theta_2$. The parameters c_1 and c_2 can be determined from (46) and (47) and satisfy the differential equations

$$\dot{c}_1 = \frac{T_0}{2i\omega} \qquad (50)$$

$$\dot{c}_2 = -\frac{T_0 e^{2i\omega t}}{2i\omega} \tag{51}$$

If Eq. (51) is solved for c_2 and the result is multiplied by θ_2, a multiple of $e^{i\omega t}$ is obtained. Such a function is a solution of the homogeneous equation corresponding to (49) and so can be neglected, since a particular solution of (49) is desired. Solving (50) gives $c_1 = T_0 t/2i\omega$ and the particular solution

$$c_1\theta_1 = \frac{T_0 t e^{i\omega t}}{2i\omega} = \frac{T_0 t(\cos \omega t + i \sin \omega t)}{2i\omega}$$

of (49). Hence, a particular solution of (48) is $\text{Re}(c_1\theta_1)$, the real part of $c_1\theta_1$:

$$\text{Re}(c_1\theta_1) = \frac{T_0 t \sin \omega t}{2\omega}$$

Consequently, every solution of (48) is of the form

$$\theta = c_1 \cos \omega t + c_2 \sin \omega t + \frac{T_0 t \sin \omega t}{2\omega} \tag{52}$$

for some choice of constants c_1 and c_2.

The sum of the first two terms in (52) is a periodic function of time. The third term represents an oscillation with increasing amplitude. Thus the forcing term $T_0 \cos \omega t$, if it is in resonance with the natural frequency of the system, will always cause unbounded oscillations. Such oscillations are not normally observed when walking or in other limb movements, such as swinging an arm. There are probably two reasons why wild oscillations do not occur. First, there is some damping. Second, and more important, is that normally there is neuromuscular feedback, which automatically changes the time-varying muscular torque if large oscillations occur. However, in certain neuromuscular disorders* large oscillations in limb movements can occur due to a defect in neuromuscular feedback. Certain muscles may also be much more relaxed than normally, which would tend to decrease the resistance to motion.

Resonance occurs frequently in mechanical systems and has been responsible for many catastrophes. Currents of air have served as driving forces setting off destructive oscillations in airplanes, steel factory chimneys, and other structures. A famous example of destructive resonance was the collapse of the Tacoma Narrows Bridge at Puget Sound in the state of Washington on November 7, 1940, about 5 months after it had been constructed.

The phenomenon of resonance was also responsible for the collapse of the Broughton suspension bridge near Manchester, England, in 1831. This occurred when a column of soldiers marched in time over the bridge, thereby setting up a periodic force of rather large amplitude. The frequency of this force was equal to the natural frequency of the bridge. Thus very large oscillations were induced and the bridge collapsed. It is for this reason that soldiers are ordered to break step when crossing a bridge.

* For example, in ballism, a symptom of a disease of the extrapyramidal nervous system.

EXERCISES

1. Find the general solution of the following differential equations:
   (a) $\ddot{y} - 3\dot{y} - 10y = 0$
   (b) $\ddot{y} - 6\dot{y} + 13y = 0$
   (c) $\dddot{y} - 3\ddot{y} + 4y = 0$
   (d) $\dddot{y} + 12\ddot{y} + 48\dot{y} + 256y = 0$
2. Solve the differential equation

$$\ddot{y} - 2\dot{y} - 35y = 0$$

   subject to the initial conditions $y(0) = \dot{y}(0) = 4$.
3. Solve the initial value problem

$$\ddot{y} - 4\dot{y} + 3y = e^{2t} \qquad y(0) = 1,\ \dot{y}(0) = 0$$

4. The method of variation of parameters often leads to difficult integrations. A particular solution can often be found by judicious guessing. Find a particular solution of the following differential equations by the indicated methods.
   (a) $\ddot{y} - 3\dot{y} + 2y = 1 + t + 3t^2$. Substitute $y_p(t) = a_0 + a_1t + a_2t^2$ and determine the unknown coefficients.
   (b) $\ddot{y} - 3\dot{y} = 1 + t + 3t^2$. Try $y_p(t) = a_0 + a_1t + a_2t^2 + a_3t^3$.
   (c) $\ddot{y} - 3\dot{y} + 2y = e^t(1 + t + 3t^2)$. Set $y = e^t v$ and obtain an equation of the form (a) for v and then solve. (Note that equations involving $\cos \omega t$ or $\sin \omega t$ instead of $e^{\omega t}$ can be treated by similar methods using the fact that $\cos \omega t = R_e e^{i\omega t}$ and $\sin \omega t = I_m e^{i\omega t}$ and using complex exponentials.)
5. The initial size of a certain fish population is 1000 and its specific growth rate is 0.75. The harvesting rate is $c(2 + \cos \pi t/4)$. If time is measured in months, what value of c would lead to a zero change in population size during the first year ($t = 0$ to $t = 12$)?
6. The initial size of a herd of cattle is $y(0)$. If its specific growth rate is k and the harvesting is he^{kt}, find the size of the population at item t.
7. The glucose concentration, $c(t)$, of the blood at time t is increased at a constant rate by infusion. Simultaneously, glucose is excreted at a rate that is proportional to the present concentration of glucose with proportionality constant $k > 0$.
   (a) Verify that $dc/dt = r - kc$.
   (b) Solve the differential equation in (a).
8. In humans, the semicircular canals are small-bore circuits containing fluid (endolymph) that rotate slightly relative to the skull when the latter rotates in space. This small relative motion deflects a watertight flap, the *cupula,* in a special compartment within the circuit, the *ampulla*. The cupula is a hair-cell transducer whose deflection produces appropriate neural signals.

   The deflection of the cupula, ϕ, per unit relative angular displacement between skull and fluid satisfies the differential equation

$$J\ddot{\phi} + b\dot{\phi} + k\phi = (aJ)\ddot{\psi}$$

   where a, b, J, and k are constants and ψ is the absolute position of the skull. Find a relation satisfied by the constants so that the system is overdamped. (To sense position changes efficiently, the oscillations of the cupula should die out rapidly.)
9. A patient's temperature is measured using a mercury-in-glass thermometer. Let
   T_b = body temperature of the patient

T_m = temperature of the mercury
S = specific heat of mercury
C = heat transfer coefficient between body and mercury
M = mass of mercury

According to Newton's law of cooling, the heat flow through the glass is

$$\dot{Q} = C(T_b - T_m)$$

(a) Show that the temperature of the mercury satisfies the differential equation

$$\frac{dT_m}{dt} + aT_m = aT_b$$

where $a = C/MS$. {*Hint:* The heat passing through the glass is absorbed by the mercury. Hence $Q = MS[T_m - T_m(0)]$. Find $\dot{Q}$.}

(b) Find T_m.

(c) Show that the temperature of the mercury eventually approaches the temperature of the body.

10. A simple model of the cardiovascular system represents the blood vessels as elastic reservoirs of volume V and constant *compliance* $C = dV/dP$, where P is the pressure in the reservoir. Let the ventricular output be $\dot{Q}_1$. The flow into the arteriole is $\dot{Q}_2$ against a fixed peripheral resistance R. Assume that $\dot{Q}_2 = P/R$.

(a) Verify that $\dot{V} = \dot{Q}_1 - \dot{Q}_2$.

(b) Show that the differential equation relating P to the output $\dot{Q}_1$ is

$$C\frac{dP}{dt} + \frac{P}{R} = \dot{Q}_1$$

(Compare this problem with Exercise 8 of Section 4.2.)

4.6 GLUCOSE TOLERANCE TEST AND THE INDICATOR-DILUTION METHOD

In this section the theory of linear differential equations will be applied to the glucose tolerance test and the indicator-dilution method. The models for these studies use the notion of a *compartment,* which is defined as follows. If a substance is present in a biological system in several distinguishable forms or locations, and if it passes from one form to another form or location at a measurable rate, then each form or location constitutes a separate compartment for the substance.

Example 17 (Glucose Tolerance Test)

In the glucose tolerance test, 50 g of glucose are administered orally. Subsequently, the blood sugar concentration level is determined from samples of blood taken at regular intervals. The blood glucose of a normal person rises from approximately 90 mg%* to approximately 140 mg% and falls back to below normal within 3 hr. A diabetic usually has a blood glucose concentration at a control value that is above 120 mg%. Upon ingestion of 50 g of glucose, a diabetic exhibits a progressive, slow rise in his or her blood glucose level for 2 to 3 hr and then the glucose level falls back to the control value

* mg% = mg/100 ml.

after about 5 or 6 hr. It never falls below the control value. This behavior indicates that the normal increase in insulin* secretion following glucose ingestion does not occur, which is an indication of diabetes mellitus.

The glucose tolerance test is sometimes performed by administering glucose intravenously at a continuous rate. This form of the test is used when the patient cannot tolerate glucose orally or suffers from the postgastrectomy dumping syndrome. The continuous infusion method avoids variation in results due to variations in the rate of intestinal absorption of sugar.

Jokipii and Turpeinen [5] obtained the average blood glucose concentration as a function of time from 24 normal subjects (persons not suffering from diabetes or other diseases known to affect carbohydrate metabolism). In this experiment, the continuous infusion rate of glucose into the blood was

$$I = 297 \text{ mg/min}$$

and the mean level of blood glucose concentration in the fasting state was $c(0) =$ 84.9 mg/dl (dl = deciliter). The blood glucose concentration in milligrams per deciliter was fitted by the formula

$$c = 138.7 - 53.8e^{-0.0519t} \tag{53}$$

where the time t is measured in minutes.

This result can be modeled using the notion of a compartment. Let V be the volume in deciliters of distribution of the glucose. Let $y(t)$ be the blood glucose concentration in milligrams per deciliter at time t, measured in minutes. Note that the amount of glucose in the glucose compartment at time t is $Vy(t)$ mg. The fractional rate (in mg/min) at which glucose leaves or is cleared from the blood will be denoted by L. Glucose is cleared from its compartment only when its concentration exceeds the normal blood concentration. Hence, the rate of glucose clearance is $LV(y - 84.9)$ mg/min. The following conservation principle,

rate of change of mass of glucose

= rate of outflow of glucose + rate of inflow of glucose

when translated into mathematical terms, yields

$$\frac{d(Vy)}{dt} = -LV(y - 84.9) + I$$

or, equivalently,

$$\frac{dy}{dt} = -Ly + 84.9L + \frac{I}{V} \tag{54}$$

The solution of (54) is

$$y = Ae^{-Lt} + 84.9 + \frac{I}{LV} \tag{55}$$

where A is a constant. In order for (53) and (55) to coincide,

$$L = 0.0519$$

*Insulin is a hormone produced by the β cells in the islet of Langerhans of the spleen that helps regulate the blood glucose level.

$$A = -54.8$$

Finally, letting $t \to \infty$ in both (53) and (55) and equating the asymptotic values yields

$$138.7 = 84.9 + \frac{297}{0.0519V}$$

Solving this last relation gives

$$V = 10.6 \text{ liters}$$

Note that the volume of distribution V is about twice the mean blood volume in an adult, which is approximately 5 liters. This difference suggests that the compartment represents both the blood and an extravascular volume.

Experiments of stopping the glucose infusion, or alternately giving a single loading dose, and allowing the system to settle to its equilibrium value have also been performed. It has been observed that the blood glucose level first decays to a value that is less than the normal fasting level but eventually reaches the normal level. Such behavior indicates that the simple one-compartment model is not adequate to explain glucose metabolism. More complicated models of glucose metabolism have been proposed. These utilize insulin as an additional variable. One such model will be discussed when the theory of systems of differential equations is presented.

Example 18 (Stewart-Hamilton Method [7, 11, 18])

The indicator-dilution method is used to determine the volume of blood or mean blood flow rate through the cardiovascular system or a portion of it. In this example it will be shown how to find the volume of the left ventricle and the rate at which the heart pumps blood, the *cardiac output*. The cardiac output can be altered significantly by certain diseases of the cardiovascular and nervous systems. The normal cardiac output is about 5 liters/min in a resting person, but can increase to more than 30 liters/min during strenuous exercise.

First the experimental method will be described. At time $t = 0$, a known amount D mg of a dye is injected into a main vein near the heart. After a short while the blood in the vein will enter the left ventricle and be pumped from there into the aorta. The concentration of the dye in the aorta is monitored at fixed time intervals. The dye concentration in the aorta is plotted as a function of time. A typical smooth curve fitted to the observations is shown in Fig. 4.5.

The experimental curve, $c(t)$, displays a time delay of about 5 sec, which is the time it takes for the injected dye to reach the heart. Note that the concentration does not reach a maximum instantaneously, because of the finite time interval for injection, the time required for the dye to diffuse and be conveyed from the injection site to the heart, and so on. Note the second local maximum at about $t = 20$ sec. It is known from physiology that about 20 sec are required for some blood passing through the aorta to make a round trip and reappear in the aorta. Thus, at about 20 sec, a little dye, having recirculated, is added to what is left of the "first-time-through" dye to cause the second local peak.

The part of the dye concentration curve after about 20 sec can be determined under the following assumption. Assume that the "first-time-through" dye concentration continues to decrease at the same rate as just before the second peak, as shown by the dashed curve in Fig. 4.5. Notice that the experimental curve $c(t)$, just after the first peak ($t > t_0$)

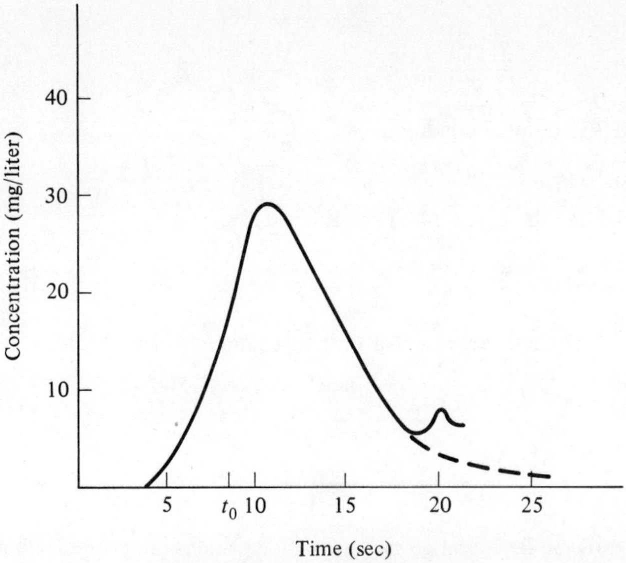

Figure 4.5 Dye concentration in the aorta as a function of time.

and its continuation by the dashed curve appears to be an exponential. This conjecture is valid, since a semilog plot of the tail of the curve $c(t)$ (disregarding the recirculation portion) is a straight line with negative slope λ. Thus this portion of the curve can be fitted by the exponential curve

$$c(t) = Ke^{\lambda t} \tag{56}$$

where K and $\lambda < 0$ are known constants determined from the experimental data. This representation is appropriate only after a certain time t_0 (see Fig. 4.5), which will be determined later.

The volume, V, of the left ventricle, the cardiac output, R, and t_0 will be determined from (56) by constructing a single-compartment model based on the following assumptions:

1. The ventricle is assumed to be a reservoir of constant volume with a constant rate of inflow and outflow.
2. A negligible amount of dye is lost through the capillary walls during the observation period.
3. All of the dye enters the ventricle instantaneously at time t_0 following its injection into thc vcin at timc $t = 0$.
4. All of the dye leaving the left ventricle must pass by the selected point of observation in the aorta.
5. The time that it takes the dye to reach the point of observation is negligibly small.

The theoretical concentration curve will be derived using the above assumptions. Let $y(t)$ be the concentration of the dye at the point of observation at time t in the aorta. By assumptions 4 and 5, $y(t)$ is also the concentration of the dye at time t in the aorta. According to assumption 1, for $t > t_0$,

$$\text{rate of change of mass of dye} = -\text{outflow rate of dye}$$

or, mathematically,

$$\frac{d(yV)}{dt} = -Ry \tag{57}$$

From assumptions 2 and 3, the initial conditions are

$$y(t_0)V = D$$

The solution of (57) satisfying the given initial conditions is

$$y(t) = \frac{D}{V}e^{-R(t-t_0)/V} \qquad (t \geq t_0) \tag{58}$$

Comparing the theoretical (58) and experimental (56) concentration curves yields

$$\frac{R}{V} = -\lambda \tag{59}$$

$$\frac{D}{V} = Ke^{\lambda t_0} \tag{60}$$

Since there are three unknowns, R, V, and t_0, and only two equations, another equation is required to solve for the unknowns.

An additional relation can be found by deriving an equation for the total amount of dye D. Consider the time interval t to $t + dt$. At the beginning of this interval the concentration of the dye is $c(t)$. The volume of blood flowing past the observation point during dt is $R\ dt$. The total amount of dye flowing past this observation point during this interval is approximately $c(t)R\ dt$. Hence the total amount of injected dye is

$$D = R\int_0^\infty c(t)\,dt$$

where the portion of the curve representing recirculation is replaced by the extrapolated value of the curve given by (56). The equation for D can be solved for the cardiac output,

$$R \approx \frac{D}{A} = \frac{D}{\displaystyle\int_0^\infty c(t)\,dt} \tag{61}$$

where A is the area under the experimental curve using the extrapolated portion. An alternative method of obtaining (61) appears in Exercise 5. Equation (61) is known as *Hamilton's formula*. Solving (59) for V in terms of R and replacing R by its value from (61) yields

$$V = \frac{-D}{\lambda A} \tag{62}$$

Replacing V in (60) by its value from (62) and solving the resulting equation for t_0 gives

$$t_0 = -\frac{1}{\lambda}\ln\left(\frac{-K}{\lambda A}\right) \tag{63}$$

The theoretical curve (58) based on an instantaneous injection of the dye into the compartment at time t_0 is shown in Fig. 4.6. The time t_0, given by (62), ensures that the areas under the theoretical and experimental curves are equal. The fact that the experi-

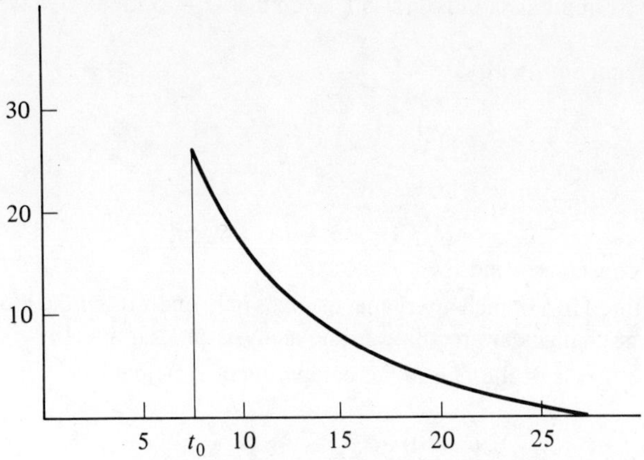

Figure 4.6 Theoretical dye concentration curve given by (58).

mental curve does not rise sharply to its maximum value is due to the lack of instantaneity of dye injection into the ventricle. A more realistic theoretical curve can be obtained by using convective diffusion equations to model indicator-dilution curves [12].

The area A under the experimental "first-time-through" dye curve cannot be found analytically. A numerical method such as approximating Riemann sums or the trapezoidal or parabolic rule [6] must be used. There are mechanical devices, called *planimeters*, that automatically yield an estimate of the area of a region by merely tracing the boundary of the region with one of the movable arms of the device. The area can also be approximated by counting squares on graph paper.

Finally, the assumptions for the model will be examined. Assumption 1 neglects the pulsatile flow of the blood due to the beating of the heart. Assumption 2 will hold if the dye is selected to bind to albumen or other serum proteins. Hypothesis 3 is not valid, but t_0 can be small. As mentioned previously, more sophisticated models do not use this assumption. Assumption 4 neglects the loss of dye in the coronary artery and other upstream arterial branches through which the dye can escape before reaching the point of observation. If the point of observation is close to the origin of the aorta, both 4 and 5 are good approximations.

EXERCISES

1. Typical data for the dye-dilution technique are shown below:

Time (sec)	0	1	2	3	4	5	6	7	8	9	10	11	12
Concentration (mg/liter)	0	0	.1	.2	.7	.9	1.3	1.9	2.7	3.0	3.7	4.0	4.2

Time (sec)	13	14	15	16	17	18	19	20	21	22	23	24
Concentration (mg/liter)	4.0	3.8	3.7	2.8	2.2	1.5	1.1	1.0	0.9	0.9	0.9	0.8

(a) Graph these observations.
(b) When does recirculation begin?
(c) Extrapolate the curve to obtain the first-time-through dye curve.

2. Calculate the cardiac output from the data in Exercise 1 given that $D = 5$ mg of dye were injected at $t = 0$ using
   (a) An approximating Riemann sum with $\Delta t = 1$
   (b) The parabolic rule
   (c) The trapezoidal rule
3. Suppose that the dye concentration is

$$c(t) = -at(t - b)$$

   where a and b are positive constants. Find R.
4. Assume that the dye is monitored in a branch artery that receives only one-fifth of the blood coming from the heart. What changes are required in the analysis in Example 18?
5. The mean lifetime or *transit time* $\bar{t}$ of the dye in the compartment is defined as

$$\bar{t} = \frac{\int_0^{\infty} tc(t)\, dt}{\int_0^{\infty} c(t)\, dt}$$

   (a) Using the theoretical concentration curve (58), show that

$$\bar{t} = t_0 + \frac{V}{R}$$

   (b) Show how V can be determined without utilizing (62), which requires a curve-fitting procedure to determine λ.
6. Substitute the theoretical concentration from (58) into the formula

$$A = \int_0^{\infty} c(t)\, dt$$

   to show that $R = D/A$. (*Hint:* The limits for finding the area under the theoretical curve are 0 and ∞.)

4.7 CONCLUSION

The emphasis in this chapter was on the direct solution of differential equations or the quantitative theory of differential equations. The reader interested in pursuing this topic should consult [1, 13, 14]. In many instances an analytic solution cannot be found. Sometimes only certain properties of the solution are of interest. For example, how does a solution behave when it is displaced from a position of rest? Such questions may be difficult to answer using only numerical methods. Some analytic techniques for studying certain properties of the solution without actually solving the differential equation will be presented in Volume II, Chapter 4. This theory is known as the qualitative theory of differential equations [2].

REFERENCES

1. Braun, M. *Differential Equations and Their Applications*. Springer-Verlag, Berlin, 1978.
2. Cronin, J. *Differential Equations: Introduction and Qualitative Theory*, Marcel Dekker, New York, 1980.

3. Eisen, M., and Schiller, J. Stability analysis of normal and neoplastic growth. *Bull. Math. Biol. 66,* 799–809, 1977.
4. Gause, G. F. *The Struggle for Existence*. Hafner, New York, 1969.
5. Jokipii, S. G., and Turpeinen, O. Kinetics of elimination of glucose from the blood during and after a continuous intravenous injection. *J. Clin. Invest. 34,* 452–454, 1954.
6. Kellison, S. G. *Fundamentals of Numerical Analysis*. Richard D. Irwin, Homewood, IL, 1975.
7. Kinsman, J. M., Moore, J. W., and Hamilton, W. F. Studies on the circulation. I. Injection method: physical and mathematical considerations. *Am. J. Physiol. 89,* 322–330, 1929.
8. Laird, A. K. Dynamics of tumor growth. *Br. J. Cancer 18,* 490–502, 1964.
9. Lotka, A. J. *Elements of Mathematical Biology*. Dover, New York, 1956.
10. Monod, J. *Recherches sur la croissance des cultures bacteriennes*. Hermann, Paris, 1942.
11. Moore, J. W., Kinsman, J. M., Hamilton, W. F., and Spurling, R. G. Studies on the circulation. II. Cardiac output determinations; Comparison of the injection method with the direct Fick procedure. *Am. J. Physiol. 89,* 331–339, 1929.
12. Norwich, K. H., and Zelin, S. The dispersion of indicator in the cardio-pulmonary system. *Bull Math. Biophys. 32,* 25–43, 1970.
13. Piaggio, H. T. H. *An Elementary Treatise on Differential Equations and Their Applications*. G. Bell and Sons, London, 1952.
14. Rabenstein, A. L. *Introduction to Ordinary Differential Equations*. Academic Press, New York, 1972.
15. Rainey, R. H. Natural displacement of pollution from the Great Lakes. *Science 155,* 1242–1243, 1967.
16. Ralph, E. K., and Michael, H. A. Twenty-five years of radiocarbon dating. *Am. Sci. 62,* 553–560, 1974.
17. Ralph, E. K., Michael, H. N., and Hahn, M. C. Radioactive dates and reality. *MASCA Newslett. 9,* 1–20, 1973.
18. Stewart, J. N. The pulmonary circulation time, the quantity of blood in the lungs and the output of the heart. *Am. J. Physiol. 58,* 20–44, 1921.

5

Convolution Representation; Properties of Linear Systems

5.1 INTRODUCTION

Certain aspects of the behavior of linear systems can be studied more efficiently by using other mathematical representations of systems besides linear differential or difference equations. The following alternative mathematical description will be developed:

1. Convolution representation
2. Transfer-function representation
3. State-space representation

These representations are equivalent in the sense that for each representation, a given input generates the same output. The main purpose for studying these alternative representations is that each has its own particular advantage for extracting particular types of information about the dynamical behavior of a linear system.

In this chapter the convolution representation will be studied. The main advantage of the convolution representation is the ability of expressing y exclusively as a function of the input signals. This representation enables one easily to prove a number of significant properties of linear systems. Representations 2 and 3 will be studied later.

5.2 CONVOLUTION REPRESENTATION OF LINEAR SYSTEMS [1]

A linear discrete system with constant coefficients is usually described by a linear difference equation of the form

$$y(k) = b_0u(k) + b_1u(k-1) + \cdots + b_mu(k-m) - a_1y(k-1) - a_2y(k-2) - \cdots - a_ny(k-n) \tag{1}$$

where $b_0, \ldots, b_m, a_1, \ldots, a_n$ are constants. The time index k is referred to as the *present time,* while any index $k - i$ ($i = 1, 2, \ldots$) is referred to as *past time*. The signal is applied at $t = 0$ and so k is a nonnegative integer. Note that the present output is expressed as a linear combination of past outputs and past and present inputs.

It is interesting to compare the difference equation description of a system with the block diagram representation or the equivalent analytic representation [c.f. Eq. (1) of Chapter 1].

$$y(k) = f(u[0, k]) \tag{2}$$

Here the present output depends only on the past and present inputs $u[0, k] = \{u(s) \mid 0 \le s \le k\}$. Is there an equivalent representation of the system described by (1) in the form (2)?

The last question can be answered by taking advantage of the linearity of f and writing the input in the form

$$u(k) = \sum_{s=0}^{k} u(s)\, \delta(k - s) \tag{3}$$

by using the definition of the delta function,

$$\delta(k) = \begin{cases} 1 & k = 0 \\ 0 & k = \pm 1, \pm 2, \pm 3, \ldots \end{cases} \tag{4}$$

Recalling the definition of $\Sigma_{i=0}^{n} c_i v_i[0, k] = \{\Sigma_{i=0}^{n} c_i v_i(s) \mid 0 \le s \le k\}$, it follows from (3) that

$$u[0, k] = \sum_{s=0}^{k} u(s)\, \delta[0, k - s] \tag{5}$$

Using the linearity of f,

$$f\left(\sum_{i=0}^{n} c_i v_i[0, k]\right) = \sum_{i=0}^{n} c_i f(v_i[0, k]) \tag{6}$$

and (5), it follows that

$$y(k) = f(u[0, k]) = f\left(\sum_{s=0}^{k} u(s)\, \delta[0, k - s]\right) = \sum_{s=0}^{k} u(s) f(\delta[0, k - s]) \tag{7}$$

where

$$\delta[0, k - s] = \{\delta(t) \mid 0 \le t \le k - s\}$$

Define the *weighting function, h,* as the system response to the delta function input,

$$h(k) = f(\delta[0, k]) \tag{8}$$

It follows from (7) that

$$y(k) = \sum_{s=0}^{k} u(s) h(k - s) = \sum_{s=0}^{k} h(s) u(k - s) \tag{9}$$

The mathematical representation (9) is of the form (2). Note that a zero-level response is required before the impulse is applied. If energy were already stored in the system from previous inputs, then the response to any new input would not be determined solely by the inherent nature of the system itself but in part by its particular past history.

The mathematical representation (9) is called the *convolution representation* of the linear system. The name originates from the following mathematical terminology. The *convolution* of two functions f and g, denoted $f * g$, is defined as

$$(f * g)(k) = \sum_{s=0}^{\infty} f(s)g(k - s) = \sum_{s=0}^{\infty} g(s)f(k - s) \tag{10}$$

for discrete-valued functions that are zero on the negative integers, and by

$$(f * g)(t) = \int_0^{\infty} f(s)g(t - s)\, ds \tag{11}$$

for functions defined on $-\infty < t < \infty$ and zero on $-\infty < t < 0$.

To determine the form of the weighting function for the convolution representation (9) for the system described by (1), let $u(s) = \delta(s)$. Then $y(k) = \sum_{s=0}^{k} \delta(s)h(k - s) = h(k)$, that is, the weighting function is the output of the system when the input is the delta function.

The above discussion motivates the next theorem, which shows the equivalence of the representations (1) and (9).

Theorem 1. The convolution representation of the system described by (1) is given by (9), where $h(k)$ is the solution (or output) of (1) for the input $u(k) = \delta(k)$ and the initial conditions

$$y(n - 1) = y(n - 2) = \dots = y(0) = 0$$

Proof. By definition,

$$h(k - i) = b_0\, \delta(k - i) + \cdots + b_m\, \delta(k - m - i) \\ - a_1 h(k - 1 - i) - \cdots - a_n h(k - n - i)$$

Multiplying this last equation by $u(i)$ and summing from $i = 0$ to $i = k$ gives

$$\begin{aligned} \sum_{i=0}^{k} u(i)h(k - i) &= b_0 \sum_{i=0}^{k} u(i)\, \delta(k - i) + \cdots + b_m \sum_{i=0}^{k} u(i)\, \delta(k - m - i) \\ &\quad - a_1 \sum_{i=0}^{k} u(i)h(k - 1 - i) - \cdots - a_n \sum_{i=0}^{k} u(i)h(k - n - i) \\ &= b_0 \sum_{i=0}^{k} u(i)\, \delta(k - i) + \cdots + b_m \sum_{i=0}^{k-m} u(i)\, \delta(k - m - i) \\ &\quad - a_1 \sum_{i=0}^{k-1} u(i)h(k - 1 - i) - \cdots - a_n \sum_{i=0}^{k-n} u(i)h(k - n - i) \end{aligned} \tag{12}$$

The left-hand side of (12) can be simplified by using (9), while the right-hand side of (12) can be simplified by using (3) and (9). These simplifications reduce (12) to (1).

An application of Theorem 1 is illustrated in the next example.

Example 1

Consider the system described by the linear difference equation

$$y(k) - 3y(k - 1) + 2y(k - 2) = u(k - 2) \tag{13}$$

The weighting function h in the convolution representation of (13) is obtained by solving

$$y(k) - 3y(k - 1) + 2y(k - 2) = \delta(k - 2) \tag{14}$$

with the initial conditions

$$y(1) = y(0) = 0$$

The characteristic equation of (14) is

$$0 = r^2 - 3r + 2 = (r - 2)(r - 1)$$

and so a particular solution of (14) can be found by the method of variation of parameters by letting

$$y(k) = c_1(k) + c_2(k)2^k$$

Then

$$\Delta c_1(k - 2) = -\delta(k - 2) \quad \text{or} \quad \Delta c_1(k) = -\delta(k)$$

$$\Delta c_2(k - 2) = \frac{\delta(k - 2)}{2^{k-1}} \quad \text{or} \quad \Delta c_2(k) = \frac{\delta(k)}{2^{k+1}}$$

The solutions of these equations are

$$c_1(k) = \begin{cases} 0 & k = 0 \\ -1 & k = 1, 2, 3, \ldots \end{cases}$$

$$c_2(k) = \begin{cases} 0 & k = 0 \\ \frac{1}{2} & k = 1, 2, 3, \ldots \end{cases}$$

Hence a particular solution of (14) is

$$y(k) = \begin{cases} 0 & k = 0 \\ -1 + 2^{k-1} & k = 1, 2, 3, \ldots \end{cases}$$

Thus the general solution of (14) is

$$y(k) = \begin{cases} c_1 + c_2 2^k + 2^{k-1} - 1 & k = 1, 2, 3, \ldots \\ c_1 + c_2 & k = 0 \end{cases}$$

Using the given initial conditions, the weighting function is

$$h(k) = \begin{cases} 0 & k = 0 \\ -1 + 2^{k-1} & k = 1, 2, 3, \ldots \end{cases}$$

Therefore the convolution representation of the system described by (13) is

$$y(k) = \sum_{s=1}^{k} u(s)(-1 + 2^{k-s-1})$$

Another method for finding the weighting function is to find the convolution representation of a system directly without solving the difference equation with input $u(k) = \delta(k)$. This can be done by successively replacing the value of $y(i)$, $i = 0, 1,$

$2, \ldots, k-1$ by $u(i)$, thereby obtaining $y(k)$ as a function of the $u(i)$. This procedure is illustrated in the next example.

Example 2 (Animal Population Size as a Function of Immigration)

Suppose that the size of a certain animal population at time k, $y(k)$, depends on the size of the population at time $k-1$ and the number of animals immigrated to the population at time k, $u(k)$, according to the difference equation

$$y(k) = (1+i)y(k-1) + u(k)$$

where i is a fixed constant. Letting $k = 0, 1, 2, \ldots$ yields the following equations:

$$\begin{aligned}
y(0) &= u(0) \\
y(1) &= (1+i)y(0) + u(1) = (1+i)u(0) + u(1) \\
y(2) &= (1+i)y(1) + u(2) = (1+i)^2u(0) + (1+i)u(1) + u(2) \\
&\vdots \\
y(k) &= (1+i)^k u(0) + (1+i)^{k-1}u(1) + \cdots + (1+i)u(k-1) + u(k)
\end{aligned}$$

Hence the weighting function $h(k) = (1+i)^k$.

Both of the methods described in Examples 1 and 2 can be difficult to apply for more complicated difference equations. A third method, which is often easier to apply than the previous methods, is based on the transfer-function representation, described in Chapter 8. In essence, the convolution equation

$$y = h * u \tag{15}$$

is transformed into

$$Y = HU \tag{16}$$

where each function in (15) has been replaced by its transform and the operation of convolution becomes ordinary multiplication. Equation (16) can be easily solved for $H = YU^{-1}$. Then h is the inverse transform of H. The details appear in Chapters 6 and 7.

In some experiments the input to a system or subsystem must be determined from the weighting function and the output, which are known. If h and y are available in analytic form, then u can be determined. Equation (16) can be solved for $U = H^{-1}Y$. Then u is the inverse transform of U. Frequently, however, the functions h and y are not available in analytic form, and a numerical method must be adopted. This is illustrated in the next example, which is a method for determining the rate of synthesis of albumin.

Example 3 (Rate of Albumin Synthesis)

Dynamic studies involving radioactive tracer materials can be used to collect data relating to the synthesis of albumin. A simplified model of the process appears in Fig. 5.1.

The following assumptions describe the process of albumin synthesis. The injection of ^{14}C-labeled carbonate into the bloodstream occurs only at time 0, T, $2T$, $3T$, . . . and is described by the input function $u(k)$. Some of this label passes into the liver, where one of its fates is synthesis into ^{14}C-labeled albumin. The output rate of newly syn-

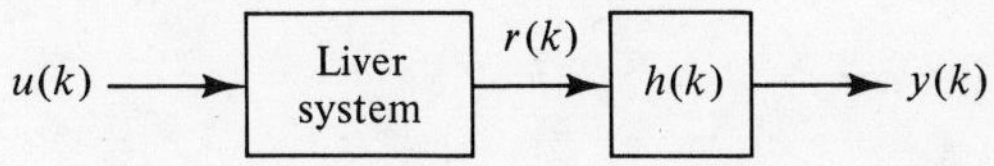

Figure 5.1 Simplified model for albumin synthesis

thesized ^{14}C-labeled albumin into the plasma also occurs only at times $0, T, 2T, 3T, \ldots$, and is denoted by $r(k)$. It is desired to estimate this function. However, the values of this function are not directly observable, since the synthesized albumin passes to the plasma compartment, where it is assumed to be uniformly distributed. The synthesized albumin is assumed to be able to pass freely to and from extravascular sites. It may also be catabolized. What can be observed is the specific activity of the synthesized albumin that has subsequently undergone the processes of distribution and breakdown. This is the output variable y. It is assumed that the output variable can also be measured at times $0, T, 2T, 3T, \ldots$.

In order to estimate r from y and u, a second tracer is used to find the weighting function, $h(k)$, of the distribution and breakdown system. A unit impulse input of ^{125}I-labeled albumin is injected into the plasma compartment and allowed to undergo the distribution and breakdown processes. It is assumed that, using an appropriate sampling procedure, the response can be measured at times $0, T, 2T, \ldots$. This determines the weighting function h at these times.

The function r can now be determined at times $0, T, 2T, \ldots$ or, equivalently, at $k = 0, 1, 2, \ldots$ as follows. Equation (9) gives

$$y(n) = \sum_{k=0}^{n} r(n-k)h(k)$$

which for $n = 0, 1, 2, \ldots$ can be written as

$$\begin{aligned} y(0) &= r(0)h(0) \\ y(1) &= r(1)h(0) + r(0)h(1) \\ y(2) &= r(2)h(0) + r(1)h(1) + r(0)h(2) \\ &\ldots \end{aligned}$$

The above system of equations can be solved successively for $r(0), r(1), r(2), \ldots$ as shown below:

$$r(0) = \frac{y(0)}{h(0)}$$

$$r(1) = \frac{y(1)}{h(0)} - \frac{r(0)h(1)}{h(0)}$$

$$r(2) = \frac{y(2)}{h(0)} - \frac{r(1)h(1)}{h(0)} - \frac{r(0)h(2)}{h(0)}$$

and so on. Hence $r(0)$ is determined, since $y(0)$ and $h(0)$ are known experimentally. Next, $r(1)$ can be determined from its equation, since $y(1)$, $h(1)$, and $h(0)$ are known experimentally and $r(0)$ has just been calculated. Similarly, $r(2)$, $r(3)$, etc., can be calculated sequentially.

In a linear time-varying discrete system, the coefficients a_i and b_i appearing in (1) will depend on k. It can be shown that a linear time-varying system has the convolution representation

$$y(k) = \sum_{s=0}^{k} h(k, s)u(k - s) \tag{17}$$

where now the weighting function h depends on k and so varies with time. This is in sharp contrast to the weighting sequence of a time-invariant system, which is a function of the single argument s. For a fixed value of k, the weighting function can be found by finding the response of the system to the inputs

$$u(k) = \delta(k - p) \quad \text{or equivalently} \quad u(k - s) = \delta(k - s - p)$$

for $p = 0, 1, 2, \ldots, k$. Inserting the delta input, for a fixed value of p, into the representation (17) gives

$$y(k) = \sum_{s=0}^{k} h(k, s)\, \delta(k - s - p) = h(k, k - p)$$

Letting $p = 0, 1, \ldots, k$ gives $h(k, k)$, $h(k, k - 1)$, $\ldots$, $h(k, 1)$, $h(k, 0)$. In general, an analytic expression for $h(k, i)$ is difficult to find, and the output of the difference equation for the input $\delta(k - s - p)$ is found numerically.

Systems whose output depends on past outputs and past, present, and *future* inputs will not be considered in the sequel. Such systems are called *anticipatory systems*, while the systems studied up to this point whose output does not depend on future intputs are called *nonanticipatory systems*. There is no loss in generality in not considering anticipatory systems in which future inputs depend on future times that are bounded. Such anticipatory systems can be converted into equivalent nonanticipatory systems. For example, the anticipatory system described by

$$y(k) = u(k + 1) + u(k + 2) + ay(k - 1)$$

can be converted to a nonanticipatory system by setting $y(k) = y(k - 2)$, which gives

$$y(k + 2) = u(k + 1) + u(k + 2) + y(k + 1)$$

or, equivalently,

$$y(k) = u(k - 1) + u(k) + y(k - 1)$$

EXERCISES

1. Show that

$$\sum_{s=0}^{k} u(s)h(k - s) = \sum_{s=0}^{k} u(k - s)h(s)$$

2. Verify that the weighting function of

$$y(k) = b_0 u(k) + b_1 u(k - 1) - a_1 y(k - 1)$$

is

$$h(0) = b_0 \quad \text{and} \quad h(i) = (b_1 - a_1 b_0)(-a_1)^{i-1} \qquad (i = 1, 2, 3, \ldots)$$

3. Find the convolution representation of the difference equation representing a hypothetical rabbit population with immigration:

$$y(k) = u(k) + y(k-2) + y(k-1)$$

4. The partial pressure after n breaths, p_n, in a plastic bag into which a subject breathes is determined by

$$p_n = ap_{n-1} + b$$

Find the convolution representation of this system.

5.3 CONVOLUTION REPRESENTATION FOR CONTINUOUS TIME-INVARIANT LINEAR SYSTEMS

In this section the analog of the theory of convolution representation for systems described by linear difference equations will be extended to systems described by a linear differential equation of the form

$$\frac{d^n y}{dt^n} + a_{n-1}\frac{d^{n-1}y}{dt^{n-1}} + \cdots + a_1\frac{dy}{dt} + a_0 y = u \tag{18}$$

where $a_0, a_1, \ldots, a_{n-1}$ are constants. Once again, t will be referred to as the present time and $t - a$, where $a > 0$, as past time. All signals are assumed to start at $t = 0$ and are zero if $t < 0$. The methods for finding the convolution representation are still applicable if the right hand of (18) depends on the past and higher-order derivatives of u. However, such equations are usually simpler to treat using the transform representation described in Chapters 6 and 7.

The block diagram or analytic description of a linear system that is time-invariant is of the form

$$y(t) = f(u[0, t]) \tag{19}$$

where f is a linear function satisfying (6). Again, the question arises if there is an equivalent representation of the system described by (18) in the form (19).

The last question can be answered by imposing certain continuity conditions on f so that if

$$u(t) = \int_0^t c(s)u_s(t)\, ds \quad \text{or equivalently} \quad u[0, t] = \int_0^t c(s)u_s[0, t]\, ds \tag{20}$$

then

$$f\left\{\int_0^t c(s)u_s[0, t]\, ds\right\} = \int_0^t c(s)f(u_s[0, t])\, ds \tag{21}$$

To take advantage of this generalized linearity of f, the delta function, δ, will be required. The delta function is described in Appendix B and has the following properties:

$$\int_0^{\infty} \delta(s)\, ds = 1 \qquad \int_t^{\infty} \delta(s)\, ds = 0 \qquad (t > 0) \tag{22}$$

$$\int_0^t u(s)\, \delta(t - s)\, ds = u(t) = \int_0^t u(t - s)\, \delta(s)\, ds \tag{23}$$

$$\int_0^t u(s) \frac{d^n \delta(t - s)}{dt^n}\, ds = \frac{d^n u(t)}{dt^n} \tag{24}$$

It follows from (21) and (23) that

$$y(t) = f\left\{\int_0^t u(s)\, \delta[0, t - s]\, ds\right\} = \int_0^t u(s) f(\delta[0, t - s])\, ds \tag{25}$$

Define the *weighting function, h,* as the system response to the delta function input

$$h(t) = f(\delta[0, t]) \tag{26}$$

It follows from (25) that

$$y(t) = \int_0^t u(s) h(t - s)\, ds = \int_0^t h(s) u(t - s)\, ds \tag{27}$$

The mathematical representation (27) is called the *convolution representation* of the linear continuous system. A zero-level response is required before the impulse is applied.

The next theorem shows that the representations (18) and (27) are equivalent. It is the analog of Theorem 1 for continuous systems.

Theorem 2. The convolution representation of the system described by (18) is given by (27), where h is the solution or output of (18) for the input $u(t) = \delta(t)$ and the initial conditions

$$\frac{d^{n-1} y(0+)}{dt^{n-1}} = \cdots = \frac{dy(0+)}{dt} = y(0+) = 0$$

Theorem 2 involves generalized functions (see Appendix B). The following equivalent version of Theorem 2 involves only real functions.

Theorem 2′. The convolution representation of the system described by (18) is given by (27), where h is the solution or output of (18) for the input $u(t) = 0$ for $t \geq 0$ and the initial conditions

$$\frac{d^{n-2} y(0+)}{dt^{n-2}} = \cdots = \frac{dy(0+)}{dt} = y(0+) = 0$$

while

$$\frac{d^{n-1} y(0+)}{dt^{n-1}} = 1$$

The equivalence of these two theorems is illustrated in the next example. The general proof of their equivalence will be given in Example 23 of Chapter 7.

Example 4

According to Theorem 2, the convolution representation of the system described by $\ddot{y} - y = u(t)$ is obtained by solving the equation

$$\frac{d^2y}{dt^2} - y = \delta(t) \tag{28}$$

with the initial conditions $\dot{y}(0+) = y(0+) = 0$. The characteristic equation of the homogeneous equation corresponding to (29) is

$$0 = \lambda^2 - 1 = (\lambda - 1)(\lambda + 1)$$

A particular solution of (29) is of the form

$$y(t) = c_1(t)e^t + c_2(t)e^{-t}$$

where c_1 and c_2 will be determined by the method of variation of parameters. From (36) and (37) of Chapter 4, the differential equations for determining c_1 and c_2 are

$$\dot{c}_1 = \frac{\delta(t)e^{-t}}{2} \tag{29}$$

$$\dot{c}_2 = \frac{-\delta(t)e^t}{2} \tag{30}$$

Hence, $c_1 = \frac{1}{2}$ and $c_2 = -\frac{1}{2}$ and so $h(t) = e^t/2 - e^{-t}/2$. Therefore the convolution representation of the given system is

$$y(t) = \frac{\int_0^t [e^{(t-s)} - e^{-(t-s)}]u(s)\,ds}{2}$$

Next, Theorem 2′ will be applied. According to this result, the function h is the solution of the corresponding homogeneous equation $\ddot{y} - y = 0$ with the initial conditions $y(0) = 0$ and $\dot{y}(0+) = 1$. It is left as an exercise to verify that, once again, $h(t) = e^t/2 - e^{-t}/2$.

Remark. The reader may be puzzled that $h(t) = e^t/2 - e^{-t}/2$ does not satisfy (28). Since the solution is defined only for $t \geq 0$, it is actually of the form

$$y(t) = h(t)H(t) \tag{31}$$

where H is the Heaviside function, defined as

$$H(t) = \begin{cases} 1 & t \geq 0 \\ 0 & t < 0 \end{cases}$$

Using some results from the theory of generalized functions (see Appendix B), it can be shown that y defined by (31) is the generalized solution of (28).

By the chain rule,

$$\dot{y} = \dot{h}H + h\dot{H} = \dot{h}H + h\delta$$

since the generalized derivative of H is the delta function. The generalized function $h\delta$ is equivalent to zero, since $h(0) = 0$. Hence $\dot{y} = \dot{h}H$ and

$$\ddot{y} = \ddot{h}H + \dot{h}\delta$$

Since $\dot{h}(0+) = 1$, $\dot{h}\delta = \delta$ in the sense of generalized functions and so

$$\ddot{y} = \ddot{h}H + \delta$$

Therefore

$$\ddot{y} - y = \ddot{h}H - hH + \delta = \delta$$

since $\ddot{h} = h$. Consequently, $y = hH$ satisfies (28).

EXERCISES

1. Finding the weighting function corresponding to the differential equation $\dot{y} = u$. (*Hint*: Integrate.)
2. Many biological processes (e.g., population growth, infusion of glucose into the bloodstream, aortic pressure) can be described by a differential equation of the form

$$\dot{y} = ay + b$$

   Find the convolution representation of the above equation.
3. Find the convolution representation of the following equations:
   (a) $\ddot{y} + \dot{y} - 2y = f(t)$
   (b) $\ddot{y} + y = f(t)$
   (c) $\ddot{y} - 2\dot{y} + y = f(t)$
4. The discrete analog of the differential equation in Exercise 2 can be written in the form

$$y_k = ay_{k-1} + b$$

   Find the convolution representation of this system.
5. Find the convolution representation of the following equations:
   (a) $y(k + 2) + y(k + 1) - 2y(k) = u(k)$
   (b) $y(k + 2) + y(k) = u(k)$
   (c) $y(k + 2) - 2y(k + 1) + y(k) = u(k)$

5.4 SOME PROPERTIES OF LINEAR SYSTEMS

In this section some properties of linear systems will be studied by using the convolution representation of the systems.

A. Linearity

The basic property of a linear system is the *principle of superposition* or the *principle of linearity*, stated below.

Principle of Linearity. If a linear system's response to the input signals u_1 and u_2 applied separately are y_1 and y_2, respectively, then the system's response to the input signal

$$u = c_1u_1 + c_2u_2$$

is

$$y = c_1y_1 + c_2y_2$$

where c_1 and c_2 are arbitrary constants.

The above result can be proved using the convolution representation as follows. If $y_1 = u_1 * h$ and $y_2 = u_2 * h$, then

$$(c_1u_1 + c_2u_2) * h = c_1(u_1 * h) + c_2(u_2 * h) = c_1y_1 + c_2y_2 \tag{32}$$

The reader should verify that (32) is valid for discrete systems, by using (9), and for continuous systems, by using (27).

The following example illustrates the principle of linearity.

Example 5

(a) Particular solutions of the difference equation

$$y(k + 1) - 2y(k) = u(k)$$

for $u_1(k) = 1$, $u_2(k) = k$, and $u_3(k) = k^2$ are $y_1 = -1$, $y_2 = -k - 1$, and $y_3 = -(k^2 + 2k + 3)$, respectively. Thus, by the superposition principle, a particular solution of the difference equation

$$y(k + 1) - 2y(k) = 6u_3 - 2u_2 - 5u_1 = 6k^2 - 2k - 5$$

is

$$y = 6y_3 - 2y_2 - 5y_1 = -(6k^2 + 10k + 11)$$

(b) The particular solutions of the differential equation

$$\dot{y} - y = u(t)$$

corresponding to the inputs $u_1 = 2$, $u_2 = t$, and $u_3 = t^2$ are $y_1 = -2$, $y_2 = -(t + 1)$, and $y_3 = -(t_2 + 2t + 2)$, respectively. Thus a particular solution of the differential equation

$$\dot{y} - y = 3u_3 + 2u_2 + 3u_1 = 3t^2 + 2t + 6$$

is

$$y = 3y_3 + 2y_2 + 3y_1 = -(3t^2 + 8t + 14)$$

B. Time-Invariant Property

Consider the continuous time-invariant system described by the convolution representation

$$y(t) = \int_0^t h(s)u(t - s)\, ds \tag{33}$$

Suppose that the input is shifted by T time units; specifically, the new input is

$$\overline{u}(t) = u(t - T)$$

The response of the system to $\overline{u}$ is

$$\int_0^t u(t - T - s)h(s)\, ds = \int_0^{t-T} u(t - T - s)h(s)\, ds = y(t - T)$$

where the last equality follows from (33) by replacing t by $t - T$. A similar result can be proved for discrete systems. This time-invariant property is summarized below.

Time-Invariant Property. If the response of a linear time-invariant system to the input $u(t)$ is $y(t)$, then the response of the system to the shifted input $u(t - T)$ is simply $y(t - T)$.

The next example illustrates how the above property can be applied.

Example 6

The response of the system

$$y(k) = 5y(k - 1) + u(k) \tag{34}$$

to the input

$$u(k) = \begin{cases} 2 & k = 0 \\ 4 & k = 1 \\ 6 & k = 2 \\ 0 & k = 3, 4, 5, \ldots \end{cases}$$

will be found by using the time-invariant property. An alternative way of writing the input is

$$u(k) = 2\,\delta(k) + 4\,\delta(k - 1) + 6\,\delta(k - 2) \tag{35}$$

Using the results of Example 2, the weighting function corresponding to (34) is $h(k) = 5^k$. The time-invariant property implies that the following input–output response pairs are generated:

Input	Output
$\delta(k)$	$h(k)$
$\delta(k - 1)$	$h(k - 1)$
$\delta(k - 2)$	$h(k - 2)$

Hence, by the linearity principle, the response of the system (34) to (35) is simply

$$y(k) = 2h(k) + 4h(k - 1) + 6h(k - 2) = 2 \cdot 5^k + 4 \cdot 5^{k-1} + 6 \cdot 5^{k-2}$$

EXERCISES

1. The response of a linear system to the inputs u_1, u_2, and u_3 is t, $t^2 + 2$, and $t^3 + t^2$, respectively. What is the response of this system to the input $u = 5u_1 - 3u_2 + 2u_3$?
2. The response of a linear system to the input $u(t)$ is $y(t) = 3t^2 + 2t - 1$. What is the response of the system to the input $u(t - 3)$?
3. **(a)** Find the response of the system

$$\ddot{y} + \dot{y} - 6y = u(t)$$

to the inputs $u_1 = 1$, $u_2 = t$, and $u_3 = t^2$.

(b) What is the response of the given system to the input:

$$u(t) = 3 - 5t + 7t^2$$

(c) Find the response of the given system to the input:

$$3 - 5(t - 1) + 7(t - 1)^2$$

4. (a) Find the response of the system

$$y(k + 2) - 4y(k + 1) + 3y(k) = u(k)$$

to the inputs $u_1 = 1$, $u_2 = k$, and $u_3 = k^2$.

(b) What is the response of the given system to the input:

$$u(k) = -2 + 3k + 4k^2$$

(c) Find the response of the given system to the input:

$$-2 + 3(k - 4) + 4(k - 4)^2$$

5. Find the response of the system

$$y(k) - 3y(k - 1) = u(k)$$

to the input

$$u(k) = \begin{cases} 1 & k = 0 \\ 0 & k = 1 \\ 3 & k = 2 \\ 5 & k = 3 \\ 0 & k = 4, 5, 6, \ldots \end{cases}$$

5.5 RESPONSE TO SPECIAL TEST INPUTS

To determine whether a given system has desirable dynamical characteristics, a special test input is applied to the system and the resultant response is observed. Depending on the criterion of performance used, the system is deemed satisfactory if its response to the test input has desirable features. The response of a linear system to three important test inputs will be discussed.

A. Kronecker Delta Input

Recall that the response of a linear time-invariant system to a delta input is the weighting function h. The weighting function h plays a significant role in determining a system's memory. Intuitively the system's memory is a measure of how significant is the effect on the present output, $y(k)$, of inputs applied in the past. For example, considêr two discrete systems with weighting functions h_1 and h_2. Suppose that $h_1(k) = 0$ for $k = 1, 2, 3, 4, 5, 6$ and zero otherwise, while $h_2(k) = 0$ for $k = 1, 2, \ldots, 27$ and zero otherwise. Using the convolution representation for system 1, it is apparent that its present output depends only on the six most recent inputs. On the other hand, the present output of system 2 depends on inputs applied 27 iterations ago. System 1 is said to have a short memory relative to system 2.

The last example suggests that a measure of a system's memory is obtained by noting how quickly the system's weighting function approaches zero. The more quickly a system's weighting function goes to zero, the shorter the system memory and vice versa. The fact that this measure of system memory is valid for general linear systems follows from the convolution representation of such systems. The effect on the present output made by an input applied i time units ago is given by $h(i)u(k - i)$.

Short memory systems are readily influenced by the most recent behavior of the input signal. Such systems are said to be *fast responding*, since the output signal responds quickly to changes in the input signal. Long memory systems do not respond as readily to changes in the recent behavior of the input and are therefore said to be *sluggish*.

The following example contains a system with a very short memory and a system with a very long memory.

Example 7

(a) Memory of an Integrator. The system $\dot{y} = u$ or $y = \int_0^t u(s)\,ds$ has a weighting function $h(t) = 1$. Thus it has an infinite memory.

(b) Memory of a Differentiator. The system $y = \dot{u}$ can be written in the equivalent form $y = \int_0^t \dot{\delta}(t - s)u(s)\,ds$ by using (24). Hence the weighting function $h(t) = \dot{\delta}(t)$ and is zero on an interval $\epsilon < t < \infty$ for any $\epsilon > 0$. Therefore the system has an extremely short memory.

B. Unit Step Input

One of the most frequently used test signals is the unit step input

$$u(t) = \begin{cases} 1 & t \geq 0 \\ 0 & t < 0 \end{cases}$$

This function is used to test whether a linear system is *stable*, that is, whether a bounded* input produces a bounded output. Another equivalent characterization of a linear time-invariant stable system is that a unit step input produces an output that approaches a steady-state value after an initial transient period. Stability will be studied in Chapter 10. The reader familiar with the theory presented in Chapters 3 and 4 might try to prove the equivalence of the two definitions of stability given above. (See Exercise 4.)

The unit step function can also be used to test whether a stable system is overdamped or underdamped. The *overdamped* response is characterized by the response approaching its steady-state value monotonically. The *underdamped* response oscillates about its steady-state value, with the amplitude of the oscillation decaying to zero as $t \to \infty$.

The type of response of a stable, linear, time-invariant system can be characterized in terms of the behavior of its weighting function h. If the system has a positive weighting function, then the system is overdamped. The proof of this last assertion will be given for a discrete system. Since the output can be written in the form

$$y(k) = \sum_{s=0}^{k} h(s)u(k - s)$$

* A function f is bounded if $|f(t)| \leq M$, where M is a fixed constant, for every t for which $f(t)$ is defined.

the response to the unit step response is

$$y(k) = \sum_{s=0}^{k} h(s)$$

Hence

$$y(k) - y(k-1) = h(k) \qquad (k = 1, 2, 3, \ldots)$$

Thus $y(k) > y(k-1)$, since $h(k) > 0$, and so y increases monotonically. Since the system is stable, y is bounded. Hence $y(k)$ approaches a limit by a well-known theorem in analysis which states that a monotonically increasing sequence bounded above approaches a limit. Note that this limit is the steady-state value $y_{ss} = \sum_{s=0}^{\infty} h(s)$. Thus the rate at which $y(k)$ approaches y_{ss} depends on how quickly h goes to zero. This could be used as a measure of the system's response time. It follows that an underdamped response can occur only if h alternates in sign as $t \to \infty$.

The unit step is also used to measure a system's speed of response. The system is subjected to a unit step input, and the following parameters are determined:

1. *Rise time:* the time required for the unit step to reach 90% of its steady-state value.
2. *Settling time:* the time required until the unit step response remains within 5% of its steady-state value.

Either of these two parameters could be defined to be the system's *response time*. A system whose response time is small is said to be a *fast responding system*. Conversely, a *sluggish system* is one that has a long response time. The particular application determines what values are considered small or large.

C. Sinusoidal Response

The test input $u(t) = \sin(t\omega T)$ for $t \geq 0$ plays an important role in determining the behavior of linear time-invariant systems. However, its use will not be investigated until transform methods have been studied in order to simplify the response.

EXERCISES

1. Using the convolution representation, find the response of
(a) A discrete system
(b) A continuous system
to the sinusoidal input $u(t) = \sin(t\omega T)$.

2. The numerical approximation $y(k)$ to the integral of a function $u(t)$ at the sampling instants kT is determined from

$$y(k) = Tu(k) + y(k-1) \qquad y(-1) = 0$$

(a) Show that the weighting function $h(k) = T$ for all k.
(b) Discuss the memory of the system.

3. An approximation to the derivative of a function $u(t)$ at the sampling instants kT is given by

$$y(k) = \frac{u(k) - u(k-1)}{T}$$

(a) Show that

$$h(k) = \frac{\delta(k) - \delta(k-1)}{T}$$

(b) Discuss the memory of this system.

4. Show that the two notions of stability described in this section are equivalent for a
   (a) Continuous system (*Hint*: The system is stable if, and only if, all characteristic roots are negative except 0, which is a simple root.)
   (b) Discrete system (*Hint*: All characteristic roots are less than 1 in absolute value except 1, which is a simple root.)

5. Consider the overdamped, stable system

$$\ddot{y} + 3\dot{y} + 2y = u(t)$$

with initial conditions $y(0) = 1$ and $\dot{y}(0) = 0$. Show that the weighting function $h > 0$.

6. Consider the underdamped, stable system

$$\ddot{y} + 2\dot{y} + 2y = u(t)$$

with initial conditions $y(0) = 0$ and $\dot{y}(0) = 1$. Show that the weighting function $h(t)$ is not positive for all values of t as $t \to \infty$.

7. Consider the system

$$\dot{y} + 2y = u(t)$$

with initial condition $y(0) = 0$.
(a) Find the rise time.
(b) Find the settling time.

5.6 INTERCONNECTIONS OF LINEAR SYSTEMS

Two or more linear systems are often interconnected in order to achieve certain objectives. Three common methods of connecting linear systems will be studied. It will be shown that the resultant system is linear and that its weighting function is related directly to the weighting functions of the two linear systems in the given configuration.

A. Parallel Configuration

Two systems are connected in *parallel* if each system is driven by the same input signal and the resultant response is the sum of the individual system responses. Such a configuration is depicted in Fig. 5.2(a).

To obtain the weighting function h of this configuration, let $u = \delta$. Then the response of each system is $y_1 = h_1$ and $y_2 = h_2$. Hence the total system response to the delta input is given by $y = y_1 + y_2 = h_1 + h_2$. Thus the equivalent weighting function of the parallel configuration is $h = h_1 + h_2$.

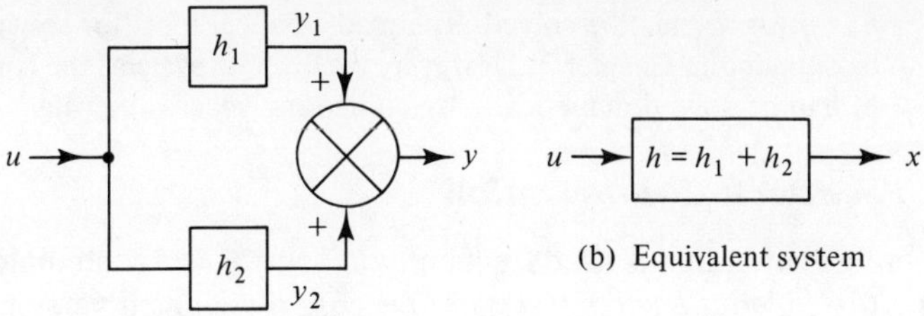

(a) Systems connected in parallel

(b) Equivalent system

Figure 5.2 Two systems connected in parallel and their equivalent.

B. Series (Cascade) Configuration

Two systems are connected in series if the output sequence of one system serves as the input sequence to the other system. A block diagram representation of a series configuration appears in Fig. 5.3(a). To find the equivalent configuration, let $u = \delta$. Then $y_1 = u * h_1$ and $y = y_1 * h_2 = (u * h_1) * h_2 = u * (h_1 * h_2)$. Hence the weighting function for the series configuration is simply $h = h_1 * h_2$.

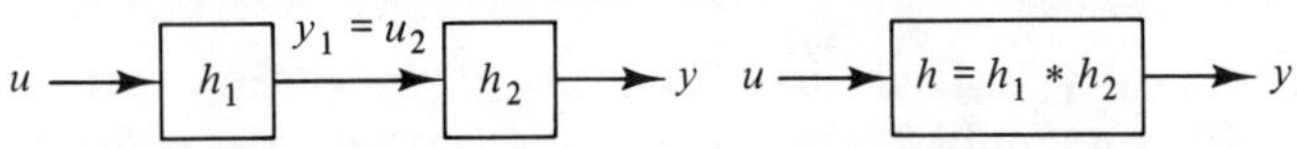

(a) Series configuration

(b) Equivalent system

Figure 5.3 Two systems connected in series and their equivalent.

Example 8

The weighting function of the system

$$y_1(k) = 0.2u_1(k) + 0.7y_1(k - 1)$$

is $h_1 = (0.2)(0.7)^k$, and the weighting function of

$$y_2(k) = 0.1u_2(k) + 0.8y_2(k - 1)$$

is $h_2(k) = (0.1)(0.8)^k$. If the two systems are connected in series, the weighting function of the resultant series configuration is

$$h = h_1 * h_2 = \sum_{i=0}^{k}(0.2)(0.7)^i(0.1)(0.8)^{k-i} = (0.02)(0.8)^k \sum_{i=0}^{k}(\tfrac{7}{8})^i$$
$$= (0.16)(0.8)^k[1 - (\tfrac{7}{8})^{k+1}]$$

Definition 1. Two systems connected in series are *inverses* of one another if the equivalent weighting function of the series is $h = \delta$.

This means that system 2 unscrambles the transformation of the input signal made by system 1 so that the overall response of the series configuration is $y = u$.

Example 9

The inverse of the system $y = \dot{u}$ is $\dot{y} = u$ with $y(0) = u(0)$. The last assertion can be verified directly by using the output of system 1, $y = \dot{u}$, as the input to system 2, yielding $\dot{y} = \dot{u}$. Taking the initial conditions into account, $y = u$ implies $y = u$.

The inverses of continuous systems are difficult to find, since the equation

$h_1 * h_2 = \delta$ must be solved. The mathematical tools for solving such an equation will be supplied in Chapter 7. However, for discrete systems the convolution equation given h_1 can be solved numerically by computing successive values of h_2.

C. Feedback Configuration

A configuration that is used frequently in control and communication is shown in Fig. 5.4. It is called a *feedback system,* because a weighted version of the output signal y is fed back and subtracted from the input signal to form the error signal e. The equations that govern this configuration are

$$s = h_2 * y \tag{36}$$

$$e = u - s \tag{37}$$

$$y = h_1 * e \tag{38}$$

In the discrete situation Eqs. (36), (37), and (38) could be solved numerically by substituting values of s from (36) into (37) and then substituting the resulting expression for e into (38), which could be solved for y. For a continuous system the above procedure leads to

$$y = h_1 * u - h_1 * h_2 * y$$

which must be solved analytically for y. In general, solving such an equation is difficult using only the techniques developed up to this point. Thus further investigation of this feedback configuration will be postponed until transform methods are developed, which replace the convolution operation by ordinary multiplication.

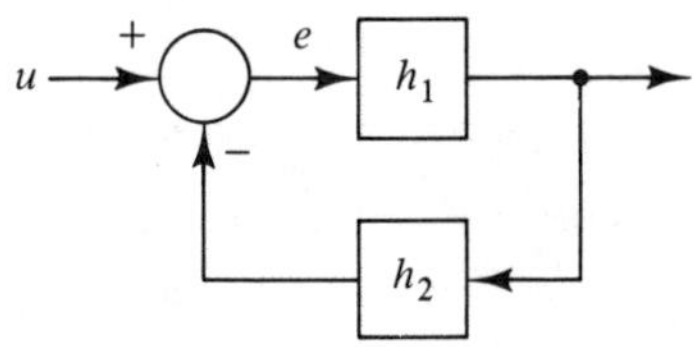

Figure 5.4 Feedback configuration.

EXERCISES

1. **(a)** Find the weighting function of the systems

$$y_1 = 0.3u_1(k) + 0.6y_1(k - 1)$$

$$y_2 = 0.4u_2(k) + 0.5y_2(k - 1)$$

(b) Determine the weighting function if these two systems are connected in series.

(c) Repeat (b) for a parallel configuration.

2. Find the inverses of the systems in (a) of Exercise 1.

3. Suppose that the weighting functions of two systems are $h_1(k) = a^k$ and $h_2(k) = b^k$, where $0 < a < 1, 0 < b < 1$.

(a) Find the weighting function for the series configuration of the given systems.

(b) Repeat (a) for the parallel configuration of the given systems.

REFERENCES

1. Cadzow, J. A. *Discrete-Time Systems*. Prentice-Hall, Englewood Cliffs, NJ, 1973.

6

The z-Transform

6.1 INTRODUCTION

The direct study of linear systems based on difference or differential equations can be difficult, as shown by the following examples: Finding the weighting function for higher-order equations, although theoretically possible, is impractical. A knowledge of distributions is required in order to understand the manipulations of the delta function. In general, the weighting function of a series of interconnected linear systems cannot be determined explicitly because of the appearance of the convolution.

Transformation methods have been used in mathematics for a long time to study linear difference and differential equations. As the name implies, a transformation is a rule by which an entity is changed uniquely into another form. For example, the z-transform is a rule by which a sequence of numbers is converted into a function of the complex variable z. The z-transform changes linear difference equations into ordinary algebraic equations and replaces the convolution operation by ordinary multiplication. These properties greatly simplify the study of linear discrete systems.

The z-transform and its properties will be studied in this chapter. The continuous analog of the z-transform, the Laplace transform, will be studied in the next chapter.

6.2 DEFINITION OF THE z-TRANSFORM [2, 3]

The following notation will be used. The letter z will denote the complex variable $x + iy$, where $i = \sqrt{-1}$ and x and y are real numbers, called Re z, the real part of z, and Im z, the imaginary part of z, respectively. Recall that $f(k) = 0$ for $k < 0$. For convenience, the sequence $\{f(k)\}$ will be denoted by $f(k)$.

Definition 1. The *z-transform* of the sequence of numbers $f(k)$ is defined by

$$Z[f] = F(z) = \sum_{k=0}^{\infty} f(k)z^{-k} \tag{1}$$

The sequence $f(k)$ is called the *generating sequence* of the z-transform $F(z)$.

If the z-transform is evaluated at a specific value of the complex variable z, the complex number given by the infinite summation (1) is generated. By definition,

$$F(z) = \lim_{n\to\infty} F_n(z)$$

where

$$F_n(z) = \sum_{k=0}^{n-1} f(k)z^{-k}$$

The set of z in the complex z-plane for which $\lim F_n(z)$ exists is called the *region of convergence* of $F(z)$, while the set of z for which the $\lim_{n\to\infty} F_n(z)$ does not exist is called the *region of divergence* of $F(z)$.

Example 1

The z-transform of the *geometric sequence*

$$f(k) = a^k \qquad (k = 0, 1, 2, \ldots) \tag{2}$$

is obtained from (1) by replacing $f(k)$ by (2). This gives

$$F(z) = \sum_{k=0}^{\infty} a^k z^{-k} = \sum_{k=0}^{\infty} \left(\frac{a}{z}\right)^k \tag{3}$$

By definition, the infinite summation in (3) converges if

$$F_n(z) = \sum_{k=0}^{n-1} \left(\frac{a}{z}\right)^k = \frac{1 - (az^{-1})^n}{1 - az^{-1}} \tag{4}$$

approaches a limit as $n \to \infty$. The only term in the last equality in (4) that depends on n is $(az^{-1})^n$. It is well known that as n increases in magnitude,

$$|(az^{-1})^n| = (|az^{-1}|)^n$$

approaches zero if $|az^{-1}| < 1$. This last inequality can be written as $|a||z|^{-1} < 1$ or $|a| < |z|$. Hence

$$F(z) = \lim_{n\to\infty} F_n(z) = \frac{1}{1 - az^{-1}} = \frac{z}{z - a} \qquad (|z| > |a|) \tag{5}$$

On the other hand, for those values of z for which $|az^{-1}| > 1$, the complex number $(az^{-1})^n$ becomes unbounded as n becomes arbitrarily large and so $F(z) = \lim F_n(z)$ is unbounded. Summarizing,

$$Z[a^k] = a^k z^{-k} = \begin{cases} \dfrac{z}{z - a} & \text{for } |z| > |a| \\ \text{unbounded} & \text{for } |z| < |a| \end{cases} \tag{6}$$

It is readily shown that the set of z in the complex z-plane that satisfies $|z| > |a|$ lies outside the circle of radius $|a|$ centered at the origin, while the set of z satisfying $|z| < |a|$ lies inside this circle. Figure 6.1 shows the region of convergence and divergence for the z-transform of the geometric sequence.

On the circle separating the convergence and divergence regions, the z-transform of the geometric sequence may or may not converge. Each z on the boundary must be tested separately. For example, at $z = a$, $F_n(a) = n$, which becomes unbounded as

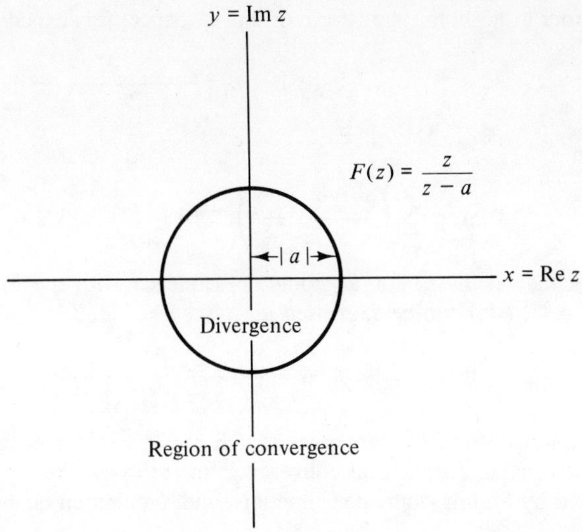

Figure 6.1 Convergence region of z-transform of the geometric sequence.

$n \to \infty$. Therefore $z = a$ lies in the region of divergence of the geometric sequence's z-transform.

Regions of convergence and divergence for the general z-transform have similar properties to those described in Example 1. The region of convergence is specified by $|z| > R$, where R is the *radius of convergence,* which depends on the generating sequence $f(k)$ [4]. This region is simply the set of z in the complex z-plane located outside the circle of radius R, centered at the origin. Inside this circle $F(z)$ becomes unbounded, so the interior of this circle constitutes the region of divergence of $F(z)$. The power series for $F(z)$ may be either bounded (convergent or divergent because of oscillations) or unbounded in absolute value for points on the circle $z = R$.

The radius of convergence, R, for the z-transform may take on the following values:

1. $R = 0$. This implies that the z-transform converges everywhere except possibly at the origin.
2. R is finite. The z-transform converges outside the circle of radius R centered at the origin.
3. R is infinite. The z-transform converges nowhere.

The sinusoidal sequence is an important test function. The z-transform of this sequence will be found in the next example.

Example 2

Consider the sinusoidal sequence defined by

$$f(k) = \sin k\omega \qquad (k = 0, 1, 2, \ldots)$$

In order to find the z-transform of this sequence, it is expedient to use Euler's identity*

$$\sin k\omega = \frac{e^{ik\omega} - e^{-ik\omega}}{2i} \tag{7}$$

Then

$$F(z) = \sum_{k=0}^{\infty}\left(\frac{e^{ik\omega} - e^{-ik\omega}}{2i}\right)z^{-k} = \frac{1}{2i}\left(\sum_{k=0}^{\infty} e^{ik\omega}z^{-k} - \sum_{k=0}^{\infty} e^{-ik\omega}z^{-k}\right)$$

using the result (6) for a geometric sequence with the first term $a = e^{i\omega}$ and then $a = e^{-i\omega}$, $F(z)$ can be expressed as

$$F(z) = \frac{1}{2i}\left(\frac{z}{z - e^{i\omega}} - \frac{z}{z - e^{-i\omega}}\right)$$

The z-transform $F(z)$ converges for all $|z| > 1$, since each of the two geometric sequences has a z-transform converging for $|z| > 1$. The right side of $F(z)$ can be subtracted by finding equivalent fractions with a common denominator. This gives

$$F(z) = \frac{1}{2i}\left[\frac{z(e^{i\omega} - e^{-i\omega})}{z^2 - (e^{i\omega} + e^{-i\omega})z + 1}\right]$$

Next using (7) and also the identity

$$\cos \omega = \frac{e^{i\omega} + e^{-i\omega}}{2} \tag{8}$$

reduces $F(z)$ to the form

$$F(z) = \frac{z \sin \omega}{z^2 - 2z \cos \omega + 1} \qquad (|z| > 1) \tag{9}$$

Applying the same technique as used in determining the z-transform of the sinusoidal sequence, it can be shown that the z-transform of the cosine sequence

$$f(k) = \cos k\omega \qquad (k = 0, 1, 2, \ldots)$$

is given by

$$Z[\cos k\omega] = \frac{z(z - \cos \omega)}{z^2 - 2z \cos \omega + 1} \qquad (|z| > 1) \tag{10}$$

The examples of z-transforms up to this point have been a simple ratio of polynomials in the variable z. Many of the sequences that will be studied later will also have z-transforms of the form

$$F(z) = \frac{b_m z^m + b_{m-1}z^{m-1} + \cdots + b_0}{z^n + a_{n-1}z^{n-1} + \cdots + a_0} \tag{11}$$

These polynomials can be factored and (11) becomes

$$F(z) = \frac{b_m(z - z_1)(z - z_2)\cdots(z - z_m)}{(z - p_1)(z - p_2)\cdots(z - p_n)} \tag{12}$$

* This identity follows from De Moivre's theorem: $e^{i\theta} = \cos \theta + i \sin \theta$ [4].

The numbers $z_1, \ldots, z_m$ are called *zeroes* of the z-transform $F(z)$. The numbers $p_1, \ldots, p_n$ are called *poles* of $F(z)$. Note that $F(z)$ is completely specified, within the multiplicative constant b_m, if all its zeroes and poles are known.

Example 3

(a) From (6), the z-transform of the geometric sequence has the zero $z_1 = 0$ and the pole $p_1 = a$.

(b) The poles of the sinusoidal sequence can be found from (9) by factoring the denominator of $F(z)$ by replacing *cos* ω by its value from (8). This gives

$$Z[\sin k\omega] = \frac{z \sin \omega}{z^2 - (e^{i\omega} + e^{-i\omega})z + 1} = \frac{z \sin \omega}{(z - e^{i\omega})(z - e^{-i\omega})}$$

Hence the poles of the z-transform are $p_1 = e^{i\omega}$ and $p_2 = e^{-i\omega}$, while the zero is $z_1 = 0$.

(c) Similarly, the poles of the z-transform of the cosine sequence, given by (10), are $p_1 = e^{i\omega}$ and $p_2 = e^{-i\omega}$. Its zeroes are $z_1 = 0$ and $z_2 = \cos \omega$.

The fact that the geometric, sinusoidal, and cosine sequences all have poles located on $|z| = R$, where R is the radius of convergence of their z-transforms, can be generalized. In general, any z-transform that may be put into the form of a ratio of polynomials in z will have one of its poles on the boundary separating its regions of convergence and divergence. This pole will be the pole of the z-transform that is located the farthest distance from the origin. The generalization is a direct consequence of the theory of power series for a complex variable [4].

EXERCISES

1. Find the z-transform of the following sequences:
   (a) $\delta(k)$
   (b) $\delta(k - m)$
   (c) 1
   (d) k
2. Verify that $z(z + 1)/(z - 1)^3$ is the z-transform of the sequence k^2 by expanding the given z-transform in powers of z^{-1}.
3. Show that the generating sequence of the z-transform $az/(z - a)^2$ is ka^k.
4. Show that the generating sequence of the z-transform $z/(z - 1)^3$ is $k(k - 1)/2!$
5. Given that the z-transform of $k(k - 1)(k - 2)/3!$ is $z(z - 1)^{-4}$, find the z-transform of k^3. {*Hint:* $k^3 = 6[k(k - 1)(k - 2)/3!] + 6[k(k - 1)]$. Use the results of 1 (d) and 4.}
6. Consider the z-transform

$$F(z) = \frac{z(z - 1)(z - 2)(z - i)(z + i)}{(z - 3)(z - 5)(z - 1 + i)(z - 1 - i)}$$

   (a) Find the poles and zeroes of $F(z)$.
   (b) What is the region of convergence of F?

7. Rewrite the following functions as a ratio of polynomials in the variable z. Find the corresponding zeroes and poles.

(a) $\dfrac{8}{8 + 6z^{-2} + z^{-4}}$ (b) $\dfrac{5 + 2z^{-2}}{1 + 6z^{-1} + 3z^{-2}}$ (c) $\dfrac{1 + 2z^{-1}}{3 + 4z^{-1} + z^{-2}}$

8. Determine the region of convergence for the z-transforms in Exercise 1.

6.3 PROPERTIES OF THE z-TRANSFORM

Finding z-transforms and the generating sequences of z-transforms can often be simplified by using certain properties of z-transforms. Some of the important properties of the z-transform are discussed below.

A. Linearity

If $f(k) = af_1(k) + bf_2(k)$ for $k = 0, 1, 2, \ldots$, where a and b are constants, then

$$F(z) = aF_1(z) + bF_2(z) \qquad (|z| > \max(R_1, R_2))$$

or in operator notation

$$Z[af_1(k) + bf_2(k)] = aZ[f_1(k)] + bZ[f_2(k)] \tag{13}$$

Here, R_1 and R_2 denote the radii of convergence for the z-transforms $F_1(z)$ and $F_2(z)$, respectively.

Example 4

It follows from the linearity property, (10) and (5) that

$$Z[3 \sin 4k + 5 \cdot 2^k] = 3Z[\sin 4k] + 5Z[2^k]$$

$$= \frac{3z \sin 4}{z^2 - 2z \cos 4 + 1} + \frac{5z}{z - 2}$$

The z-transform converges for $|z| > \max(1, 2) = 2$.

B. Right Shifting

Suppose that the input, f, of a linear system is *identically zero for negative time*. If the system consists of m delay units connected in series, then the output of the system is

$$y(k) = f(k - m) \qquad (k = 0, 1, 2, \ldots)$$

and its z-transform is

$$Z[f(k - m)] = z^{-m}F(z) = z^{-m}Z[f(k)] \tag{14}$$

Here m may be any nonnegative integer. The radius of convergence of this z-transform is the radius of convergence of $F(z)$. A block diagram representation of the right-shifting property is shown in Fig. 6.2.

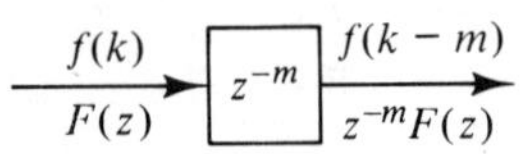

Figure 6.2 The right-shifting property.

Example 5

Consider the system characterized by

$$y(k) = 5u(k) + 3y(k - 1) \qquad (k = 0, 1, 2, \ldots) \tag{15}$$

By the linearity of the z-transform (13),

$$Z[y(k)] = 5Z[u(k)] + 3Z[y(k-1)]$$

Using the right-shifting property, the last equality becomes

$$Y(z) = 5U(z) + 3z^{-1}Y(z)$$

which when solved for Y gives

$$Y(z) = 5z(z-3)^{-1}U(z)$$

Here it is assumed that $y(-1) = 0$.

The procedure used in Example 5 is the basis of an alternative method for solving linear difference equations. First the z-transform Y of the output is determined. Then the solution y of the difference equation is determined by finding the generating sequence of Y. The exact procedure for doing this will be given in Section 6.5. The usefulness of this alternative method will become apparent in the sequel.

C. Convolution-Summation

Suppose that

$$y = h * u$$

Then

$$Y = H \cdot U$$

or, in operator notation,

$$Z[h * u] = Z[h]Z[u] \tag{16}$$

The convolution-summation property is depicted in Fig. 6.3.

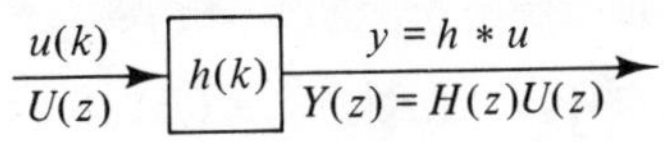

Figure 6.3 The convolution-summation property.

Example 6

The z-transform of the unit step response of the system (15) can be determined by utilizing the convolution-summation property. The weighting sequence of system (15) is the geometric sequence

$$h(k) = 5 \cdot 2^k \qquad (k = 0, 1, 2, \ldots)$$

whose z-transform is

$$H(z) = 5z(z-2)^{-1} \qquad (|z| > 2)$$

The output according to the convolution representation is $y = h * u$, which by the convolution-summation property transforms into

$$Y(z) = H(z)U(z) = 5z(z-2)^{-1}U(z)$$

D. Left Shifting

The left-shifting property is

$$Z[f(k+1)] = zZ[f(k)] - zf(0) \qquad (|z| > R) \tag{17}$$

where R is the radius of convergence of $F(z)$. More generally, for $m > 0$,

$$Z[f(k+m)] = z^m Z[f(k)] - f(k)z^{m-k} \qquad (|z| > R) \tag{18}$$

E. Summation

Suppose that

$$g(k) = \sum_{i=0}^{k} f(i) \qquad (k = 0, 1, 2, \ldots)$$

then

$$G(z) = z(z-1)^{-1}F(z) \qquad [|z| > \max(1, R)] \tag{19}$$

where R is the radius of convergence of $F(z)$.

Example 7

The z-transform of the sequence $g(k) = k$ can be obtained by defining the sequence

$$f(k) = \begin{cases} 0 & k = 0 \\ 1 & k = 1, 2, 3, \ldots \end{cases}$$

and writing g as a sum:

$$g(k) = \sum_{i=0}^{k} f(i) = k$$

It follows from the summation property that $G(z) = z(z-1)^{-1}F(z)$. To determine $F(z)$, note that $f(k)$ is the discrete step shifted one unit. Hence, from (14) with $m = 1$,

$$F(z) = z^{-1}z(z-1)^{-1} = (z-1)^{-1} \qquad (|z| > 1)$$

Hence $G(z) = z(z-1)^{-2}$ for $|z| > 1$.

F. Multiplication by a^k

The multiplication property is

$$Z[a^k f(k)] = F(a^{-1}z) \qquad (|z| > |a|R) \tag{20}$$

where R is the radius of convergence of the z-transform, $F(z)$, of $f(k)$.

Example 8

From (9) and (20),

$$Z[a^k \sin k\omega] = \frac{a^{-1}z \sin \omega}{(a^{-1}z)^2 - 2a^{-1}z \cos \omega + 1} = \frac{az \sin \omega}{z^2 - 2az \cos \omega + a^2}$$

In some situations the generating sequence may be difficult to determine from its z-transform. However, if only the initial value of f [that is, $f(0)$] or its final value [$\lim_{k\to\infty} f(k)$] is required, then its generating sequence need not be found. The required values can be determined directly from the z-transform as shown in the next two results.

G. Initial Value Theorem

The initial value theorem states that

$$f(0) = \lim_{|z| \mapsto \infty} F(z) \tag{21}$$

Example 9

The initial value of the cosine sequence can be determined directly from its transform (10) as

$$\cos 0 = \lim_{|z| \mapsto \infty} \frac{z(z - \cos \omega)}{z^2 - 2z \cos \omega + 1} = 1$$

H. Final Value Theorem

The following definition is required in order to state the final value theorem. A function, which is the ratio of two polynomials of a complex variable, is *analytic* in a region if it has no poles in the region. The final value theorem is

If $(z - 1)F(z)$ is analytic for $|z| \geq 1$, then

$$f(\infty) = \lim_{k \to \infty} f(k) = \lim_{z \to 1} (z - 1)F(z) \tag{22}$$

Example 10

The final value of the geometric sequence $f(k) = a^k$ can be determined from its z-transform $F(z) = z(z - a)^{-1}$ for $|z| > a$. To apply the final value theorem, it is necessary that $(z - 1)F(z)$ is analytic for $|z| \geq 1$ and so $|a| \leq 1$. If $a = 1$, then $f(\infty) = \lim_{z \to 1}(z - 1)z(z - 1)^{-1} = 1$. If $|a| < 1$, then $f(\infty) = \lim_{z \to 1}(z - 1)z(z - a)^{-1} = 0$.

I. Periodic Sequences

A number sequence that is repeated every N discrete time units is said to be *periodic* with *period* N. Analytically, the last definition means that

$$f(k) = f(k + N) \qquad (k = 0, 1, 2, \ldots)$$

The *first period* of the above sequence is the set of values $f(0), f(1), \ldots, f(N - 1)$. The z-transform of the first period will be denoted by

$$F_1(z) = \sum_{k=0}^{N-1} f(k)z^k$$

The z-transform of a periodic sequence of period N is

$$F(z) = \left(\frac{z^N}{z^N - 1}\right)F_1(z) \qquad (|z| > 1) \tag{23}$$

Example 11

Consider the periodic sequence $f(k + 5) = f(k)$ for $k = 0, 1, 2, \ldots,$ where the first period is defined by $f(0) = f(1) = f(2) = 1$ and $f(3) = f(4) = -1$. Hence

$$F_1(z) = 1 + z^{-1} + z^{-2} - z^{-3} - z^{-4} = \frac{z^4 + z^3 + z^2 - z - 1}{z^4}$$

The z-transform of the periodic sequence is

$$F(z) = \frac{z^5}{z^5 - 1}F_1(z) = \frac{z(z^4 + z^3 + z^2 - z - 1)}{z^5 - 1} \qquad (|z| > 1)$$

J. Differentiation

If the z-transform of a sequence is known, the z-transform of more complicated sequences can be found by repeatedly applying the following result:

$$Z[kf(k)] = -z\frac{dF(z)}{dz} \qquad (|z| > R) \tag{24}$$

where $F(z) = Z[f(k)]$ and R is the radius of convergence of $F(z)$.

Example 12

Since

$$Z[a^k] = \frac{z}{z - a} \qquad (|z| > a)$$

the differentiation property implies that

$$Z[ka^k] = -z\frac{d}{dz}\left(\frac{z}{z - a}\right) = \frac{az}{(z - a)^2} \qquad (|z| > a)$$

Another application of the differentiation property gives

$$Z[k^2a^k] = -z\frac{d}{dz}\left(\frac{az}{(z - a)^2}\right) = \frac{az(z + a)}{(z - a)^2}$$

Table 6.1 summarizes the properties of the z-transform.

TABLE 6.1 PROPERTIES OF THE z-TRANSFORM

Property	Sequence	z-Transform
1. Linearity	$af(k) + bf(k)$	$aF(z) + bG(z)$
2. Right shifting	$f(k - m)$	$z^{-m}F(z)$
3. Left shifting	$f(k + m)$	$z^mF(z) - \sum_{i=0}^{m-1} f(i)z^{m-i}$
4. Convolution	$\sum_{i=0}^{k} f_1(k - i)f_2(i)$	$F_1(z)F_2(z)$
5. Summation	$\sum_{i=0}^{k} f(i)$	$\left(\frac{z}{z - 1}\right)F(z)$
6. Multiplication by a^k	$a^kf(k)$	$F(a^{-1}z)$
7. Initial value theorem	$f(0) = \lim_{\lvert z\rvert\to\infty} F(z)$	
8. Final value theorem	$f(\infty) = \lim_{z\to 1}(z - 1)F(z)$ if $(z - 1)F(z)$ is analytic for $\lvert z\rvert \geq 1$	
9. Periodic sequences	$f(k) = f(k + N)$	$F(z) = \frac{z^N}{z^N - 1}F_1(z)$
10. Differentiation	$kf(k)$	$-z\frac{dF(z)}{dz}$

EXERCISES

1. If the z-transform of $1/k$ is $\ln[z/(z-1)]$ for $k = 1, 2, 3, \ldots$ and the z-transform of $a^k/k!$ is $e^{a/z}$, what is the z-transform of

$$\frac{3a^k}{k!} + \frac{4}{k}$$

2. Find the z-transforms of the following functions:
   (a) $(k-4)^{-1}$
   (b) $(k+4)^{-1}$
   (c) $\dfrac{1}{k} * \dfrac{a^k}{k}$
   (d) $\displaystyle\sum_{i=0}^{k} \frac{a^i}{i!}$
   (e) $\dfrac{1}{k}$
3. Apply the final value theorem to the z-transform of $1/k$ given in Exercise 1 and verify the result directly.
4. Find the initial value of the generating sequence of $z(z-1)^{-m}$.
5. What is the generating function of the periodic sequence $f(k+4) = f(k)$ if $f(0) = 1$ and $f(2) = f(3) = f(4) = 3$?
6. Prove that the generating function of k^3a^k is $az(z^2 + 4az + a^2)(z-a)^{-4}$.

6.4 TABLE OF z-TRANSFORM PAIRS

The table of z-transform pairs, Table 6.2, is used in two ways. First, given a sequence of numbers, its z-transform may be extracted directly from the table rather than going

TABLE 6.2 z-TRANSFORM PAIRS

$f(k)$	$Z[f(k)]$
1. $\delta(k)$	1
2. 1	$z(z-1)^{-1}$
3. k	$z(z-1)^{-2}$
4. a^k	$z(z-a)^{-1}$
5. $(k+1)(k+2)\cdots(k+m)a^k/m!$	$z^{m+1}/(z-a)^{m+1}$
6. $\sin k\theta$	$z\sin\theta/(z^2 - 2z\cos\theta + 1)$
7. $\cos k\theta$	$z(z-\cos\theta)/(z^2 - 2z\cos\theta + 1)$
8. $a^k \sin k\theta$	$az\sin\theta/(z^2 - 2az\cos\theta + a^2)$
9. $a^k \cos k\theta$	$z(z - a\cos\theta)/(z^2 - 2az\cos\theta + a^2)$
10. $\sinh k\theta$	$z\sinh k\theta/(z^2 - 2z\cosh\theta + 1)$
11. $\cosh k\theta$	$z(z-\cosh\theta)/(z^2 - 2z\cosh\theta + 1)$
12. $a^k/k!$	$e^{a/z}$
13. $(\ln a)^k/k!$	$a^{1/z}$
14. $n!a^kb^{n-k}/(n-k)!k!$	$(bz+a)^n z^{-n}$
15. k^{-1} $(k = 0, 1, 2, \ldots)$	$\ln[z/(z-1)]$
16. $k!/(k-m)!m!$	$z(z-1)^{-m}$

through the more formal procedure of evaluating an infinite summation. Second, if z-transform theory is to play a significant role in the study of linear discrete systems (c.f. Example 5), it is essential to be able to find the generating sequence of a given z-transform. This can be accomplished by decomposing the given z-transform, which is a ratio of polynomials, into equivalent simpler expressions whose z-transforms appear in the table. This process will be described in the next section.

A more extensive table of z-transforms can be found in [1, 2, 3].

EXERCISES

1. Verify the entries in Table 6.2.

2. Use the result of Exercise 6 in Section 6.3 to prove that the z-transform of k^5a^k is

$$az(z^4 + 26az^3 + 66a^2z^2 + 26a^3z + a^4)(z - a)^{-6}$$

3. What is the z-transform of $(k - 4)^5$?

4. Prove that the z-transform of $ka^k \sin k\theta$ is

$$\frac{z(z - a)(z + a)a \sin \theta}{(z^2 - 2az \cos \theta + a^2)^2}$$

5. What is the z-transform of $k(k - 1)(k - 2)/3!$?

6. Find the z-transforms of the following functions:

(a) $(k + 1)(k + 2)a^k/2$

(b) $\delta(k - m)$

(c) a^{k-n}

7. Show that the z-transform of $\sin(k\theta + \omega)$ is

$$\frac{z[z \sin \omega - \sin(\theta - \omega)]}{z^2 - 2z \cos \theta + 1}$$

(*Hint:* Use the addition formula

$$\sin(A + B) = \sin A \cos B + \sin B \cos A$$

the linearity of the z-transform, and Table 6.2.)

6.5 PARTIAL FRACTIONS AND INVERSE z-TRANSFORMS

In this section the partial fraction method for decomposing a ratio of polynomials into simpler fractions will be described. Each fraction will be a z-transform, $F(z)$, whose generating sequence, $f(k)$, appears in a table of z-transform pairs.

The process of determining the sequence that generates a given z-transform is called finding the *inverse z-transform* and is denoted by

$$f(k) = Z^{-1}[F(z)]$$

The inverse of a sum of fractions is found by using the linearity of the inverse

z-transform—namely,

$$Z^{-1}[c_1 F_1(z) + c_2 F_2(z) + \cdots + c_n F_n(z)]$$
$$= c_1 Z^{-1}[F_1(z)] + c_2 Z^{-1}[F_2(z)] + \cdots + c_n Z^{-1}[F_n(z)] \qquad (25)$$

The result in (25) follows from the linearity of the z-transform. The generating sequence of each $F_i(z)$ is determined by merely looking it up in a table of z-transform pairs. This procedure is illustrated in the following example.

Example 13

Using the result of Example 5, it can be shown that the z-transform of the output of system (15) to the unit step input is

$$Y(z) = 5z(z-3)^{-1}(z-1)^{-1}$$

It will be shown subsequently that the partial fraction expansion of Y is

$$Y(z) = c_1 + c_2\left(\frac{z}{z-3}\right) + c_3\left(\frac{z}{z-1}\right)$$

where c_1, c_2, and c_3 are constants. From Table 6.2 and (25),

$$y(k) = c_1 Z^{-1}[1] + c_2 Z^{-1}\left[\frac{z}{z-3}\right] + c_3 Z^{-1}\left[\frac{z}{z-1}\right]$$
$$= c_1\delta(k) + c_2 3^k + c_3 1^k$$

The inversion procedure hinges on the expansion of a ratio of polynomials into fractions, which will now be studied. The z-transform of a sequence will be assumed to be of the form

$$F(z) = \frac{b_n z^n + \cdots + b_1 z + b_0}{z^n + a_{n-1}z^{n-1} + \cdots + a_1 z + a_0}$$

The degree of the numerator is assumed to be less than or equal to the degree of the denominator. This form can be achieved by dividing the denominator into the numerator if the degree of the numerator is greater than the degree of the denominator. To form the partial fraction expansion of F, it is first necessary to factor the denominator polynomial. This gives

$$F(z) = \frac{b_n z^n + \cdots + b_1 z + b_0}{(z-p_1)^{m_1}(z-p_2)^{m_2}\cdots(z-p_r)^{m_r}}$$

Where $p_1, p_2, \ldots, p_r$ are the distinct roots of the denominator of multiplicity $m_1, m_2, \ldots, m_r$, where

$$m_1 + m_2 + \cdots + m_r = n$$

The terms in the partial fraction expansion of $F(z)$ are determined by its poles and their multiplicity according to the following rules:

A. There is always a constant A_0. If zero is not a pole, then $A_0 = F(0)$. If zero is a pole, then the evaluation of A_0 is more complicated and is described in D below.
B. Each simple ($m_i = 1$) pole $p_i \neq 0$ gives rise to the term

$$\frac{A_i z}{z - p_i} \tag{26}$$

where the constant

$$A_i = \left(\frac{z}{z - p_i}\right)^{-1} F(z)\bigg|_{z=p_i} \tag{27}$$

The above notation indicates that $z^{-1}(z - p_i)F(z)$ is evaluated at $z = p_i$.

The simple pole $p_i = 0$ gives rise to the term

$$\frac{B_i}{z} \tag{28}$$

where

$$B_i = zF(z)\,|_{z=0}$$

C. Each multiple ($m_i > 1$) pole $p_i \neq 0$ gives rise to the terms

$$\frac{A_1 z}{z - p_i} + \frac{A_2 z^2}{(z - p_1)^2} + \cdots + \frac{A_{m_i} z^{m_i}}{(z - p_i)^{m_i}} \tag{29}$$

The constant

$$A_{m_i} = z^{-m_i}(z - p_i)^{m_i} F(z)\,|_{z=p_i} \tag{30}$$

Unfortunately, there is no simple procedure for finding the remaining constants. Some methods for finding the remaining constants will be described in D below.

The multiple pole $p_i = 0$ gives rise to the terms

$$\frac{B_1}{z} + \frac{B_2}{z^2} + \cdots + \frac{B_{m_i}}{z^{m_i}} \tag{31}$$

where only B_{m_i} can be determined easily as

$$B_{m_i} = z^{m_i} F(z)\,|_{z=0} \tag{32}$$

The method of determining the remaining constants will be described in D below.

Example 14

(a) The partial fraction expansion of

$$F(z) = \frac{z^2 + 6z - 4}{(z - 1)(z + 2)}$$

is of the form

$$\frac{z^2 + 6z - 4}{(z - 1)(z + 2)} = A_0 + \frac{Bz}{z - 1} + \frac{Cz}{z + 2}$$

The constant

$$A_0 = F(0) = \frac{z^2 + 6z - 4}{(z - 1)(z + 2)}\bigg|_{z=0} = \frac{-4}{-2} = 2$$

From (27),

$$B = \frac{z - 1}{z} F(z)\bigg|_{z=1} = \frac{z^2 + 6z - 4}{z(z + 2)}\bigg|_{z=1} = \frac{1 + 6 - 4}{3} = 1$$

Similarly,

$$C = \frac{z+2}{z}F(z)\bigg|_{z=-2} = -2$$

(b) The partial fraction expansion of

$$F(z) = \frac{z^2 + 4z - 1}{z^3(z-1)^2(z+1)}$$

is of the form

$$\frac{z^2 + 4z - 1}{z^3(z-1)^2(z+1)} = A_0 + \frac{A}{z} + \frac{B}{z^2} + \frac{C}{z^3} + \frac{Dz}{z-1} + \frac{Ez^2}{(z-1)^2} + \frac{Fz}{z+1}$$

where

$$C = z^3F(z)|_{z=0} = \frac{z^2+4z-1}{(z-1)^2(z+1)}\bigg|_{z=0} = -1$$

$$E = \frac{(z-1)^2}{z^2}F(z)\bigg|_{z=1} = \frac{z^2+4z-1}{z^5(z+1)}\bigg|_{z=1} = 2$$

$$F = \frac{z+1}{z}F(z)\bigg|_{z=-1} = \frac{z^3+4z-1}{z^4(z-1)^2}\bigg|_{z=-1} = -1$$

The remaining constants will be determined below, since there is no simple formula for them.

D. Methods of Determining Coefficients in the Partial Fraction Expansion of a z-Transform with Multiple Roots

The following two methods can be used.

1. After the coefficients of the highest-order poles (terms of the form $z^{m_i}/(z - p_i)^{m_i}$) have been found by the simple method described above, transfer these terms to the left side of the partial fraction identity, obtaining

$$G(z) = F(z) - \frac{c_1 z^{m_1}}{(z-p_1)^{m_1}} - \frac{c_2 z^{m_2}}{(z-p_2)^{m_2}} - \cdots - \frac{d_i}{z^{m_i}}$$

where the constants $c_1, c_2, \ldots, d_i$ have already been determined, The right side contains unknown coefficients in terms of the form $A_{m_i-1}z^{m_i-1}/(z-p_i)^{m_i-1}$ and B_{m_i-1}/z^{m_i-1}, which are now the poles of highest order. The A_{m_i-1} and the B_{m_i-1} can be obtained from (30) and (32) with m_i replaced by $m_i - 1$ and $F(z)$ by $G(z)$. Once the A_{m_i-1} and the B_{m_i-1} have been determined, the corresponding terms can be transferred to the right side of the new partial fraction identity and the above process can be repeated to determine A_{m_i-2} and B_{m_i-2}. The procedure is repeated until all the coefficients have been determined.

2. Add all the fractions on the right side of the partial fraction expansion. The resulting numerator is a polynomial in z whose coefficients are a linear combination of the unknown constants. Since the resulting denominator is the same as the one in $F(z)$, the resulting numerator must be identical to the numerator of $F(z)$. Equating coefficients of like powers of z produces a system of linear equations for the unknown coefficients in the partial fraction expansion. The solution of these equations yields the unknown constants.

An alternative method of obtaining a system of linear equations for the unknown constants is to evaluate the partial fraction identity at as many different values of z as there are unknown constants. This method can be shortened by a proper selection of the values of z depending on the particular example.

Example 15

(a) The first method will be applied to obtain the remaining constants in Example 14b.

$$\frac{z^2 + 4z - 1}{z^3(z-1)^2(z+1)} = A_0 + \frac{A}{z} + \frac{B}{z^2} - \frac{1}{z^3} + \frac{Dz}{z-1} + \frac{2z^2}{(z-1)^2} - \frac{z}{z+1} \tag{33}$$

Transferring the terms not involving the unknown constants to the left side gives

$$\frac{z^2 + 4z - 1}{z^3(z-1)^2(z+1)} + \frac{1}{z^3} - \frac{2z^2}{(z-1)^2} + \frac{z}{z+1} = A_0 + \frac{A}{z} + \frac{B}{z^2} + \frac{Dz}{z-1}$$

Carrying out the indicated operations on the left side, the above identity reduces to

$$-\frac{z^3 + 4z^2 + 3}{z^2(z-1)} = A_0 + \frac{A}{z} + \frac{B}{z^2} + \frac{Dz}{z-1} \tag{34}$$

Hence, from (32) and (30),

$$B = -\frac{z^3 + 4z^2 + 3}{z-1}\bigg|_{z=0} = 3$$

$$D = -\frac{z^3 + 4z^2 + 3}{z^3}\bigg|_{z=1} = -8$$

According to the first method for determining the remaining constants, (34) is rewritten as

$$-\frac{z^3 + 4z^2 + 3}{z^2(z-1)} - \frac{3}{z^2} + \frac{8z}{z-1} = A_0 + \frac{A}{z}$$

This last expression simplifies to

$$7 + \frac{3}{z} = A_0 + \frac{A}{z}$$

Hence $A_0 = 7$ and $A = 3$.

(b) The second method will be applied to the problem discussed in (a). Adding the right side of (33) using the common denominator $z^3(z-1)^2(z+1)$ gives

$$\frac{z^2 + 4z - 1}{z^3(z-1)^2(z+1)}$$
$$= \frac{A_0 z^3(z-1)^2(z+1) + Az^2(z-1)^2(z+1) + Bz(z-1)^2(z+1)}{z^3(z-1)^2(z+1)}$$
$$+ \frac{-(z-1)^2(z+1) + Dz^4(z-1)(z+1) + 2z^5(z+1) - z^4(z-1)^2}{z^3(z-1)^2(z+1)}$$

Rewriting the right-hand side in powers of z gives

$$\frac{z^2 + 4z - 1}{z^3(z-1)^2(z+1)}$$

$$= \frac{(A_0 + D + 1)z^6 + (-A_0 + A + 4)z^5 + (-A_0 - A + B - D - 1)z^4 + (A_0 - A - B - 1)z^3}{z^3(z-1)^2(z+1)}$$

$$+ \frac{(A - B + 1)z^2 + (B + 1)z - 1}{z^3(z-1)^2(z+1)}$$

Equating coefficients of powers of $z^6, z^5, \ldots, z$ in the numerator of the above identity yields the following system of equations:

$$\begin{aligned} A_0 + D + 1 &= 0 \\ -A_0 + A + 4 &= 0 \\ -A_0 - A + B - D - 1 &= 0 \\ -A_0 - A - B - 1 &= 0 \\ A - B + 1 &= 1 \\ B + 1 &= 4 \end{aligned}$$

The reader can verify that these equations have the same solution as given in part (a).

The method for finding the inverse z-transform is outlined below.

Procedure for Determining the Inverse z-Transform Using Partial Fractions

1. Factor the denominator polynomial of the z-transform $F(z)$.
2. Decompose the factored $F(z)$ by the partial fraction expansion into a linear combination of elementary fractions and evaluate the coefficients as described above.
3. Use the linearity of the inverse z-transform and the following result to find the inverse z-transform of elementary fractions:

$$Z^{-1}\left[\frac{z^{m+1}}{(z-a)^{m+1}}\right] = \frac{(k+1)(k+2)\cdots(k+m)a^k}{m!} \tag{35}$$

Here k and m are nonnegative integers.

Example 16

Determine the inverse z-transform of the function

$$F(z) = \frac{z^3 + 2z^2 + z + 1}{z^3 - z^2 - 8z + 12}$$

Following the procedure outlined above, factor the denominator polynomial to obtain

$$z^3 - z^2 - 8z + 12 = (z-2)^2(z+3)$$

The partial fraction expansion must be of the form

$$\frac{z^3 + 2z + z + 1}{(z-2)^2(z+3)} = A_0 + \frac{A_1 z}{z-2} + \frac{A_2 z^2}{(z-2)^2} + \frac{B_1 z}{z+3} \tag{36}$$

From rules 1, 2, and 3,

$$A_0 = F(0) = \frac{1}{12}$$

$$A_2 = \left(\frac{z-2}{z}\right)^2 F(z)\big|_{z=2} = \frac{19}{20}$$

$$B_1 = \frac{z+3}{z} F(z)\big|_{z=-3} = \frac{11}{75}$$

The remaining coefficient A_1 will be evaluated using the method described in the second paragraph of method 2 in D. Replace z by a value not previously selected in (36) and solve the resulting equation for A_1. For example, setting $z = 1$ in (36) yields

$$\frac{5}{4} = A_0 - A_1 + A_2 + \frac{B_1}{4}$$

Substituting the known values for A_0, A_2, and B_1 implies that $A_1 = -\frac{9}{50}$. Thus the partial fraction expansion is

$$F(z) = \frac{1}{12} - \frac{9}{50}\left(\frac{z}{z-2}\right) + \frac{19}{20}\left[\frac{z^2}{(z-2)^2}\right] + \frac{11}{75}\left(\frac{z}{z+3}\right) \tag{37}$$

The inverse z-transform can now be found by using the linearity of Z^{-1} as expressed in (25). Hence, from (37),

$$f(k) = \frac{1}{12}Z^{-1}[1] - \frac{9}{50}Z^{-1}\left(\frac{z}{z-2}\right) + \frac{19}{20}Z^{-1}\left(\frac{z^2}{(z-2)^2}\right) + \frac{11}{75}Z^{-1}\left(\frac{z}{z+3}\right)$$

Looking up the inverse z-transform in Table 6.2 and using (3) gives

$$f(k) = \frac{1}{12}\delta(k) + \frac{77}{100}(2^k) + \frac{19}{20}(k+1)2^k + \frac{11}{75}(-3)^k \qquad (k = 0, 1, 2, \ldots)$$

EXERCISES

1. Find the inverse z-transforms of

(a) $F(z) = \dfrac{z+1}{z^2-4}$

(b) $F(z) = \dfrac{z+3}{z-1}$

2. **(a)** Determine the partial fraction expansion of

$$F(z) = \frac{z+1}{z^2(z-1)}$$

(b) Determine the inverse z-transform of $F(z)$.

3. Find the inverse z-transforms of

(a) $F(z) = \dfrac{2z - 7}{(z-2)(z-5)}$

(b) $F(z) = \dfrac{(2z - 7)z}{(z-2)(z-5)}$

4. Determine the inverse z-transforms of

(a) $F(z) = \dfrac{6z^3}{(z-2)^2(z+1)}$

(b) $F(z) = \dfrac{z+3}{z(z-1)^3}$

5. Find the inverse z-transforms of

(a) $F(z) = 3 + 2z^{-1} + \dfrac{3z}{(z-1)(z-2)}$

(b) $F(z) = 5 + 6z^{-1} + 10z^{-2} + 3z^{-4}$

6.6 DIRECT METHOD OF FINDING THE INVERSE z-TRANSFORM

The z-transform can be inverted directly by division. This method is based on the fact that the z-transform is a power series in z^{-1}. In the situation when the function $F(z)$ to be inverted is a rational function of z, the denominator is simply divided into the numerator in such a way as to yield a power series in the variable z^{-1}. The corresponding terms of this division are compared with

$$F(z) = f(0) + f(1)z^{-1} + f(2)z^{-2} + \cdots$$

to determine the generating sequence of $F(z)$. This inversion method is useful when only the first few terms of the sequence $f(k)$ are required. Otherwise, the partial fraction expansion method is used.

Example 17

Determine the first three terms of the sequence $f(k)$ that generates the z-transform

$$F(z) = \frac{z^2 + 6z - 4}{z^2 + z - 2}$$

Performing the indicated division yields

$$\begin{array}{r}
1 + 5z^{-1} - 7z^{-2} + \cdots \\
z^2 + z - 2 \,\overline{\smash{)}\, z^2 + 6z - 4} \\
\underline{z^2 + z - 2} \\
5z - 2 \\
\underline{5z + 5 - 10z^{-1}} \\
-7 + 10z^{-1} \\
\underline{-7 - 7z^{-1} + 14z^{-2}} \\
17z^{-1} - 14z^{-2}
\end{array}$$

Therefore

$$F(z) = 1 + 5z^{-1} - 7z^{-2} + \ldots$$

and

$$f(0) = 1, f(1) = 5, f(2) = -7, \ldots$$

which is in agreement with the results of Example 14a.

EXERCISES

1. Find the first four terms of the generating sequences of the z-transforms appearing in Exercises 1–4 in Section 6.5 by the direct division method and compare your answers with the answers obtained previously by the partial fraction method.

REFERENCES

1. Beightler, C. S., Mitten, L. G., and Nemhauser, G. L. A short table of z-transforms and generating functions. *Oper. Res.* 9(4), 574–578, 1961.
2. Bishop, A. B. *Introduction to Discrete Controls Theory and Application*. Academic Press, New York, 1975.
3. Cadzow, J. A. *Discrete-Time Systems,* Prentice-Hall, Englewood Cliffs, NJ, 1973.
4. Churchill, R. V., and Brown, J. W. *Complex Variables and Applications*. McGraw-Hill, New York, 1974.

7

The Laplace Transform

7.1 INTRODUCTION

The Laplace transform is the continuous analog of the z-transform. It is a transformation by which a function of time is converted into a function of a complex variable s. The Laplace transform changes a linear differential equation into an ordinary algebraic equation and replaces the convolution of two functions by ordinary multiplication of their transforms. Nearly all the properties of the z-transform studied in the last chapter have Laplace transform counterparts. It will be seen that these properties simplify the study of systems of linear differential equations.

7.2 DEFINITION OF THE LAPLACE TRANSFORM

Real-valued functions f of a real variable t defined for $0 \leq t < \infty$ with the following two properties will be considered in this chapter.

1. The function f is *piecewise continuous*. This means that f has at most a finite number of discontinuities on any interval of finite length and both the left and right limits exist at each discontinuity point.
2. The function f is of *exponential order;* that is, there exist constants c and M, depending on f, such that

$$|f(t)| < Me^{ct} \qquad (0 \leq t < \infty)$$

Definition 1. The Laplace transform of f is the function

$$\mathcal{L}[f] = F(s) = \int_0^\infty f(t)e^{-st}\,dt \tag{1}$$

where s is a complex variable.

Readers not familiar with complex variables can regard s as real.

Recall that the definition of the integral appearing in (1) is

$$\int_0^\infty f(t)e^{-st}\,dt = \lim_{T\to\infty}\int_0^T f(t)e^{-st}\,dt$$

The function $f(t) = e^{t^2}$ provides an example of a function for which the integral appearing in (1) fails to exist for any value of s (see Exercise 4). Properties 1 and 2 are required in order that this integral exists, as shown in the following theorem.

Theorem 1. If f is a piecewise continuous function of exponential order, then its Laplace transform exists for $\text{Re}(s) > c$.

Proof. A standard result on the convergence of improper integrals is that absolute convergence implies convergence. Hence, it is sufficient to prove that $\int_0^\infty |e^{-st}f(t)|\,dt$ exists for $\text{Re}(s) > c$ or, equivalently, that the value of $\int_0^b |e^{-st}f(t)|\,dt$ remains bounded as $b \to \infty$. From the assumption $|f(t)| \le Me^{ct}$ for all $t \ge 0$, it follows that

$$\int_0^b |e^{-st}f(t)|\,dt \le \int_0^b |e^{-st}||Me^{ct}|\,dt = M\int_0^b e^{-[\text{Re}(s)-c]t}\,dt$$

$$\le M\int_0^\infty e^{-[\text{Re}(s)-c]t}\,dt = \frac{M}{\text{Re}(s) - c}$$

if $\text{Re}(s) > c$.

Corollary 1. If f is a function satisfying the conditions of the theorem, then

$$|F(s)| < \frac{M}{\text{Re}(s) - c} \tag{2}$$

Corollary 2.

$$\lim_{s\to\infty} F(s) = 0 \tag{3}$$

Corollary 1 was proved during the proof of Theorem 1. Corollary 2 follows from Corollary 1.

The form of the Laplace transform of a piecewise continuous function of exponential order is limited by the property (3). For instance, $3s/(2s + 5)$ cannot be the Laplace transform of a piecewise continuous function of exponential order, since its limit is $\frac{3}{2}$ as $s \to \infty$. More generally, a quotient of two polynomials can be a Laplace transform only if the degree of its denominator is greater than the degree of its numerator. It will be seen that such Laplace transforms can be decomposed into partial fractions. This enables one to determine the function that was originally transformed.

The hypotheses of Theorem 1 are sufficient conditions for the existence of a Laplace transform. They are not necessary conditions since it can be shown, by using the gamma function (see Exercise 2), that

$$\mathcal{L}[t^{-1/2}] = \left(\frac{\pi}{s}\right)^{1/2}$$

and $f(t) = t^{-1/2}$ is not piecewise continuous at zero. Note that Corollary 1 does not hold for this function.

The following examples illustrate the application of Definition 1.

Example 1

(a) The Laplace transform of $f(t) = 1$ is

$$F(s) = \lim_{T\to\infty} \int_0^T e^{-st}\, dt = \lim_{T\to\infty} \left(\frac{1 - e^{-sT}}{s}\right) = \begin{cases} \dfrac{1}{s} & \operatorname{Re}(s) > 0 \\ \infty & \operatorname{Re}(s) \le 0 \end{cases}$$

(b) The Laplace transform of $f(t) = t$ is

$$F(s) = \lim_{T\to\infty} \int_0^T te^{-st}\, dt = \lim_{T\to\infty} -\frac{1}{s}\int_0^T t\, de^{-st} = \lim_{T\to\infty}\left(-\frac{Te^{-sT}}{s} + \frac{1}{s}\int_0^T e^{-st}\, dt\right)$$

$$= \frac{1}{s^2} \qquad (\operatorname{Re}(s) > 0)$$

using integration by parts and the results of (a). In the sequel only the finite part of the Laplace transform will be stated.

(c) The delta function, δ, has the property that

$$\int_0^\infty \delta(t - a)g(t)\, dt = g(a)$$

(see Appendix B). Setting $g(t) = e^{-st}$, it follows that its Laplace transform is e^{-sa} (alternatively, see Exercise 5).

Given f, its Laplace transform is uniquely determined by (1). Conversely, $F(s)$ uniquely determines $f(t)$, as shown by the next theorem, whose proof can be found in [2].

Theorem 2. If f_1 and f_2 are piecewise continuous functions of exponential order having the same Laplace transform F, then $f_1(t) = f_2(t)$ wherever f_1 and f_2 are both continuous.

Hence two piecewise continuous functions of exponential order with the same Laplace transform can differ only at their isolated discontinuity points. In many practical applications, this difference is inconsequential or does not exist. For example, two functions that satisfy a given differential equation must both be continuous; hence they must be the same solution if they have the same Laplace transform.

The unique inverse transform f of F is written symbolically as

$$f = \mathcal{L}^{-1}[F]$$

The operator $\mathcal{L}^{-1}$ is called the *inverse Laplace operator,* and f is called the *inverse* of F. Now, just as $F(s)$ is expressed explicitly in terms of $f(t)$ by (1), an explicit formula can be written for f that involves complex integration of F[6]. In practice, this formula is rarely used. Instead, a similar approach to that used in inverting z-transforms is adopted. This involves using a table of Laplace transforms, partial fractions, and certain properties of the Laplace transform. These properties appear in the next section.

EXERCISES

1. Determine the Laplace transform of the following functions:

(a) t^2 **(b)** t^n **(c)** e^{at} **(d)** $e^{at} \sin bt$ **(e)** $e^{at} \cos bt$

2. Given that $\int_0^\infty e^{-x^2}\,dx = \sqrt{\pi}/2$, find $\mathcal{L}[t^{-1/2}]$. [*Hint*: Make the change of variables $u = t^{1/2}$ in (1).]

3. Prove that each of the following functions is of exponential order:

(a) t^n

(b) $\sin at$

(c) e^t

4. Show that e^{t^2} does not have a Laplace transform. [*Hint*: Use the fact that $e^{t^2-st} > e^t$ for $t > \operatorname{Re}(s) + 1$.]

5. (a) Find the Laplace transform of $f(t) = \epsilon^{-1}$ for $a < t < a + \epsilon$ and 0 otherwise.

(b) Find the Laplace transform of the delta function $\delta(t - a)$ by letting $\epsilon \to 0$ in the answer obtained in (a).

7.3 PROPERTIES OF THE LAPLACE TRANSFORM [2,6]

A. Real-Time Translation

If h is a real number, then

$$\mathcal{L}[f(t - h)] = e^{-sh}F(s)$$

where $f(t)$ is defined to be 0 for $t < 0$.

Example 2

The Laplace transform of the function $g(t) = 1$ for $t \geq 5$ and 0 otherwise can be found using real-time translation. Noting that $g(t) = f(t - 5)$, where $f(t) = 1$ for $0 \leq t < \infty$, it follows from Example 1(a) that

$$\mathcal{L}[g] = \mathcal{L}[f(t - 5)] = \frac{e^{-5s}}{s}$$

B. Scale Change

If a is a real positive number, then

$$\mathcal{L}\left[f\left(\frac{t}{a}\right)\right] = aF(as)$$

Example 3

Given that the Laplace transform of $\sin t$ is $F(s) = (s^2 + 1)^{-1}$, then by the scale change property, the Laplace transform of $\sin at$ is

$$\frac{F(s/a)}{a} = \frac{1/[(s/a)^2 + 1]}{a} = \frac{a}{s^2 + a^2}$$

C. Complex Translation

Let a be a complex number. Then

$$\mathcal{L}[e^{-at}f(t)] = F(s + a)$$

Example 4

Let $f(t) = 1$; then, from Example 1(a), $F(s) = s^{-1}$. Hence, by the complex translation property,

$$\mathcal{L}[e^{-at}] = \frac{1}{s + a}$$

D. Linearity

The Laplace transform associates with every function f a new function f by means of (1). Recall that a function defined on functions that associates with each function a unique function is called an *operator*. The Laplace operator $\mathcal{L}$ has the following property:

$$\mathcal{L}[c_1 f_1 + c_2 f_2] = c_1 \mathcal{L}[f_1] + c_2 \mathcal{L}[f_2]$$

called *linearity*. Thus $\mathcal{L}$ is an example of a *linear operator*, an operator with the linearity property.

Example 5

Since $\mathcal{L}[t^3] = 6s^{-4}$ and $\mathcal{L}[\sin 2t] = 2(s^2 + 4)^{-1}$, it follows from the linearity property that

$$\mathcal{L}[5t^3 + 7 \sin 2t] = 5\mathcal{L}[t^3] + 7\mathcal{L}[\sin 2t] = 30s^{-4} + \frac{14}{s^2 + 4}$$

The fact that $\mathcal{L}$ is a linear operator implies that $\mathcal{L}^{-1}$ is a linear operator (see Exercise 8).

E. Convolution

The *convolution* of f and g, denoted $f * g$, is defined to be the function

$$(f * g)(t) = \int_0^t f(x)g(t - x)\,dx = \int_0^t g(x)f(t - x)\,dx$$

Convolution Theorem. If the integrals defining $\mathcal{L}[f]$ and $\mathcal{L}[g]$ converge absolutely, then

$$\mathcal{L}[f * g] = \mathcal{L}[f] \cdot \mathcal{L}[g]$$

Corollary

$$\mathcal{L}^{-1}[FG] = f * g = g * f$$

The proof of the theorem can be found in [2, 6]. The corollary is a direct consequence of the theorem.

Example 6

(a) $\mathcal{L}[\int_0^t x \sin(t - x)\,dx] = \mathcal{L}[t]\mathcal{L}[\sin t] = s^{-2}(s^2 + 1)^{-1}$
(b) $\mathcal{L}^{-1}[(s + 5)^{-1}(s^2 + 1)^{-1}] = \mathcal{L}^{-1}[1/(s + 5)] * \mathcal{L}^{-1}[1/(s^2 + 1)] = e^{-5t} * \sin t$

F. Differentiation

If $f, \ldots, f^{(n-1)}$ are piecewise continuous, then

$$\mathcal{L}[f^{(n)}] = s^n F(s) - s^{n-1}f(0+) - s^{n-2}f'(0+) - \cdots - f^{(n-1)}(0+)$$

The proof of this result is left as an exercise (see Exercise 9).

Example 7

Consider the differential equation

$$\ddot{y} + \dot{y} - 2y = 0 \tag{4}$$

with the initial conditions $\dot{y}(0) = 0$ and $y(0) = 1$. Since

$$\mathcal{L}[\ddot{y}] = s^2Y(s) - sy(0) - \dot{y}(0)$$

$$\mathcal{L}[\dot{y}] = sY(s) - y(0)$$

it follows by taking the Laplace transform of both sides of (4) and using the given initial conditions that

$$s^2Y(s) - s + sY(s) - 1 - 2Y(s) = 0$$

Solving for $Y(s)$ yields

$$Y(s) = \frac{1 + s}{s^2 + s - 2} = \frac{2}{3(s - 1)} + \frac{1}{3(s + 2)}$$

Using the linearity of the inverse Laplace operator and the fact that

$$\mathcal{L}^{-1}[(s + a)^{-1}] = e^{-at}$$

it follows that

$$\mathcal{L}^{-1}[Y(s)] = \frac{2}{3}\mathcal{L}^{-1}\left[\frac{1}{s - 1}\right] + \frac{1}{3}\mathcal{L}^{-1}\left[\frac{1}{s + 2}\right]$$

or

$$y(t) = \frac{2}{3}e^{t} + \frac{1}{3}e^{-2t}$$

The reader should verify the solution by solving (4) using the method described in Chapter 4. The Laplace transform method for solving linear differential equations with constant coefficients will be discussed further in Section 7.6. The procedure for decomposing a fraction into partial fractions, in order to find the inverse transform, is presented in Section 7.5.

G. Integration

The following notation will be required:

$$f^{(-1)}(t) = \int_{a_1}^{t} f(x_1)\, dx_1$$

$$f^{(-2)}(t) = \int_{a_2}^{t} \int_{a_1}^{x_2} f(x_1)\, dx_1\, dx_2$$

$$\vdots$$

$$f^{(-p)}(t) = \int_{a_p}^{t} \int_{a_{p-1}}^{x_p} \cdots \int_{a_1}^{x_2} f(x_1)\, dx_1\, dx_2 \cdots dx_p$$

Using the above notation, the Laplace transform of an iterated integral is

$$\mathcal{L}[f^{(-p)}(t)] = \frac{1}{s^p}F(s) + \frac{1}{s^p}f^{(-1)}(0+) + \cdots + \frac{1}{s}f^{(-p)}(0+)$$

The proof is left as Exercise 10.

Example 8

The function t^2 for $t \geq 0$ can be written in the form

$$t^2 = 2\int_0^t \int_0^{x_2} dx_1\, dx_2$$

The integration property will be used to find $\mathcal{L}[t^2]$, noting that $f(x_1) = 1$, $f^{(-1)}(0+) = 0$, and $f^{(-2)}(0+) = 0$. Then

$$\mathcal{L}[t^2] = \frac{2}{s^2}\mathcal{L}[1] = \frac{2}{s^3}$$

since $\mathcal{L}\{1\} = s^{-1}$. The result should be verified by finding $\mathcal{L}[t^2]$ directly.

H. Initial Value Theorem

If f has a piecewise continuous derivative, then

$$\lim_{t \to 0+} f(t) = \lim_{s \to \infty} sF(s)$$

The proof is left as Exercise 11. This result is useful when only the initial value is required and the Laplace transform is difficult to invert.

Example 9

Suppose that

$$\mathcal{L}[f] = \frac{2s^2 + 5}{3s^3 + 2s^2 + 5}$$

From the initial value theorem,

$$\lim_{t \to 0+} f(t) = \lim_{s \to \infty} s\left(\frac{2s^2 + 5}{3s^3 + 2s^2 + 5}\right) = \frac{2}{3}$$

Note that the initial value of f has been found without inverting $F(s)$.

I. Final Value Theorem

If f has a piecewise continuous derivative, then

$$\lim_{t \to \infty} f(t) = \lim_{s \to 0+} sF(s)$$

Example 10

If

$$\mathcal{L}[f] = \frac{s^2 + 7}{5s^4 + 2s^3 + s}$$

then $$\lim_{t \to \infty} f(t) = \lim_{s \to 0+} s\mathcal{L}[f] = \lim_{s \to 0+} \frac{s^2 + 7}{5s^3 + 2s^2 + 1} = 7$$

J. Differentiation of the Laplace Transform

$$F'(s) = \mathcal{L}[-tf(t)]$$

Corollary

$$\mathcal{L}^{-1}[F'(s)] = -t\mathcal{L}^{-1}[F(s)]$$

The corollary is useful for inverting Laplace transforms.

Example 11

Since $-(s-3)^{-2} = d(s-3)^{-1}/ds$ and $(s-3)^{-1} = \mathcal{L}[e^{3t}]$,

$$\mathcal{L}^{-1}[-(s-3)^{-2}] = -t\mathcal{L}^{-1}[(s-3)^{-1}] = -te^{3t}$$

The differentiation property can be used to avoid the tedious integration in finding the Laplace transform of $tf(t)$ once the Laplace transform, $F(s)$, of $f(t)$ is known. It is only necessary to differentiate $F(s)$ and multiply by -1.

Example 12

To compute the Laplace transform of t^{10}, the differentiation property can be applied 10 times to give

$$\mathcal{L}[t^{10}] = (-1)^{10}\frac{d^{10}}{ds^{10}}\mathcal{L}[1] = \frac{d^{10}}{ds^{10}}(s^{-1}) = (10!)s^{-11}$$

K. Laplace Transform of Periodic Functions

If f is a periodic function with period T, that is, T is the smallest real number such that $f(t+T) = f(t)$ for every t, then

$$\mathcal{L}[f(t)] = \frac{1}{1-e^{-sT}}\int_0^T e^{-sT}f(t)\,dt$$

Table 7.1 summarizes the properties of the Laplace transform.

TABLE 7.1 PROPERTIES OF THE LAPLACE TRANSFORM

Property	Function	Laplace transform
1. Real time translation	$f(t-h)$	$e^{-sh}F(s)$
2. Scale change	$f(t/a)$	$aF(as)$
3. Complex translation	$e^{-at}f(t)$	$F(s+a)$
4. Linearity	$c_1f_1 + c_2f_2$	$c_1F_1 + c_2F_2$
5. Convolution	$f * g$	FG
6. Differentiation	$f^{(n)}(t)$	$s^nF(s) - s^{n-1}f(0+) - s^{n-2}f'(0+) - \cdots - f^{(n-1)}(0+)$
7. Integration	$f^{(-p)}(t)$	$s^{-p}F(s) + s^{-p}f^{(-p)}(0+) + \cdots + s^{-1}f^{(-p)}(0+)$
8. Initial value	$\lim_{t\to 0+} f(t)$	$\lim_{s\to\infty} sF(s)$
9. Final value	$\lim_{t\to\infty} f(t)$	$\lim_{s\to 0+} sF(s)$
10. Differentiation of $\mathcal{L}[f]$	$-tf(t)$	$F'(s)$
11. Periodic: $f(t) = f(t+T)$	$f(t)$	$\frac{1}{1-e^{-sT}}\int_0^T e^{-st}f(t)\,dt$

EXERCISES

1. Given that $\mathscr{L}[\sin bt] = b/(s^2 + b^2)$, find the Laplace transform of $e^{at} \sin bt$ by using complex translation.
2. Given that $\mathscr{L}[e^t] = (s - 1)^{-1}$, find $\mathscr{L}^{-1}[(s - 1)^{-2}]$.
3. (a) Find the Laplace transform of t^n.
   (b) Find $\mathscr{L}[t^n e^{at}]$.
4. Given that $\mathscr{L}[ate^{-at}] = a/(s + a)^2$, deduce the functions or values indicated below by using the properties described in the preceding section.
   (a) $\mathscr{L}[e^{-at}]$
   (b) $\mathscr{L}[\int_0^t axe^{-ax}\, dx]$
   (c) $\lim\limits_{t \to 0+} ate^{-at}$
   (d) $\lim\limits_{t \to \infty} ate^{-at}$
5. Prove the real-time translation property.
6. Prove the scale change property.
7. Prove the translation property.
8. (a) Verify the linearity property.
   (b) Prove that $\mathscr{L}^{-1}$ is a linear operator.
9. Prove the differentiation property by using induction. {*Hint*: Use integration by parts to show that
$$\mathscr{L}[f^{(n)}(t)] = -f^{(n-1)}(0+) + s\mathscr{L}[f^{(n-1)}(t)].\}$$
10. Prove the integration property by using induction. {*Hint*: Use integration by parts to show that
$$\mathscr{L}[f^{(-p)}(t)] = \frac{1}{s}f^{(-p)}(0+) + \frac{1}{s}\mathscr{L}[f^{(-p-1)}(t)].\}$$
11. Prove the initial value theorem. {*Hint*: Use integration by parts to show that $\mathscr{L}[f'] = -f(0+) + sF(s)$ and the fact that $\lim\limits_{s \to \infty} \mathscr{L}[f'] = 0$ by the corollary of Theorem 1.}
12. Prove the final theorem. [*Hint*: use the identities
$$\int_0^\infty e^{-st}f'(t)\, dt = sF(s) - f(0+)$$
$$\lim_{s \to 0+} \int_0^\infty e^{-st}f'(t)\, dt = f(\infty) - f(0+).]$$

7.4 TABLE OF LAPLACE TRANSFORMS

Laplace transforms are frequently calculated by using the properties in the last section and a table of Laplace transforms. For example, if the Laplace transforms of two functions can be found in a table, then the Laplace transform of a linear combination of these functions can be determined by using the linearity of the Laplace operator.

The inverse of a rational function is usually calculated by using algebraic identities that decompose the given function into simpler fractions whose Laplace transforms appear in a table. The procedure is illustrated in the following example.

Example 13

Find the inverse transform of

$$F(s) = \frac{1}{s(s-1)}$$

Using algebra,

$$F(s) = \frac{1}{s-1} - \frac{1}{s}$$

By the linearity of the inverse Laplace operator,

$$\mathcal{L}^{-1}[F(s)] = \mathcal{L}^{-1}\left[\frac{1}{s-1} - \frac{1}{s}\right] = \mathcal{L}^{-1}\left[\frac{1}{s-1}\right] - \mathcal{L}^{-1}\left[\frac{1}{s}\right]$$

From Table 7.2, it follows that

$$f(t) = e^t - 1$$

Table 7.2 is a short table of Laplace transforms. More extensive tables can be found in [1, 3, 5].

The essence of the method used in Example 13 was resolving a fraction into a sum of fractions $(1/(s-1) - 1/s)$ called *partial fractions*. This is the reverse process of adding fractions by the common-denominator process. Once the resolution into partial fractions has been accomplished, the inverse Laplace transform can easily

TABLE 7.2 LAPLACE TRANSFORMS

Time function, $f(t)$, $t \geq 0$	Laplace transform, $F(s)$
$\delta(t - t_0)$, unit impulse at $t = t_0$	e^{-st_0}
1	$\frac{1}{s}$
t	$\frac{1}{s^2}$
t^2	$\frac{2}{s^3}$
t^n	$\frac{n!}{s^{n+1}}$
e^{-at}	$\frac{1}{s+a}$
$\frac{e^{-at} - e^{-bt}}{b-a}$	$\frac{1}{(s+a)(s+b)}$
$\frac{1}{(n-1)!}t^{n-1}e^{-at}$	$\frac{1}{(s+a)^n}$
$\frac{1}{(b-a)}[(\alpha - a)e^{-at} - (\alpha - b)e^{-bt}]$	$\frac{s+\alpha}{(s+a)(s+b)}$

TABLE 7.2 *(cont.)*

Time function, $f(t)$, $t \geq 0$	Laplace transform, $F(s)$
$\dfrac{e^{-at}}{(b-a)(c-a)} + \dfrac{e^{-bt}}{(c-b)(a-b)} + \dfrac{e^{-ct}}{(a-c)(b-c)}$	$\dfrac{1}{(s+a)(s+b)(s+c)}$
$\sin \omega t$	$\dfrac{\omega}{s^2+\omega^2}$
$\cos \omega t$	$\dfrac{s}{s^2+\omega^2}$
$\omega t e^{-\omega t}$	$\dfrac{\omega}{(s+\omega)^2}$
$\dfrac{t^{n-1}e^{-t/T}}{T^n(n-1)!}$	$\dfrac{1}{(1+Ts)^n}$
$1 - e^{-t/T}$	$\dfrac{1}{s(1+Ts)}$
$1 - \dfrac{t+T}{T} e^{-t/T}$	$\dfrac{1}{s(1+Ts)^2}$
$e^{-at} \sin \omega t$	$\dfrac{\omega}{(s+a)^2+\omega^2}$
$e^{-at} \cos \omega t$	$\dfrac{(s+a)}{(s+a)^2+\omega^2}$
$1 + \dfrac{e^{-\zeta\omega_n t}}{\sqrt{1-\zeta^2}} \sin(\omega_n \sqrt{1-\zeta^2}\, t - \alpha)$ where $\cos \alpha = -\zeta$	$\dfrac{\omega_n^2}{s(s^2+2\zeta\omega_n s+\omega_n^2)}$
$t - \dfrac{2\zeta}{\omega_n} + \dfrac{1}{\omega_n\sqrt{1-\zeta^2}} e^{-\zeta\omega_n t} \sin(\omega_n\sqrt{1-\zeta^2}\, t - \theta)$ where $\theta = 2\tan^{-1}\dfrac{\sqrt{1-\zeta^2}}{-\zeta}$	$\dfrac{\omega_n^2}{s^2(s^2+2\zeta\omega_n s+\omega_n^2)}$
$\dfrac{-1}{(1+T^2\omega_n^2)} e^{-t/T} + \dfrac{1}{\sqrt{1+T^2\omega_n^2}} \cos(\omega_n t - \theta)$ where $\theta = \tan^{-1} \omega_n T$	$\dfrac{s}{(1+Ts)(s^2+\omega_n^2)}$
$\dfrac{1}{2\omega_n} t \sin \omega_n t$	$\dfrac{s}{(s^2+\omega_n^2)^2}$
$\dfrac{e^{-bt}}{(b-a)^2+\omega^2} + \dfrac{e^{-at}\sin(\omega t - \theta)}{\omega[(b-a)^2+\omega^2]^{1/2}}$ where $\theta = \tan^{-1}\dfrac{\omega}{b-a}$	$\dfrac{1}{(s+b)[(s+a)^2+\omega^2]}$
$\sin at \sin bt$	$\dfrac{2abs}{[s^2+(a+b)^2][s^2+(a-b)^2]}$
$t + (a - T_1 - T_2) + \dfrac{b - aT_1 + T_1^2}{T_1 - T_2} e^{-t/T_1} - \dfrac{b - aT_2 + T_2^2}{T_1 - T_2} e^{-t/T_2}$	$\dfrac{1+as+bs^2}{s^2(1+T_1 s)(1+T_2 s)}$

be determined using linearity. Linearity reduces the problem to evaluating terms of the form

$$\mathcal{L}^{-1}\left[\frac{1}{s+a}\right] \quad \text{or} \quad \mathcal{L}^{-1}\left[\frac{1}{s^2+bs+c}\right]$$

where a, b, and c are constants. These types of inverses can be found directly from a table of Laplace transforms. Hence the methods of partial fraction expansion will be studied in the next section. Occasionally the method of partial fractions is not readily applicable. The inverse is then found by using the complex variable inversion technique as described in [6].

EXERCISES

1. Using the given table find the Laplace transform of
   (a) $1 + 2t + 5t^3$
   (b) $7e^{-5t} - 3 \sin 4t$
   (c) $5 \cos t + 4 \sin 3t \sin 6t$
2. Determine the inverse Laplace transform of the following functions:
   (a) $2s^{-1} + 3s^{-2} - 6s^{-5}$
   (b) $2e^{-5s} + (1 + 3s)^{-5}$
   (c) $20/[(s + 3)^2 + 4] + 18/(s + 6)^2$
3. Verify the entry for $t^{n-1}e^{-t/T}/T^n(n - 1)!$ in Table 7.2.
4. Show that

$$\mathcal{L}[ate^{-at}] = \frac{a}{(s+a)^2}$$

5. Prove that

$$\mathcal{L}\left[\frac{t^{n-1}e^{-at}}{(n-1)!}\right] = (s+a)^{-n}$$

7.5 EXPANSION INTO PARTIAL FRACTIONS

In this section, methods for decomposing ratios of two polynomials into simpler fractions will be studied. The form of the partial fractions differs from the ones used in Section 6.5, since the form of the fractions appearing in a z-transform table is different from the form of the fractions appearing in a Laplace transform table. A fraction is first reduced, if necessary, to a proper fraction—that is, a fraction whose numerator is of lower degree than its denominator. This is accomplished by dividing the numerator by the denominator, as shown in the following example.

Example 14

Consider the following improper fraction:

$$\frac{x^4 + 2x^3 + 4x^2 + 7}{x^3 + 3x^2 + 6}$$

Dividing the numerator by the denominator yields

$$\frac{x^4 + 2x^3 + 4x^2 + 7}{x^3 + 3x^2 + 6} = x - 1 + \frac{7x^2 - 6x + 13}{x^3 + 3x^2 + 6}$$

a polynomial and a proper fraction.

In general, any improper fraction can be reduced to a sum of a polynomial and a proper fraction by a similar division. Thus, only the proper fraction in the sum need be expanded into partial fractions. In the sequel, it will be assumed that the numerator and the denominator of the given fraction have no common factors. Four situations can occur depending on the form of the denominator.

1. If the denominator contains a single linear factor of the form $ax + b$, then there will correspond to this factor a partial fraction of the form

$$\frac{A}{ax + b}$$

in which A is a nonzero constant. In fact, if the given fraction is denoted by f, then

$$A = (ax + b)f(x)\,|_{x=-b/a} \tag{5}$$

The following example illustrates the use of situation 1 in a partial fraction expansion.

Example 15

Consider the fraction

$$\frac{x^2 - 2x + 5}{(x - 1)(2 - 3x)(x + 1)}$$

The factors of the denominator are $x - 1$, $2 - 3x$, and $x + 1$. Therefore the given fraction can be resolved into the sum of the fractions

$$\frac{A}{x - 1}, \quad \frac{B}{2 - 3x}, \quad \text{and} \quad \frac{C}{x + 1}$$

where, from (5),

$$A = (x - 1)f(x)\,|_{x=1} = \frac{x^2 - 2x + 5}{(2 - 3x)(x + 1)} = \frac{4}{(-1)(2)} = -2$$

$$B = (2 - 3x)f(x)\,|_{x=2/3} = \frac{x^2 - 2x + 5}{(x - 1)(x + 1)} = \frac{\frac{4}{9} - \frac{4}{3} + 5}{(-\frac{1}{3})(\frac{5}{3})} = -\frac{37}{5}$$

$$C = (x + 1)f(x)\,|_{x=-1} = \frac{x^2 - 2x + 5}{(x - 1)(2 - 3x)} = \frac{8}{(-2)(5)} = -\frac{4}{5}$$

Hence

$$\frac{x^2 - 2x + 5}{(x - 1)(2 - 3x)(x + 1)} = -\frac{2}{x - 1} - \frac{37}{5(2 - 3x)} - \frac{4}{5(x + 1)}$$

An alternative method for determining the coefficients A, B, and C is described in Section 6.5 part D under 2. This method will be illustrated in 2 below.

2. If the denominator of the given fraction contains the linear factor $ax + b$ repeated n times, then there will correspond to this factor the sum of n partial fractions of the form

$$\frac{A_1}{ax + b} + \frac{A_2}{(ax + b)^2} + \cdots + \frac{A_n}{(ax + b)^n}$$

where

$$A_k = \frac{a^{k-n}}{(n - k)!} \frac{d^{n-k}}{dx^{n-k}} (ax + b)^n f(x) \big|_{x=-b/a} \qquad (k = 1, 2, \ldots, n) \tag{6}$$

and $A_n \neq 0$. Note that

$$A_n = (ax + b)^n f(x) \big|_{x=-b/a} \tag{7}$$

Frequently, the general formula is not used because of difficulties in carrying out the repeated differentiations. The method of undetermined coefficients is used as shown in the following example (c.f. method 2 in Section 6.5 part D)

Example 16

According to result 2, the fraction

$$\frac{-x^2 + 11x - 6}{(x + 2)(x - 2)^2} = \frac{A}{x + 2} + \frac{B}{x - 2} + \frac{C}{(x - 2)^2} \tag{8}$$

where A, B, and C are constants to be determined so that both sides of the equation are equal. This can be accomplished by obtaining a common denominator on the left-hand side, which is just the denominator on the left side, and cancelling this denominator on both sides of the equation so that only the numerators remain equated. After expanding and collecting terms, this procedure yields

$$-x^2 + 11x - 6 = (A + B)x^2 + (C - 4A)x + (4A - 4B + 2C)$$

Since this equation is an identity, the coefficients of like powers of x must be equal. Hence

$$A + B = -1$$

$$C - 4A = 11$$

$$4A - 4B + 2C = -6$$

The simultaneous solution of these linear algebraic equations yields

$$A = -2$$

$$B = 1$$

$$C = 3$$

Therefore the partial fraction expansion is

$$\frac{-x^2 + 11x - 6}{(x + 2)(x - 2)^2} = \frac{-2}{x + 2} + \frac{1}{x - 2} + \frac{3}{(x - 2)^2}$$

Alternatively, from (5),

$$A = (x + 2)f(x)\,|_{x=-2} = \frac{-(-2)^2 + 11(-2) - 6}{(-2 - 2)^2} = -2$$

In a similar manner, from (7),

$$C = (x - 2)^2 f(x)\,|_{x=2} = \frac{12}{4} = 3$$

The terms involving A and C are then transferred to the left side of (8) and the resulting fractions are simplified. The single remaining, unknown constant (B) can then be found using (5) or the method of unknown coefficients. In general, some of the coefficients can first be calculated using (5) or its modification (7) and then the corresponding fractions can be transferred to the left-hand side of the partial fraction identity. After combining the fractions on the left-hand side to form a single fraction, the remaining unknown coefficients can be determined by the method of undetermined coefficients or by repeating the previous step as many times as necessary.

Results 1 and 2 can be used to decompose any proper rational function into partial fractions whose denominators are linear polynomials. However, some of the coefficients may be complex numbers. The complex numbers occur in conjugate pairs, so that if two of these pairs are combined, the result is a real fraction whose denominator is a quadratic polynomial with real coefficients. It is often convenient to carry out this process directly and have quadratic polynomials in the denominators of some of the partial fractions. Then only real numbers appear in the partial fractions. The next two rules deal with fractions whose denominators contain factors of the form $ax^2 + bx + c$, where a, b, and c are real numbers. It is assumed that $ax^2 + bx + c$ can only be factored into linear factors with complex coefficients. This is equivalent to the fact that $ax^2 + bx + c = 0$ has complex roots or the condition that the discriminant $b^2 - 4ac < 0$.

3. If the denominator of the given fraction contains a nonrepeated quadratic factor of the form $ax^2 + bx + c$, where $b^2 - 4ac < 0$, then there will correspond to this factor the partial fraction

$$\frac{Ax + B}{ax^2 + bx + c}$$

in which A and B are real constants, at least one of which is not zero.

The constants A and B can be determined by the method of undetermined coefficients as shown in the next example.

Example 17

Consider the following fraction and its partial fraction expansion according to results 1 and 3:

$$\frac{x^2 + 2x + 3}{(x + 1)(x^2 + x + 1)} = \frac{A}{x + 1} + \frac{Bx + C}{x^2 + x + 1}$$

Adding the two fractions on the right-hand side of the above identity and equating numerators yields

$$x^2 + 2x + 3 = (A + B)x^2 + (A + B + C)x + (A + C)$$

Equating coefficients of like powers of x gives

$$A + B = 1$$

$$A + B + C = 2$$

$$A + C = 3$$

Solving these linear algebraic equations simultaneously yields

$$A = 2 \qquad B = -1 \qquad C = 1$$

Therefore the partial fraction expansion is

$$\frac{x^2 + 2x + 3}{(x + 1)(x^2 + x + 1)} = \frac{2}{x + 1} + \frac{-x + 1}{x^2 + x + 1}$$

The following alternative method, based on complex numbers, can be used to find B and C. The roots of $x^2 + x + 1 = 0$ are

$$r_1 = \frac{-1 + \sqrt{3}\, i}{2} \qquad r_2 = \frac{-1 - \sqrt{3}\, i}{2}$$

Hence $x^2 + x + 1 = (x - r_1)(x - r_2)$. Multiplying both sides of the partial fraction identity by $x - r_1$ yields

$$\frac{x^2 + 2x + 3}{(x + 1)(x - r_2)} = \frac{A(x - r_1)}{x + 1} + \frac{Bx + C}{x - r_2}$$

When $x = r_1$, the last equation reduces to

$$\frac{r_1^2 + 2r_1 + 3}{r_1 + 1} = Br_1 + C$$

or

$$\frac{3 + \sqrt{3}\, i}{1 + \sqrt{3}\, i} = \frac{B(-1 + \sqrt{3}\, i)}{2} + C$$

Writing the complex numbers in the last equation in their standard form $a + bi$, where a and b are real numbers, gives

$$\frac{3}{2} - \frac{\sqrt{3}}{2} i = \left(-\frac{1}{2}B + C\right) + \left(\frac{\sqrt{3}}{2}B\right)i$$

where the standard form of $(3 + \sqrt{3}\, i)/(1 + \sqrt{3}\, i)$ is obtained by multiplying the numerator and the denominator by $1 - \sqrt{3}\, i$. Hence

$$\frac{\sqrt{3}\, B}{2} = -\frac{\sqrt{3}}{2} \qquad \frac{-B}{2} + C = \frac{3}{2}$$

Therefore $B = -1$ and $C = 1$, in agreement with the previous method. The value of A can be easily found from (5).

4. If the quadratic expression $ax^2 + bx + c$ occurs exactly n times as a factor of

the denominator of the given fraction, then there will correspond to this factor the partial fraction

$$\frac{A_1x + B_1}{ax^2 + bx + c} + \frac{A_2x + B_2}{(ax^2 + bx + c)^2} + \cdots + \frac{A_nx + B_n}{(ax^2 + bx + c)^n}$$

in which $A_1, B_1, A_2, B_2, \ldots, A_n, B_n$ are real constants, of which at least A_n or B_n is not zero.

The constants appearing in the partial fraction can be determined by the method of undetermined coefficients, as shown in the next example.

Example 18

Consider the following fraction and its partial fraction expansion according to results 1 and 4:

$$\frac{x^4 - x^2 + 1}{x(x^2 + x - 1)^2} = \frac{A}{x} + \frac{Bx + C}{x^2 + x - 1} + \frac{Dx + E}{(x^2 + x - 1)^2}$$

Adding the fractions on the right-hand side of the above identity and then equating the numerators yields

$$x^4 - x^2 + 1 = (A + B)x^4 + (2A + B + C)x^3 + (-A - B + C + D)x^2 + (-2A - C + E)x + A$$

Equating the coefficients of like powers of x yields

$$A + B = 1$$

$$2A + B + C = 0$$

$$-A - B + C + D = -1$$

$$-2A - C + E = 0$$

$$A = 1$$

Solving these linear algebraic equations simultaneously gives

$$A = 1 \qquad B = 0 \qquad C = -2 \qquad D = 2 \qquad E = 0$$

Therefore the partial fraction expansion is

$$\frac{x^4 - x^2 + 1}{x(x^2 + x - 1)^2} = \frac{1}{x} - \frac{2}{x^2 + x - 1} + \frac{2x}{(x^2 + x - 1)^2}$$

EXERCISES

Resolve into partial fractions:

1. $\dfrac{3x}{x^2 + 2x - 8}$
2. $\dfrac{3 + 5x}{(x + 2)(2 - 3x)}$
3. $\dfrac{2x + 3}{(x + 1)^2(2x + 3)^2}$

4. $\dfrac{3x + 8}{x(x + 2)(x^2 - 2x - 1)}$

5. $\dfrac{11x - 6 - x^2}{(x - 2)^2(2x + 1)^2}$

6. $\dfrac{x^4 - 4x + 8}{2x^3 + 4x}$ (*Hint*: Use division to obtain a proper fraction.)

7. $\dfrac{x^4 + 2x^3 - 3x + 2}{x^3 + 3x^2 - 4x}$

7.6 LAPLACE TRANSFORM SOLUTION OF ORDINARY DIFFERENTIAL EQUATIONS

The following examples will indicate the method of solving differential equations by using the Laplace transform.

Example 19

Solve the differential equation $\dot{y} + 2y = e^{3t}$ subject to the initial condition $y = 2$ when $t = 0$.

Taking transforms of both sides of the given equation yields

$$\mathscr{L}[\dot{y} + 2y] = \mathscr{L}[\dot{y}] + 2\,\mathscr{L}[y] = \mathscr{L}[e^{3t}] \tag{9}$$

From Table 7.2,

$$\mathscr{L}[e^{3t}] = \frac{1}{s - 3} \tag{10}$$

From the differentiation property in Section 7.3,

$$\mathscr{L}[\dot{y}] = sY - 2 \tag{11}$$

Substituting (10) and (11) into (9) and using the notation $\mathscr{L}[y] = Y$ gives

$$sY - 2 + 2Y = \frac{1}{s - 3}$$

Solving for Y and expanding into partial fractions yields

$$Y = \frac{2s - 5}{(s + 2)(s - 3)} = \frac{9}{5(s + 2)} + \frac{1}{5(s - 3)} \tag{12}$$

Taking the inverse transform of (12) and using Table 7.2 gives

$$y = \mathscr{L}^{-1}[Y] = \mathscr{L}^{-1}\left[\frac{9}{5(s + 2)} + \frac{1}{5(s - 3)}\right] = \frac{9e^{-2t}}{5} + \frac{e^{3t}}{5}$$

Example 20

Solve the equation $\ddot{y} - 2\dot{y} - 3y = 6e^t$ with the initial conditions $y(0) = 1$ and $\dot{y}(0) = 3$.

Taking transforms of both sides of the given equation gives

$$\mathscr{L}[\ddot{y} - 2\dot{y} - 3y] = \mathscr{L}[6e^t] = \frac{6}{s - 1}$$

The differentiation property in Section 7.3 combined with the initial conditions yields

$$\mathcal{L}[\dot{y}] = sY - 1$$

$$\mathcal{L}[\ddot{y}] = s^2Y - s - 3$$

Therefore the transformed differential equation becomes

$$s^2Y - s - 3 - 2sY + 2 - 3Y = \frac{6}{s-1}$$

Solving for Y yields

$$Y = \frac{s^2 + 5}{(s-1)(s+1)(s-3)}$$

Expanding into partial fractions, taking inverse transforms, and using Table 7.2 yields

$$y = \mathcal{L}^{-1}[Y] = \frac{7e^{3t}}{4} - \frac{3e^t}{2} + \frac{3e^{-t}}{4}$$

The Laplace transform can also be used to solve certain types of integral and integrodifferential equations, as shown in the next example.

Example 21

Solve the integrodifferential equation $\dot{y} - 3y + 2\int_0^t y\,dt = 1 + t$ if $y(0) = 1$.

From the differentiation property,

$$\mathcal{L}[y] = sY - 1$$

and the integration property in Section 7.3,

$$\mathcal{L}\left[\int_0^t y\,dt\right] = \frac{Y}{s}$$

Therefore the given equation is transformed into

$$sY - 1 - 3Y + \frac{2}{s}Y = \frac{1}{s} + \frac{1}{s^2}$$

Solving this equation for Y gives

$$Y = \frac{s^2 + s + 1}{s(s-1)(s-2)}$$

Expanding the last equation into partial fractions, using the inverse transform and Table 7.2 yields

$$y = \mathcal{L}^{-1}[Y] = \mathcal{L}^{-1}\left[\frac{1}{2s} + \frac{7}{2(s-2)} - \frac{3}{s-1}\right] = \frac{1}{2} + \frac{7}{2}e^{2t} - 3e^t$$

The Laplace transform can be used to study systems of linear differential equations with more than one unknown and constant coefficients. Such systems will be studied in detail by another method in Volume II, Chapter 2.

Example 22

Solve the simultaneous differential equations in the two unknowns x and y,

$$-2x + \frac{dy}{dt} = -1$$

$$2\frac{dx}{dt} - y = t$$

in which $x(0) = \frac{1}{2}$ and $y(0) = 0$.

Laplace transforming the given system of equations yields

$$-2X + sY = -\frac{1}{s}$$

$$2sX - 1 - Y = \frac{1}{s^2}$$

Solving the above transformed equations algebraically yields

$$X = \frac{s}{2(s-1)(s+1)} = \frac{1}{4(s-1)} + \frac{1}{4(s+1)}$$

$$Y = \frac{1}{s^2 - 1} = \frac{1}{2(s-1)} - \frac{1}{2(s+1)}$$

Hence, from Table 7.1,

$$x = \frac{e^t}{4} + \frac{e^{-t}}{4}$$

$$y = \frac{e^t}{2} - \frac{e^{-t}}{2}$$

The next example uses the property of Laplace transforms to show the equivalence of Theorems 2 and 2′ of Chapter 5.

Example 23

The solution of the differential equation

$$y^{(n)} + a_{n-1}y^{(n-1)} + \cdots + a_1 y' + a_o y = u(t)$$

where the differentiation is with respect to t, the a_i are constants, $u(t) = \delta(t)$, and the initial conditions

$$y^{(n-1)}(0+) = \cdots = y'(0+) = y(0+) = 0$$

are the same as the solution of the given differential equation with $u(t) = 0$ and the initial conditions

$$y^{(n-1)}(0+) = 1 \qquad y^{(n-2)}(0+) = \cdots = y(0+) = 0$$

Let y_1 be the solution of the given differential equation with $u(t) = \delta(t)$ and the first set of initial conditions. Taking Laplace transforms of both sides of the given equation and using the differentiation property gives

$$(s^n + a_{n-1}s^{n-1} + \cdots + a_1 s + a_0)Y_1 = 1$$

where Y is the Laplace transform of y_1. Similarly, with $u(t) = 0$ and the second set of initial conditions, the Laplace transform Y_2 of the second solution y_2 satisfies

$$s^n Y_2 - 1 + (a_{n-1}s^{n-1} + \cdots + a_1 s + a_0)Y_2 = 0$$

It is evident that $Y_1 = Y_2$ or, equivalently, $y_1 = y_2$.

EXERCISES

Solve by means of Laplace transforms:

1. $\dot{y} - 2y = 1$ with $y(0) = 0$

2. $\ddot{y} + 2\dot{y} - 8y = 0$ with $y(0)$ and $\dot{y}(0) = 1$

3. $\dot{y} - y - 6\int_0^t y\,dt = 12e^{3t}$ with $y(0) = -3$

4. $\dot{y} + 2y - 3\int_0^t y\,dt = 3 + 5t$ with $y(0) = 2$

5. $\ddot{y} + 4y = 6\sin 2t$ with $y(0) = 2$ and $\dot{y}(0) = 3$

6. $x - \dot{y} = t - 2$
$\dot{x} + y = 1 + 2t$; $x = 0$, $y = 2$ when $t = 0$

7. $-x + \dot{y} = -3$
$\dot{x} - y = 2t$; $x = 1$, $y = 0$ when $t = 0$

8. A suggested model for traffic flow is

$$\ddot{x}_2 = K_2(x_1 - x_2 - \tau_2\dot{x}_2)$$

where x_1 is the position of the lead car, x_2 is the position of the second car, and K_2 and τ_2 are the natural frequency and time constant, respectively, of the car and driver in the second car. A third car is also in the single-lane traffic flow, with a model similar to the above relating the third car to the second car. Assume that:
(a) All vehicles are initially at rest.
(b) Initial spacing between cars is 20 ft.
(c) The cars have zero length.
If the lead car instantaneously drags out at 30 ft/sec and holds a straight, constant speed, will there be any intercar accidents?

$K_2 = 16 \qquad \tau_2 = 0.05$ sec (a high-strung driver)

$K_2 = 1 \qquad \tau_2 = 2$ sec (a sluggish driver)

9. A model of nanoplankton respiration has been formulated [4] that accounts reasonably for gross community photosynthesis, storage, and respiration in tropical waters. This model has been represented in terms of direct analog elements in the following electric circuit:

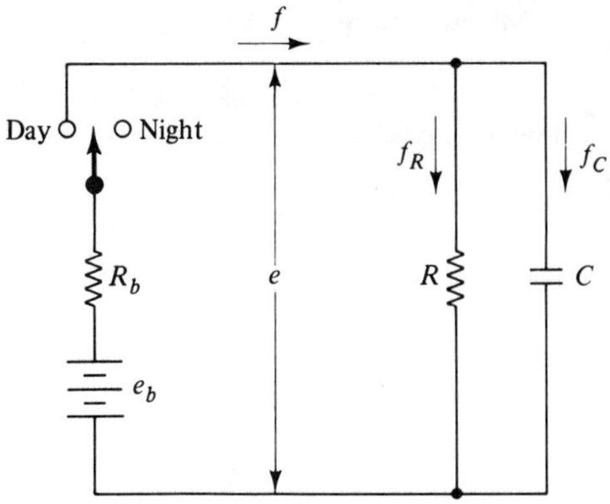

e_b = sunlight (approximated by a step function in the tropics)
f = production rate of material by photosynthesis
f_R = community respiration rate
e = backup potential of material in system
f_C = storage rate
C = community cellular storage capacity

The sunlight is the input to the system and is represented by a battery of voltage e_b. The rate f is proportional to the difference between e_b and a "backup" potential e, of material in the system, with the constant of proportionality being looked upon as the battery conductance $1/R_b$. The rate f_R is assumed proportional to the potential e, and the storage rate f_C proportional to the rate of change of potential in the community cellular storage capacity C. Finally, the total production rate f must equal the sum of respiration and storage rates.

(a) Verify that the differential equations for f and f_R are

$$\frac{C}{1/R + 1/R_b}\frac{df}{dt} + f = \frac{1}{R + R_b}\left(\frac{C}{1/R}\frac{de_b}{dt} + e_b\right)$$

$$\frac{C}{1/R + 1/R_b}\frac{df_R}{dt} + f_R = \frac{1}{R + R_b}e_b$$

(b) Solve the above equations using Laplace transforms and verify that the solutions behave as shown in the following diagram:

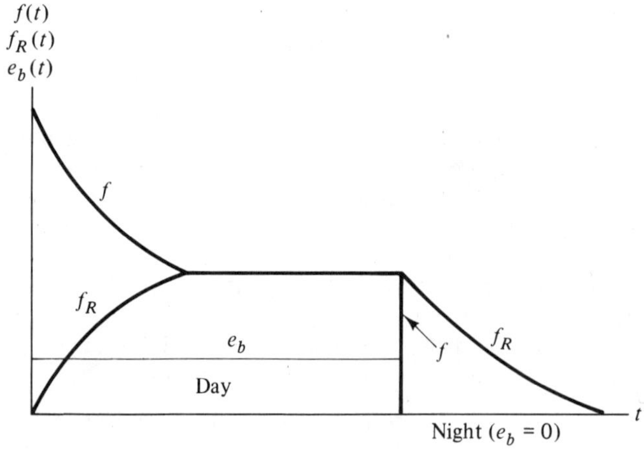

REFERENCES

1. Abramowitz, M., and Slegun, I. A. *Handbook of Mathematical Functions*. Dover, New York, 1970.
2. Churchill, R. V. *Operational Mathematics*. McGraw-Hill, New York, 1972.
3. Erdélyi, A., and Bateman, H. (eds.). *Tables of Integral Transforms,* Vols. I, II. McGraw-Hill, New York, 1954.
4. Odum, H. T., Beyers, R. J., and Armstrong, N. E. Consequences of small storage capacity in Nanoplankton pertinent to primary production in tropical waters. *J. Marine Res. 20*(3), 191–198, 1963.
5. Selby, S. M., and Weast, R. C. (eds.). *Handbook of Tables for Mathematics*. CRC Press, Cleveland, Ohio, 1975.
6. Widder, D. V. *The Laplace Transform*. Princeton University Press, Princeton, NJ, 1941.

8

Transfer Functions and Their Pictorial Representation

8.1 INTRODUCTION

In this chapter systems or subsystems will be considered in which the input is related to the output by means of a linear, time-invariant, differential, integrodifferential, or difference equation. For brevity, instead of specifying the defining equations of the system, the output will be said to be related to the input by a *linear operator*.

If such equations with zero initial conditions are Laplace transformed if time is continuous, or z-transformed if time is discrete, they are reduced to algebraic equations. These algebraic equations can be reduced to the form in which a certain function of the transform variable times the transformed input equals the transformed output. The function multiplying the transformed input is called the *transfer function*.

It will be seen that the algebraic relation between the transformed input and output can be conveniently represented by means of a block diagram. Thus a system will be represented by interconnections of blocks corresponding to its subsystems. Frequently it is desirable to reduce the interconnecting blocks to a single block that is an equivalent representation of the given system. Rules for carrying out this reduction directly and by means of signal graph theory will be given.

8.2 THE TRANSFER FUNCTION

Consider an input x_n to the nth component or subsystem of a linear system. Let L_n be the linear operator that completely specifies the nature of the component. If y_n is the output, then the general nonhomogeneous equation for the nth component will be

$$y_n(t) = L_n x_n(t) \tag{1}$$

A general expression for the time-dependent behavior of the system as a whole is desired. This can be done using Laplace transforms.

The Laplace transform approach requires *zero initial conditions for generality*. Nonzero initial conditions would imply that energy was already stored in the system from previous inputs. Thus the response to any new input would not be determined solely by the inherent nature of the component itself but would be determined in part by the particular past history.

Taking Laplace transforms of both sides of (1) yields

$$Y_n(s) = G_n(s)X_n(s)$$

where the function $G_n(s)$ arises from transforming derivatives, integrals, or differences. The function $G_n(s)$ is called the *transfer function* of the nth component. G_n is determined solely by the linear operator L_n and thus depends only on the nature of the nth component.

The Laplace transform of the output can be used as the transformed input for the next component in line, and so on. Hence, by stages, a general expression for the transformed output for the whole system can be obtained. By taking the inverse Laplace transform, the result can be translated back into the time domain. Thus the notion of the transfer function is an important concept for a linear system or a subsystem and so is formally defined in the following definition.

Definition 1. The *transfer function* G of a linear, stationary system or subsystem with zero initial conditions is defined as

$$G(s) = \frac{\text{transform of output}}{\text{transform of input}}$$

where the transform is the Laplace transform for continuous systems and the z-transform for discrete systems.

Remarks

1. The transfer function is the Laplace transform (or z-transform) of the system weighting function (see Exercise 4).
2. There is no loss in generality in using zero initial conditions, since by using the delta function as an input, arbitrary initial conditions can be obtained.
3. A transfer function may be defined only for a linear, stationary (constant-parameter) system. A nonstationary (time-varying) system has one or more time-varying parameters, and the Laplace transformation may not be used.
4. A transfer function function is an input–output description of the behavior of a system. Therefore the transfer function description does not include any information concerning the internal structure of the system and its behavior.

The following examples illustrate the application of Definition 1.

Example 1

A fundamental component of the circulatory system is a blood vessel. Let p_1 and p_2 be the pressures at the ends of the vessel. Within limits, the flow q through the vessel is

$$q = kp \tag{2}$$

where k is a parameter characteristic of the vessel and $p = p_1 - p_2$ is the pressure gradient, which is assumed to be variable.

The input to this flow system is the pressure gradient, and the output is the blood flow. By Laplace transforming (2), it follows from Definition 1 that the transfer function of the system is

$$G = \frac{Q}{P} = k$$

Example 2

In Example 12(b) of Chapter 3, a system of beavers was described by the difference equation

$$y(k) - (\tfrac{5}{4})y(k-1) = x(k)$$

where $x(k) = k - 5$ and $y(0) = 100$. Here the input is $x(k)$ and the output is the population size $y(k)$. Taking the z-transform of both sides of the given difference equation yields

$$Y - (\tfrac{5}{4})z^{-1}Y = X$$

Hence the transfer function of the given system is

$$G(z) = \frac{Y}{X} = \frac{4z}{z - 5}$$

Example 3 (Transfer Function of Leg Vibration System)

The output, θ, of the system describing the forced vibration of a leg (see Example 15 in Chapter 4) satisfies the differential equation.

$$\ddot{\theta} + b\dot{\theta} + \omega^2\theta = x(t)$$

where the input $x(t) = T_0 \cos \psi t$. Laplace transforming both sides of the given differential equation yields

$$s^2\Theta + bs\Theta + \omega^2\Theta = X$$

Therefore the transfer function of the given system is

$$G(s) = \frac{\Theta}{X} = \frac{1}{s^2 + bs + \omega^2}$$

Engineers and biologists frequently use transfer functions in two different ways. Engineers, after designing the system, usually know the form of the transfer function. Their problem is adjusting parameters in the transfer function to give a particular performance. By contrast, biologists usually have the problem of determining the transfer function, which is unknown. They do this by applying test signals to the system and observing the responses. A method for determining the transfer function from such information will be described in the next section.

EXERCISES

1. It has been shown [2, 9] that if a load of water is suddenly placed in the stomach by means of a tube, the volume of the water in the stomach decreases exponentially.
   (a) Show that this implies that

$$\text{flow out of stomach} = K_1 \times \text{stomach volume}$$

(b) Find the transfer function of this system.

2. Find the transfer function of the following discrete systems whose output is y and whose input is u:
   (a) $y(k+1) - 2y(k) = u(k)$
   (b) $y(k+2) + 4y(k+1) + y(k) - 2y(k-1) = u(k)$
   (c) $y(k+2) - 3y(k+1) + 2y(k) = u(k)$
3. Find the transfer function function of the following continuous systems whose output is y and whose input is u:
   (a) $\dot{y} + ay = u(t)$
   (b) $\ddot{y} - 3\dot{y} + 2y = u(t)$
   (c) $y^{(4)} + 3y^{(3)} - 7y^{(2)} - 15y^{(1)} + 18y = u(t)$
4. Show that the transform of the weighting function of a system is its transfer function. (*Hint:* $y = h * u$. Use the convolution property.)
5. The weighting function of

$$y(k) = b_0 u(k) + b_1 u(k-1) - a_1 y(k-1)$$

   is $h(0) = b_0$ and $h(i) = (b_1 - a_1 b_0)(-a_1)^{i-1}$ for $i = 1, 2, 3, \ldots$. Verify that the z-transform of h is the transfer function of the given system.
6. The weighting function of system 1 is G_1, and the weighting function of system 2 is G_2. If the output of system 1 is the input to system 2, what is the weighting function of the two combined systems?

8.3 USE OF IMPULSE INPUTS TO DETERMINE THE TRANSFER FUNCTIONS

The black box problem is the problem of using experimental observations to find the transfer function of the system hidden in the box. The transfer function, in turn, may enable one to formulate some plausible hypothesis about what the components inside the box are. Aside from this, once the transfer function is known, the output response to *any* input can be predicted.

By definition,

$$G(s) = \frac{\mathcal{L}[y(t)]}{\mathcal{L}[A\delta(t)]} = \frac{Y(s)}{A} \tag{3}$$

which implies the following result for $A = 1$.

Theorem 1. The transfer function of a linear system is the Laplace transform of the output corresponding to a unit impulse input.

An alternative way of stating Theorem 1 is that the inverse Laplace transform of the transfer function of a system is simply the natural response to a unit impulse input.

Before applying Theorem 1, the given system must be checked for linearity in the range of the inputs that are going to be applied. The next example illustrates the application of Theorem 1 under the assumption that the described system is linear.

Example 4 (Determination of Drug Concentrations Using Transfer Functions)

One gram of a drug was injected intravenously at $t = 0$ into a man. Thereafter, the concentration of the drug in samples of blood plasma was determined at suitable intervals. These experimental observations could be fitted satisfactorily by the exponential function

$$c(t) = 0.2e^{-0.08t} \text{ g/liter} \qquad (t \geq 0) \tag{4}$$

where c is the concentration of the drug in the plasma and t is the time in hours. The output of this man–drug system is c, and the input is defined as u, the rate of intravenous administration of the drug in grams per hour.

(a) Find the transfer function $G(s)$.

(b) Predict $c(t)$ if a single dose of 0.5 g of the drug were injected into the same subject every 2 hr for 1 day, the first such dose being given at $t = 0$.

(c) Predict $c(t)$ if the drug were infused continuously for 1 day at a rate of 0.25 g/hr starting at $t = 0$.

(a) The Laplace transform of the output c given in (4) is

$$C(s) = \frac{0.2}{s + 0.08}$$

The Laplace transform of the impulse input $1\delta(t)$ is 1. Therefore, from (3), the transfer function is

$$G(s) = \frac{C(s)}{1} = \frac{0.2}{s + 0.08}$$

(b) The input can be described by the following series of 12 impulses,

$$u(t) = 0.5[\delta(t) + \delta(t-2) + \delta(t-4) + \cdots + \delta(t-22)]$$

since the drug is administered at $t = 0, 2, 4, \ldots, 22$. Its Laplace transform is

$$U(s) = 0.5(1 + e^{-2s} + e^{-4s} + \cdots + e^{-22s})$$

Thus

$$C(s) = G(s)U(s) = \frac{0.1}{s + 0.08}(1 + e^{-2s} + e^{-4s} + \cdots + e^{-22s})$$

The inverse transform of both sides of (5) gives the output

$$c(t) = 0.1[e^{-0.08t}\sigma(t) + e^{-0.08(t-2)}\sigma(t-2) + \cdots + e^{-0.08(t-22)}\sigma(t-22)]$$

where

$$\sigma(t) = \begin{cases} 1 & (t \geq 0) \\ 0 & (t < 0) \end{cases}$$

is the unit step function.

(c) This input is

$$u(t) = 0.25[\sigma(t) - \sigma(t-24)]$$

Thus

$$U(s) = \frac{0.25}{s}(1 - e^{-24s})$$

and so the Laplace transform of the output is

$$C(s) = G(s)U(s) = \frac{0.05(1 - e^{-24s})}{s(s + 0.08)}$$

$$= \frac{0.05}{0.08}\left(\frac{1}{s} - \frac{1}{s + 0.08}\right)(1 - e^{-24s})$$

Taking the inverse of the transformed output gives

$$c(t) = 0.625(1 - e^{-0.08t})\sigma(t) - 0.625(1 - e^{-0.08(t-24)})\sigma(t - 24)$$

or

$$c(t) = \begin{cases} 0.625(1 - e^{-0.08t}) & \text{for } 0 \le t \le 24 \\ c(24)e^{-0.08(t-24)} & \text{for } t \ge 24 \end{cases}$$

Figure 8.1 shows $c(t)$ in (b) and (c) plotted against time.

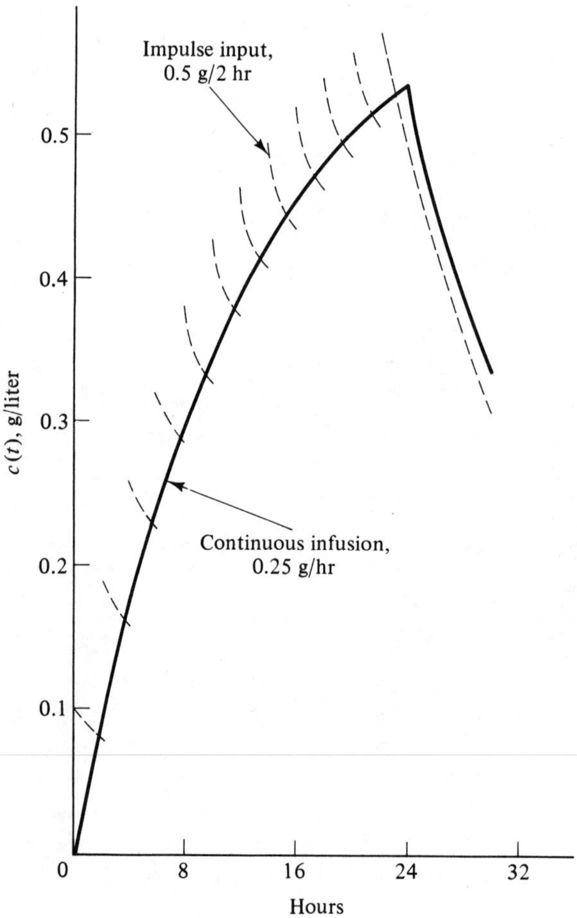

Figure 8.1 Plasma concentration of the drug in Example 4.

The solid curves in Fig. 8.1 show how the plasma concentration, $c(t)$, of the drug would behave if 0.5 g of the drug were given every 2 hr for 1 day. The dashed curves indicate how $c(t)$ would behave if the same total dose were administered continuously at the rate of 0.25 g/hr for 24 hr starting at $t = 0$. Note that at the end of about 14 hr, the mean plasma concentration is nearly the same for both types of input. However, with the intermittent impulse inputs there are fluctuations above and below the mean.

In Example 4, the system remained a black box since no information was obtained about the metabolism or catabolism of the drug. It was not known if the drug was being sequestered in the bone, or was being excreted by expiration or urination, or was being metabolized by the liver or other tissues, or was being disposed of by more than one of such pathways. Nevertheless, transfer function analysis did yield the output for a given input for a drug that disappears exponentially.

EXERCISES

1. The output of a linear system in response to a given input appears below.

	Output	Input
(a)	1	$2\delta(t)$
(b)	sin $3t$	$4\delta(t)$
(c)	t^2	$3\delta(t)$

Find the corresponding transfer function.

2. If the response of a linear system to the input $5\delta(t)$ is e^{-2t}, find the output of this system if the input is $6t^2 + e^{4t}$.
3. Answer the questions posed in Example 4 if the observations are fitted by the curve $c(t) = 0.5e^{-0.04t}$ and in (b) the single dose is 0.25 g every 4 hr for 2 days while in (c) the continuous infusion is 0.0625 g/hr for 2 days.
4. The output current i when an input voltage e is applied across a resistance and capacitance in series is described by the integral equation

$$iR + \frac{1}{C}\int_0^t i(x)\,dx = e$$

where R and C are constants. Find the transfer function of this system by letting e be a unit impulse.

5. The output current i when an input voltage e is applied across a resistance, capacitance, and inductance in series is described by the integrodifferential equation

$$iR + \frac{1}{C}\int_0^t i(x)\,dx + L\frac{di}{dt} = e$$

where R, C, and L are constants. Find the transfer function of this system.

8.4 BLOCK DIAGRAMS

Block diagrams were introduced in Sections 1.2 and 2.1. Here these diagrams will be specialized in the sense that the input and output variables will be the Laplace transforms of the input and output functions. Moreover, the relation connecting the input and output, which is written in the block, will always be a transfer function. The reason for this special form is that complicated systems composed of many such blocks can often be reduced to an equivalent single block. A detailed description of the block diagram of a single component and the reduction process appear below.

The reader might wonder what useful advantage the block diagram representation has over simply writing the equivalent equation for each component and their connections. For a complex system such as the circulatory system, a set of hundreds of equations may be required. Such a set of equations is exceedingly difficult to understand and follow in a logical order. A block diagram incorporating the same information is much easier to follow. The flow of information is clearly shown and is intuitively appealing.

The block diagrams in this section will describe stationary linear systems (i.e., those with constant coefficients) in which each component has a single input and a single output. Each such component is represented in the diagram by a rectangle (block) containing the symbolic representation of the transfer function. The Laplace transform of the input variable is denoted by a line with an arrowhead pointing to the block, and the Laplace transform of the output variable is shown by a line with an arrowhead pointing away from the block (Fig. 8.2). Such a block represents a complete subsystem in the sense that, given the transfer function in the block, the output variable is determined by the input variable. For example, for the subsystem depicted in Fig. 8.2, the output variable Y is determined from the input variable U by the relation $Y = GU$. Thus the signal flow along each pathway in a block diagram must be unidirectional.

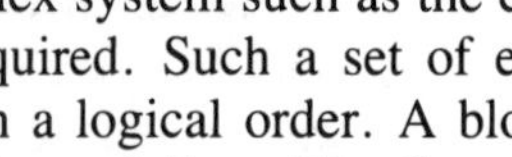

Figure 8.2 Block diagram of a component with transfer function G.

The block diagram representation of a given system may often be reduced by block-diagram reduction techniques to a simplified block diagram with fewer blocks than the original diagram. This reduction in the number of blocks often leads to a deeper insight into the behavior of the system. These diagram transformations are given in Table 8.1. All the transformations in the table can be derived by simple algebraic manipulations of the equations representing each block. For example, the equivalent diagram for combining blocks in cascade in Table 8.1 is derived as follows. By definition, $Y_2 = G_1 Y_1$ and $Y_3 = G_2 Y_2$. Hence $Y_3 = G_2 G_1 Y_1 = G_1 G_2 Y_1$, which is the relation expressed by the equivalent diagram.

As another example of block diagram transformation, the equivalence of the two diagrams for a negative feedback loop will be derived. The proofs of the remaining transformations in the table are left as exercises. The block diagram that will be used in the proof appears in Fig. 8.3. Since the output, Y, is related to the actuating signal, E, it follows by definition that

$$Y = GE \tag{5}$$

TABLE 8.1 BLOCK DIAGRAM TRANSFORMATIONS

Transformation	Original diagram	Equivalent diagram
1. Combining blocks in series (cascade)	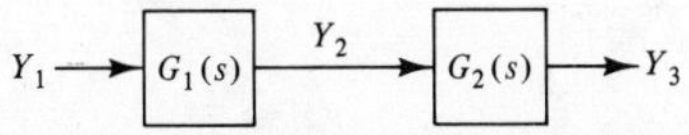	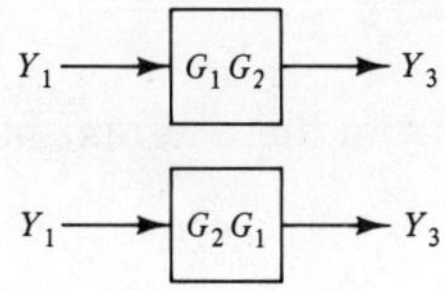
2. Moving a summing point behind a block	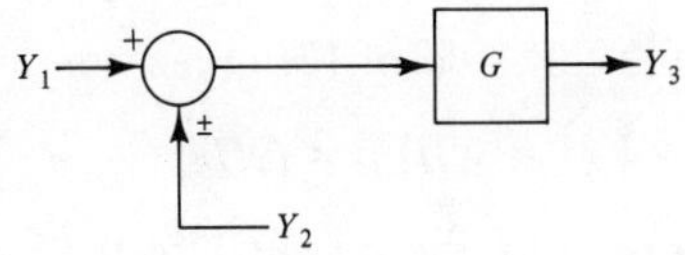	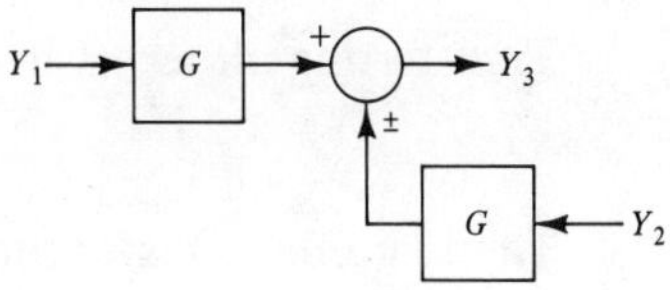
3. Moving a pick-off point ahead of a block	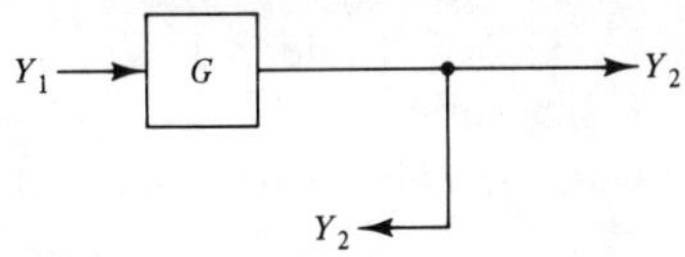	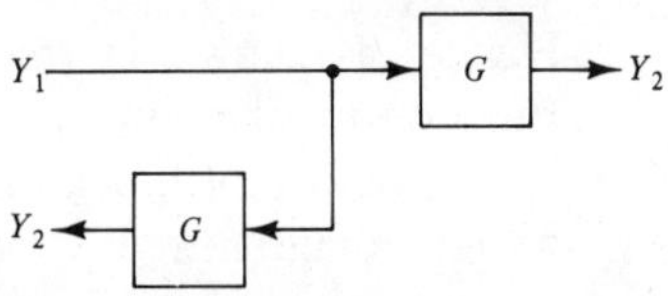
4. Moving a pick-off point behind a block	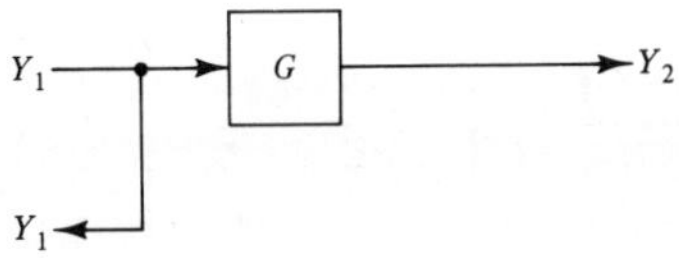	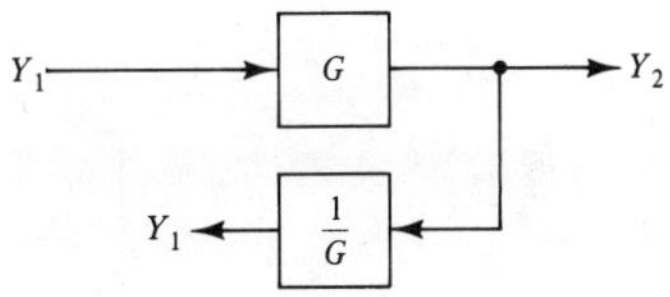
5. Moving a summing point ahead of a block	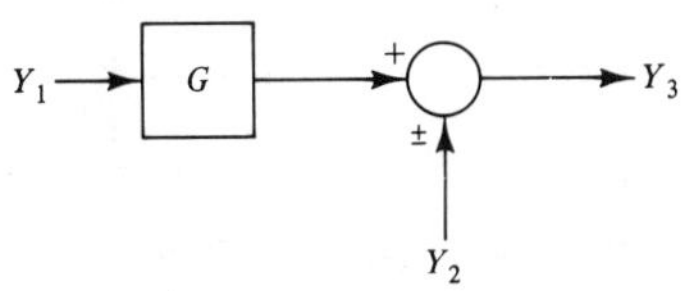	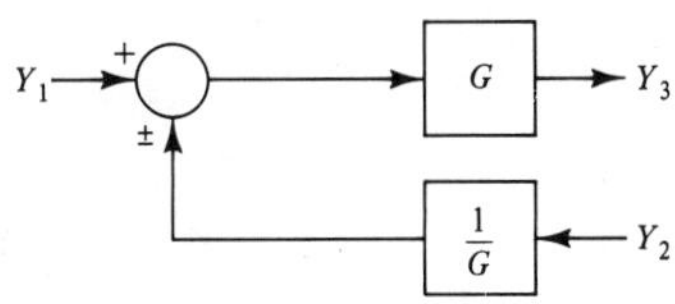
6. Eliminating a feedback loop	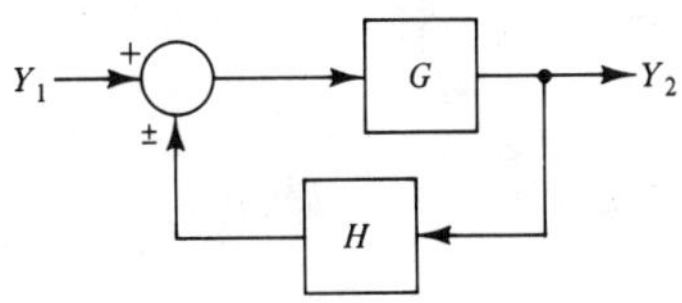	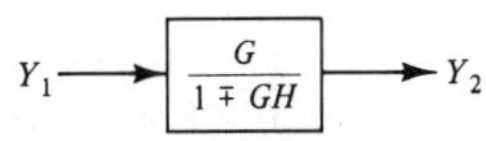

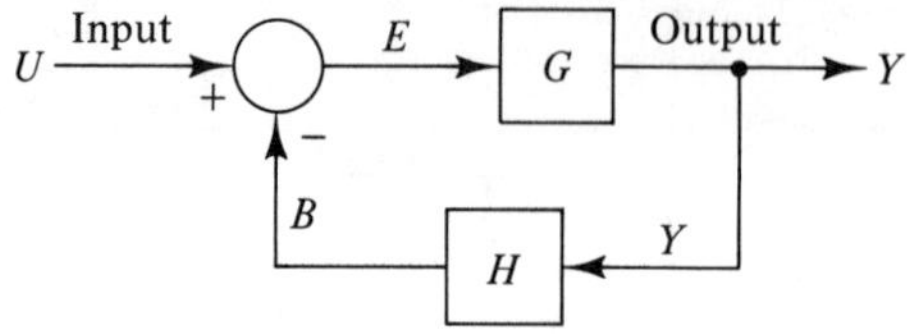

Figure 8.3 Negative feedback control system.

From the diagram in Fig. 8.3, the actuating signal is

$$E = U - B = U - HY \tag{6}$$

Substituting the value of E given in (6) into (5) yields

$$Y = G(U - HY) \tag{7}$$

Transferring all terms in Y to the left side of (7) gives

$$Y(1 + GH) = GU$$

Therefore the transfer function relating the output Y to the input U is

$$\frac{Y}{U} = \frac{G}{1 + GH}$$

Hence, as shown in transformation 6 of Table 8.1 $G(1 + GH)^{-1}$ is the transfer function for the equivalent block diagram.

The use of block diagram transformations will be illustrated by an example of a block diagram reduction.

Example 5

The block diagram transformations will be applied to the multiple loop in Fig. 8.4.

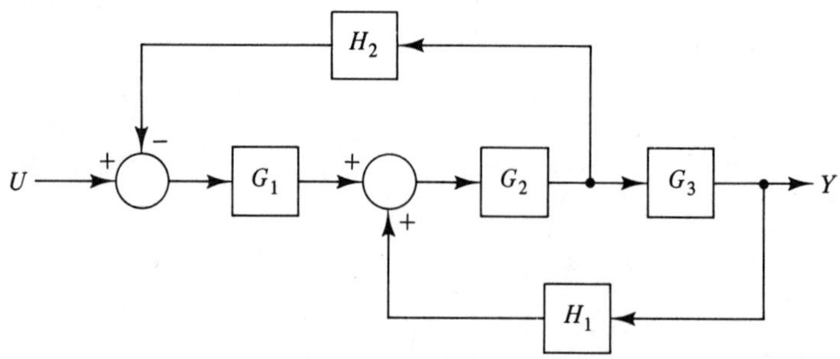

Figure 8.4 Multiple-loop feedback control system.

First, in order to eliminate the loop $G_1 G_2 H_2$, move H_2 behind G_3 by using rule 4 of Table 8.1 and thereby obtain Fig. 8.5(a). Combining the blocks G_2 and G_3 by rule 1 and then eliminating the loop $G_2 G_3 H_1$ by rule 6 gives Fig. 8.5(b). Next, combining the two blocks G_1 and $G_2 G_3/(1 - G_2 G_3 H_1)$ by using rule 1 and then eliminating the loop in Fig. 8.5(b) leads to Fig. 8.5(c) after some algebraic simplifications.

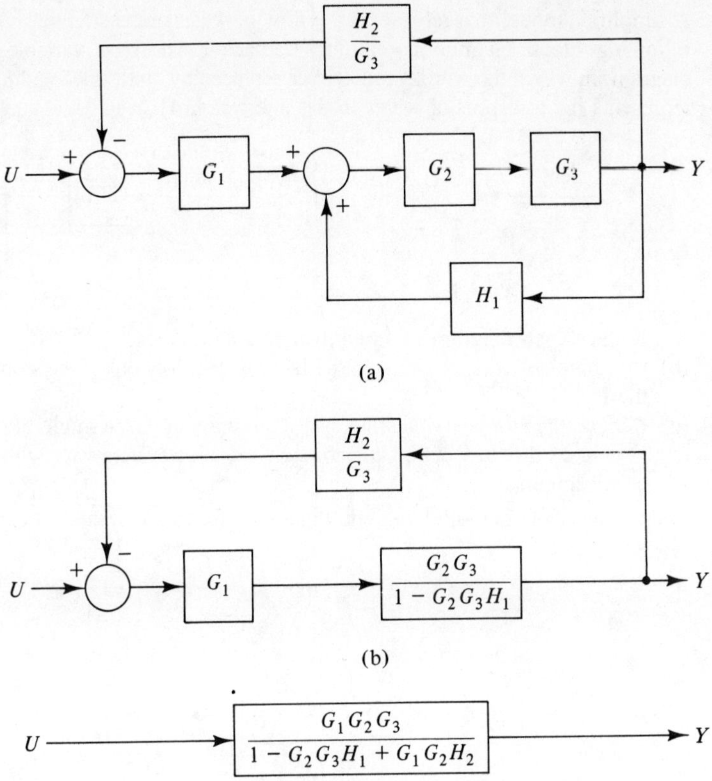

(c)

Figure 8.5 Block diagram reduction of the system of Fig. 8.4.

EXERCISES

1. Prove rule 2.
2. Prove rule 3.
3. Prove rule 4.
4. Prove rule 5.
5. Prove rule 6 for a positive feedback control system.
6. Reduce the following block diagrams to an equivalent single block diagram:

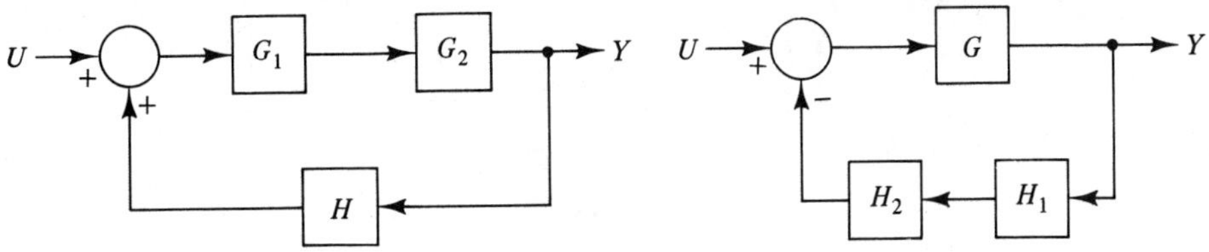

(a) (b)

7. A simplified model that represents the flow of water in the stomach [2, 9] is shown in the following block diagram containing Laplace-transformed variables: A similar block diagram, in which the word "stomach" is replaced by "intestine" and K_1 is replaced by K_2, represents the transport of water in the intestine [14].

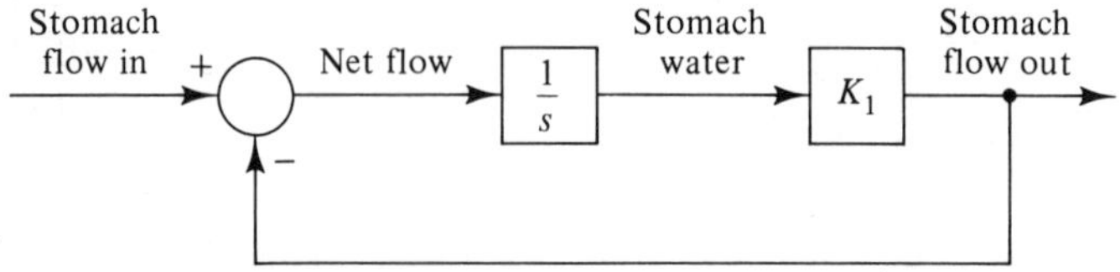

(a) Reduce each diagram to an equivalent single block.
(b) Combine both diagrams using the fact that the flow out of the stomach is the flow into the intestine.
(c) Reduce the stomach–intestine system to an equivalent single block.
(d) How does the flow out of the stomach–intestine system vary with time in response to a unit impulse input?

8. Reduce each of the following multiple-loop feedback systems to an equivalent single block.

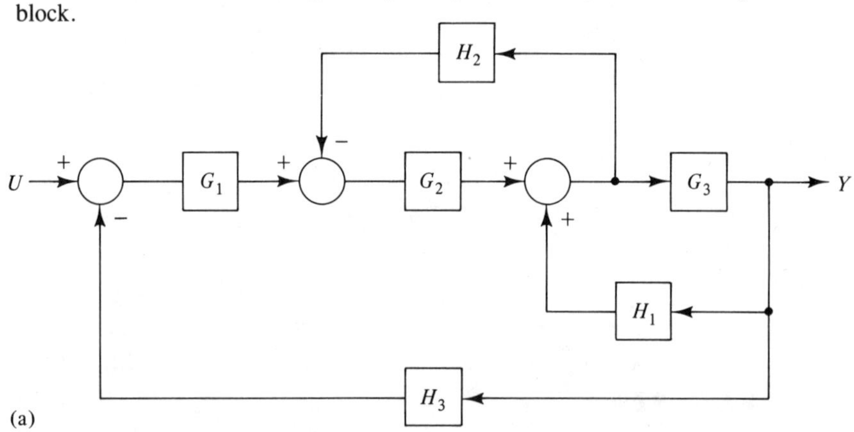

(a)

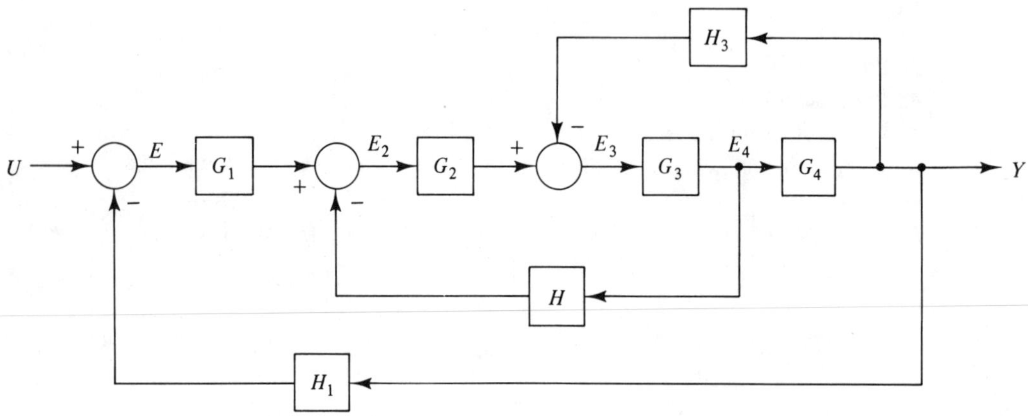

(b)

9. The blocks represent a system that enables a person, paralyzed below the neck, to drive himself in a motorized wheelchair. The headgear sensor provides an output proportional to the magnitude of the head movement. Sensors are mounted at 90° intervals so that forward, left, right, and reverse can be commanded. Find the transfer function of the given system by reducing the block diagram to a single block.

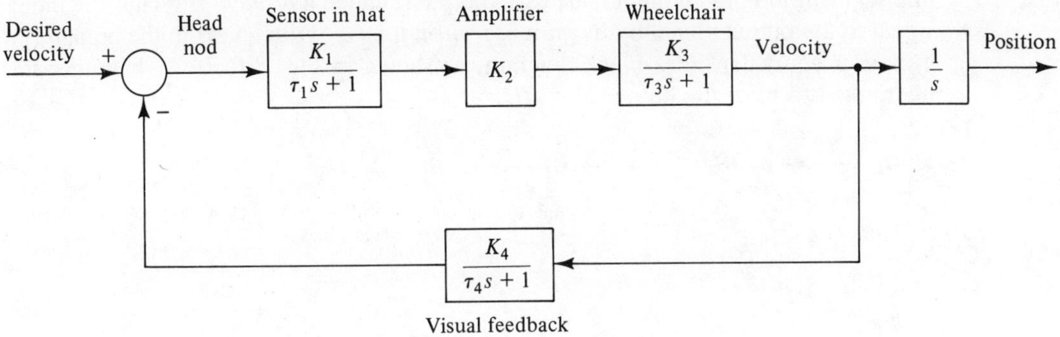

10. A simplifed model of the human speech system is shown below.

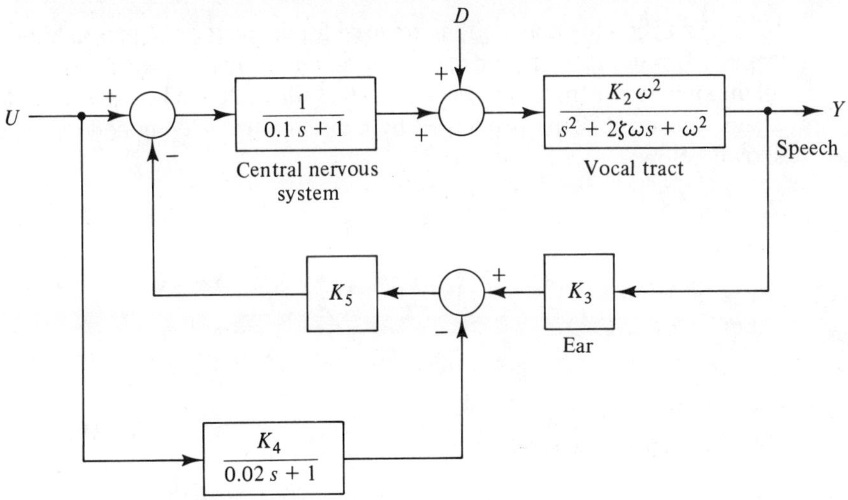

Speech control is a feedback process that occurs naturally in humans. When the system becomes unstable, a person stutters. The central nervous system is approximated by the synaptic delay of the nerves. Find the transfer function of the given system.

8.5 SIGNAL FLOW GRAPHS

Block diagram representation for a complex system can be cumbersome. An alternative method for determining relationships among system variables is based on representation of the system by line segments and is called the *signal flow graph method*. The advantage of this method is the availability of a flow graph gain formula that provides the relation between system variables without requiring any reduction procedure or manipulation of the flow graph.

A *signal flow graph* is a diagram consisting of dots (or circles) that are connected by several directed line segments and is a graphical representation of a set of linear relations. The dots are called *nodes* and represents inputs and outputs. The directed line segment joining an input node to an output is called a *branch*. It relates the input signal to the output signal by its *gain* or *transmittance,* written next to the branch, in a manner equivalent to a block diagram, as shown in Fig. 8.6. Both diagrams are interpreted to have the output $Y = HU$.

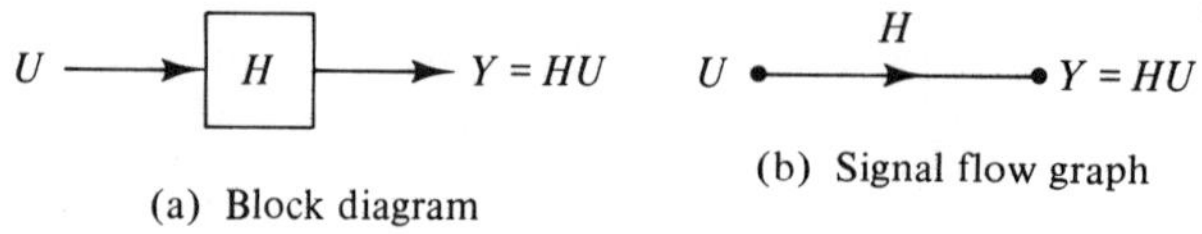

Figure 8.6 Signal flow graph of a single block.

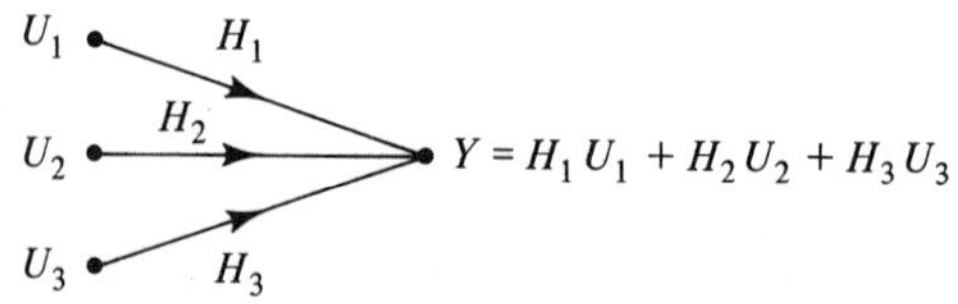

Figure 8.7 Summation of signals.

The following conventions are used for the pictorial representation of signal flow graphs. If a number of branches enter an output node, then the output is the sum of all the signals transmitted by each branch as illustrated in Fig. 8.7. A node can transmit a signal to itself. This is denoted by a circular arc leaving and entering the node as shown below:

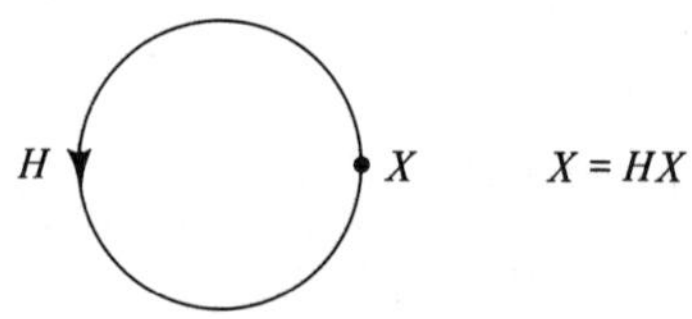

Example 6

The signal flow graph of

$$
\begin{aligned}
C_1 &= G_{11}R_1 + G_{12}R_2 \\
C_2 &= G_{21}R_1 + G_{22}R_2
\end{aligned}
\tag{8}
$$

is shown in Fig. 8.8.

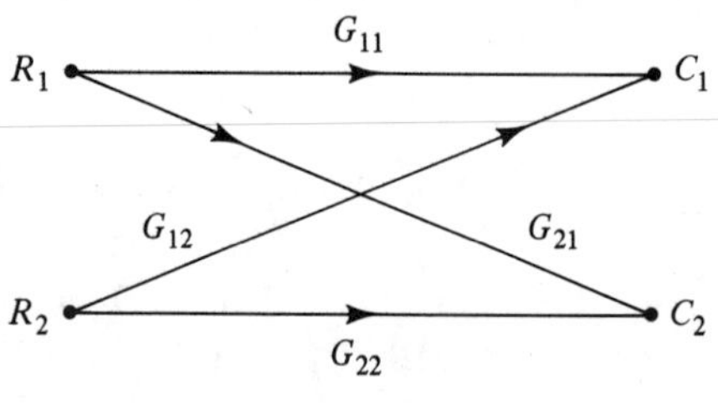

Figure 8.8 Signal flow graph of the system (8).

Example 7

The signal flow graph of

$$\begin{aligned} a_{11}x_1 + a_{12}x_2 + r_1 &= x_1 \\ a_{21}x_1 + a_{22}x_2 + r_2 &= x_2 \end{aligned} \tag{9}$$

is shown in Fig. 8.9.

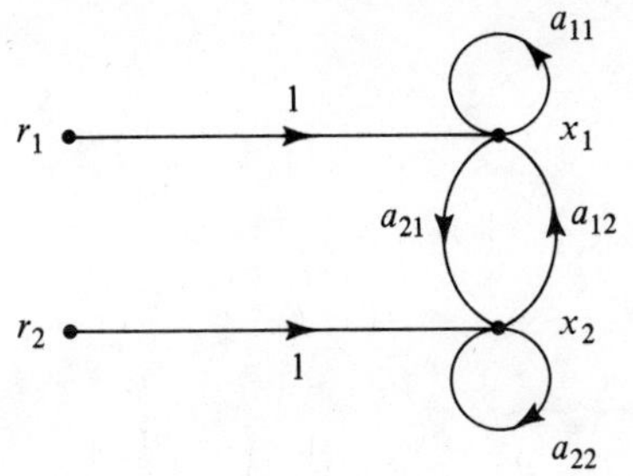

Figure 8.9 Signal flow graph of the system (9).

The following terminology is used in signal flow graph theory:

source = a node having only outgoing branches—for example, R_1 and R_2 in Fig. 8.8; r_1 and r_2 in Fig. 8.9.

sink = a node having only incoming branches—for example, C_1 and C_2 in Fig. 8.8.

path = a group of connected branches traversed from one node to another—for example, r_1x_1, $r_1x_1x_2$, $r_1x_1x_2x_1$ in Fig. 8.9.

forward path = a path that originates from a source and terminates in a sink and along which no node is encountered more than once—for example, r_1x_1 in Fig. 8.9.

path gain = the product of coefficients associated with the branches along the path—for example, in Fig. 8.9 the gain of $r_1x_1x_2$ is $1 \cdot a_{21}$.

loop = a closed path that originates and terminates on the same node and along which no node is met twice—for example, $x_1x_2x_1$ in Fig. 8.9.

loop gain = the product of coefficients associated with the branches of the loop—for example, the loop gain of $x_1x_2x_1$ in Fig. 8.9 is $a_{21}a_{12}$.

nontouching loops = loops having no common node—for example, x_1x_1 and x_2x_2 in Fig. 8.9.

touching loops = loops having at least one common node—for example, x_1x_1 and $x_1x_2x_1$ in Fig. 8.9.

A complicated flow graph can be reduced to a simpler, equivalent flow graph having fewer nodes or branches than the original graph by means of the rules in Table 8.2. This table is the analog of Table 8.1 for the simplification of block diagrams. Just as a block diagram having a single input and output can be reduced to an equivalent single block, so can a flow graph with a single input and output node be reduced to an equivalent flow graph with one branch. Frequently, the purpose of such a reduction is to find the transfer function of the given system. However, for flow graphs there is a formula for finding the transfer function that is applicable without simplifying the

TABLE 8.2 FLOW GRAPH DIAGRAM TRANSFORMATIONS

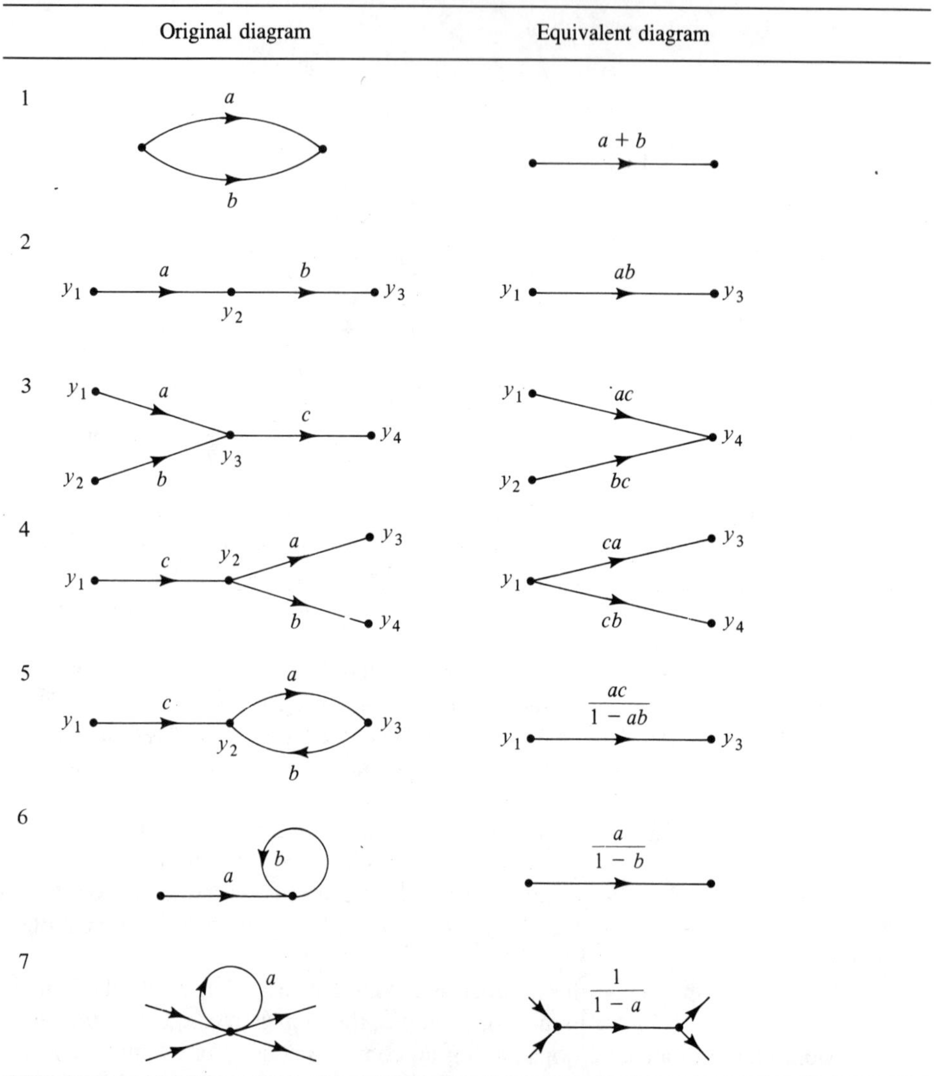

graph. Thus flow graph transformations for simplification are not used as often as the analogous block diagram transformations. In fact, block diagrams are often translated into their equivalent flow graphs in order to find the transfer functions of certain of the variables by means of the flow graph gain formula. The following example illustrates this translation process and the simplification of the resulting flow graph. The flow graph gain formula will be applied later.

Example 8

The block diagram in Fig. 8.10 is the z-transform version of the life cycle of a bacterium that undergoes binary fission with a fixed interfission interval, T. The input, u, is the flow

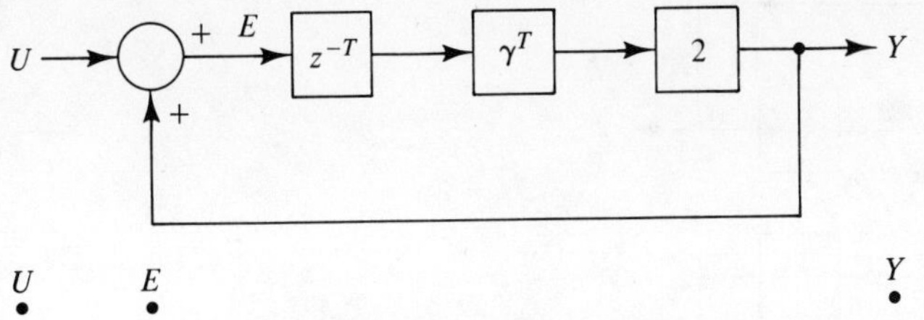

Figure 8.10 Block diagram of binary fission.

of newly produced bacteria from an external source, and the output, y, is the observed flow of newly generated bacteria.

To convert the block diagram to a flow graph, nodes for the input, E, and output variables must be drawn as shown at the bottom of Fig. 8.10. Since the variables U and Y are added to produce E, directed line segments labeled with 1 must be drawn from U and Y to E. Another branch from E to Y labeled $z^{-T}\gamma^{T}2$ must also be drawn, since the variable E is transformed into Y through the operations shown in the blocks in Fig. 8.10. The resulting flow graph is shown in Fig. 8.11.

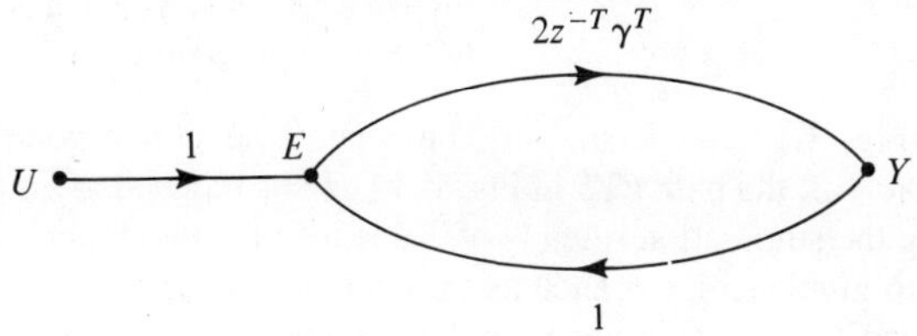

Figure 8.11 Flow graph corresponding to the block diagram in Fig. 8.10.

Using rule 5 in Table 8.2, the node E can be eliminated, leaving a single line from the input node to the output node. This results in the configuration of Fig. 8.12. The transfer function is the label of the single remaining branch.

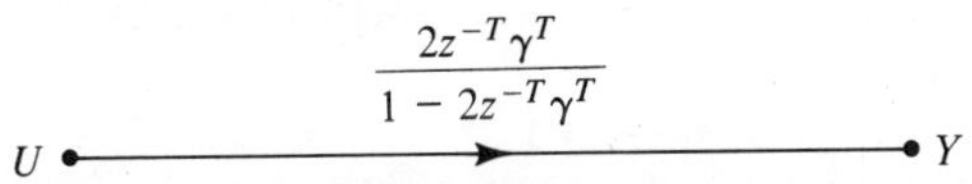

Figure 8.12 Simplified flow graph equivalent to the flow graph in Fig. 8.11.

Note that the flow graph in Fig. 8.11 is not unique. For example, another flow graph equivalent to the block diagram of Fig. 8.10 can be obtained by using the output of z^{-T} as a node.

Example 9 (Life Cycle of the Female Herring Gull)

The block diagram in Fig. 8.13 represents the z-transform of the idealized model of the life cycle of the female herring gull. These birds have a maturation time of 30 months between fledglings and the production of the first fledged brood and a nesting interval of 10 months. The input to the population is the flow of adults by migration from another colony, and the output is the observed flow of returning breeders.

The equivalent flow graph is constructed as follows. Four nodes, U, S, E, and Y, are selected to represent the variables U, the output of the adder, E, and Y, respectively. Since the variable

$$S = U + 0.8\gamma_1 z^{-30}E + Y$$

branches connecting the nodes U, E, and Y to S and labeled with 1, $0.8\gamma_1 z^{-30}$, and 1, respectively, are drawn as shown in Fig. 8.14. Finally, since $Y = \gamma_2 z^{-10}E$, the branch connecting E to Y and labeled $\gamma_2 z^{-10}$ is drawn. This results in Fig. 8.14.

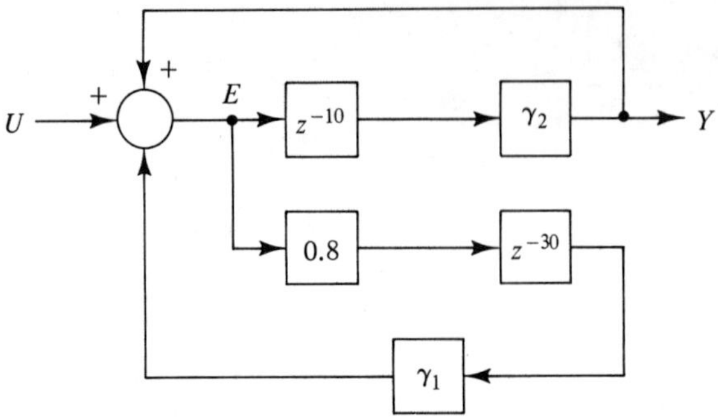

Figure 8.13 The z-transform of a block diagram of an idealized population of herring gulls.

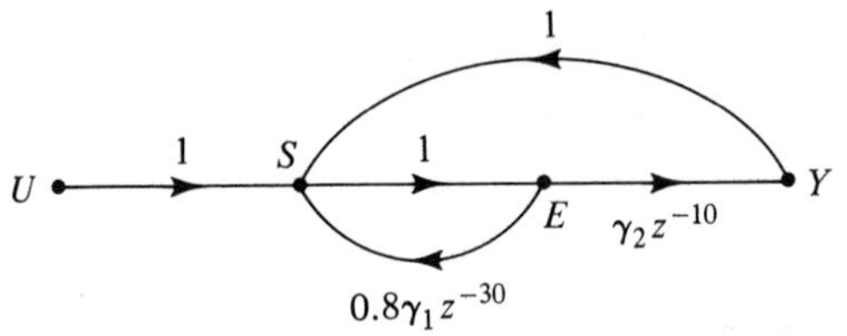

Figure 8.14 A flow graph equivalent to the block diagram of Fig. 8.13.

The flow graph in Fig. 8.14 can be simplified to an equivalent two-node graph as follows. By rule 2 in Table 8.2, the path *EYS* in Fig. 8.14 can be reduced to *ES* as shown in Fig. 8.15(a). Applying the summation rule (1) of Table 8.2, the two branches joining *E* to *S* can be combined to give a single branch as shown in Fig. 8.15(b). Since $U = S$, the node *S* can be relabeled *U*, and the node *S* can be omitted. By rule 5, the loop *ESE* can be replaced by a branch labeled $(1 - 0.8\gamma_1 z^{-30} - \gamma_2 z^{-10})^{-1}$. Finally, another application of rule 2 yields Fig. 8.15(c). The transfer function of the block diagram in Fig. 8.13 is the label of the branch *UY* in Fig. 8.15(c).

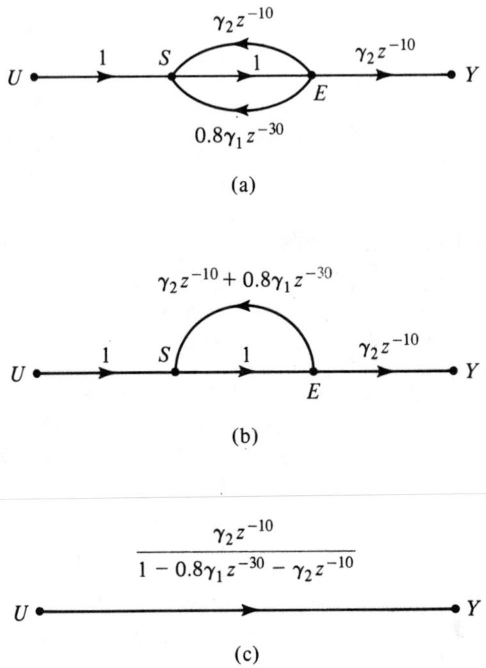

Figure 8.15 Simplification of the flow graph in Fig. 8.14.

The analytic formula for finding a transfer function will be motivated by an example. The system of equations (9) can be rewritten as

$$x_1(1 - a_{11}) + x_2(-a_{12}) = r_1$$
$$x_1(-a_{21}) + x_2(1 - a_{22}) = r_2 \tag{10}$$

The simultaneous solution of the system (10) is

$$x_1 = \frac{(1 - a_{22})r_1 + a_{12}r_2}{(1 - a_{11})(1 - a_{22}) - a_{12}a_{21}} = \frac{(1 - a_{22})}{\Delta}r_1 + \frac{a_{12}}{\Delta}r_2$$

$$x_2 = \frac{(1 - a_{11})r_2 + a_{21}r_2}{(1 - a_{11})(1 - a_{22}) - a_{12}a_{21}} = \frac{(1 - a_{11})}{\Delta}r_2 + \frac{a_{21}}{\Delta}r_1$$

where

$$\Delta = (1 - a_{11})(1 - a_{22}) - a_{12}a_{21}$$
$$= 1 - a_{11} - a_{22} + a_{11}a_{22} - a_{12}a_{21}$$

Here Δ is equal to 1 minus the gain of each self-loop a_{11}, a_{22} and the gain of the loop $a_{12}a_{21}$, plus the product of the gains of the two nontouching loops a_{11} and a_{22} (See Fig. 8.9). This result leads to the speculation that for an arbitrary flow graph the denominator in the gain formula is

$$\begin{aligned}\Delta = 1 &- \text{(sum of all different loop gains)} \\ &+ \text{(sum of gain products of all combinations of two nontouching loops)} \\ &- \text{(sum of gain products of all combinations of three nontouching loops)} \\ &+ \cdots\end{aligned} \tag{11}$$

The numerator for x_1 with the input r_1 is 1 times $(1 - a_{22})$, which can be interpreted as the gain of the path joining r_1 to x_1 times the value of Δ when all the gains of all the loops touching the path joining r_1 to x_1 have been set equal to zero. According to this last interpretation, the numerator for x_1 with the input r_2 is the gain of the forward path joining r_2 to x_1 (a_{12}) times the value of Δ when all the gains of all loops touching the path joining r_2 to x_1 have been set to zero (1, since all loops touch this path as shown in Fig. 8.9). Therefore the numerator is a_{12}, which agrees with the formula for x_1. The numerators for x_2 are symmetrical to those for x_1 and can be obtained by replacing 1 by 2 and 2 by 1 in the formula for x_1. These results lead to the speculation that for an arbitrary flow graph, the numerator of x_j for the input r_i is

$$\sum_k G_{ij}(k)\, \Delta_{ij}(k)$$

where the summation is over all possible forward paths from r_i to x_j and every such path is identified by an integer k. Here

$$G_{ij}(k) = \text{gain of the } k\text{th forward path from the variable } r_i \text{ to } x_j \tag{12}$$

$$\Delta_{ij}(k) = \text{the value of } \Delta \text{ with the gains of all loops touching the } k\text{th path set equal to zero} \tag{13}$$

The denominator Δ defined in (11) is called the *determinant* of the graph, while $\Delta_{ij}(k)$ is called the *cofactor* of the path k. Students familiar with the theory of determinants will see the reason for extending the terminology of determinants to flow graph theory.

In general, the *linear dependency*, T_{ij}, between the input r_i (often called the independent variable) and the output x_j (often called the dependent variable) is given by the following formula.

Mason's Loop Rule [6, 7]. In a linear flow graph, the linear dependency between an input r_i and an output x_j is

$$T_{ij} = \frac{\sum_k G_{ij}(k)\, \Delta_{ij}(k)}{\Delta} \tag{14}$$

where Δ, $G_{ij}(k)$, and $\Delta_{ij}(k)$ are defined by (11), (12), and (13), respectively, and the summation is over all possible forward paths from r_i to x_j.

An alternative way of writing (14) can be formulated by using the following notation:

L_i = gain of the loop i in the flow graph, $i = 1, 2, \ldots, m$
P_s = gain of the forward path s joining r_i to x_j, $s = 1, 2, \ldots, k(i, j)$
$*$ = drop all terms containing products of gains of touching loops
$**$ = drop all terms containing products of gains of touching loops or products of gains of touching loops and forward paths from r_i to x_j.

Using the above notation, (14) can be rewritten as

$$T_{ij} = \frac{[(P_1 + P_2 + \cdots + P_{k(i,j)})(1 - L_1)(1 - L_2) \cdots (1 - L_m)]^{**}}{[(1 - L_1)(1 - L_2) \cdots (1 - L_m)]^{*}} \tag{15}$$

Note that for a flow graph with a single input and output, which represents the transform of a system, the gain formula gives the transfer function of the system.

The examples below will illustrate the usefulness and ease of application of Mason's formula.

Example 10

Consider the signal flow graph in Fig. 8.16. The forward paths connecting the input U to the output Y have the following gains:

$$\text{gain of path 1} = P_1 = G_1G_2G_3G_4$$

$$\text{gain of path 2} = P_2 = G_5G_6G_7G_8$$

There are four loops, which have the following gains:

$$L_1 = G_2H_2 \qquad L_2 = G_3H_3 \qquad L_3 = G_6H_6 \qquad L_4 = G_7H_7$$

Since loops L_1 and L_2 do not touch loops L_3 and L_4, the determinant is

$$\Delta = 1 - (L_1 + L_2 + L_3 + L_4) + (L_1L_3 + L_1L_4 + L_2L_3 + L_2L_4)$$

The cofactor of the determinant along path 1 is evaluated by removing the loops that touch path 1 from Δ or, equivalently, setting the gains of these loops equal to zero. Since

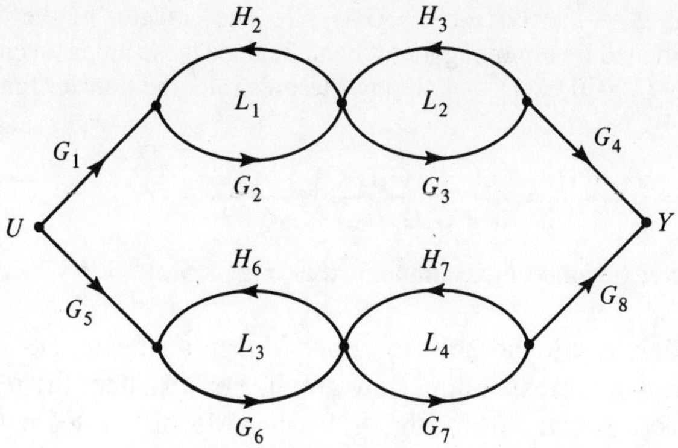

Figure 8.16 A two-path system.

loops 1 and 2 touch path 1, set $L_1 = L_2 = 0$. Therefore

$$\Delta(1) = 1 - (L_3 + L_4)$$

Similarly, the cofactor of path 2 is

$$\Delta(2) = 1 - (L_1 + L_2)$$

By formula (14), the transfer function of the system is

$$\frac{Y}{U} = T = \frac{P_1\Delta(1) + P_2\,\Delta(2)}{\Delta}$$

$$= \frac{G_1G_2G_3G_4(1 - L_3 - L_4) + G_5G_6G_7G_8(1 - L_1 - L_2)}{1 - L_1 - L_2 - L_3 - L_4 + L_1L_3 + L_1L_4 + L_2L_3 + L_2L_4}$$

Example 11

The transfer function found in Example 5 by block diagram simplification will now be found by using Mason's rule. This new approach requires a flow graph diagram equivalent to the block diagram of the given system appearing in Fig. 8.4. The reader should verify that the flow graph drawn in Fig. 8.17 is correct. Here

E = output of plus-minus junction with input U
V = output of the block labeled G_1
S = output of the second summing junction
X = output of the block labeled G_2

There is only a single path 1 in the flow graph. Its gain is $P_1 = G_1G_2G_3$. The gains of the two loops are $L_1 = G_1G_2H_2$ and $L_2 = G_2G_3H_1$. Since these two loops are

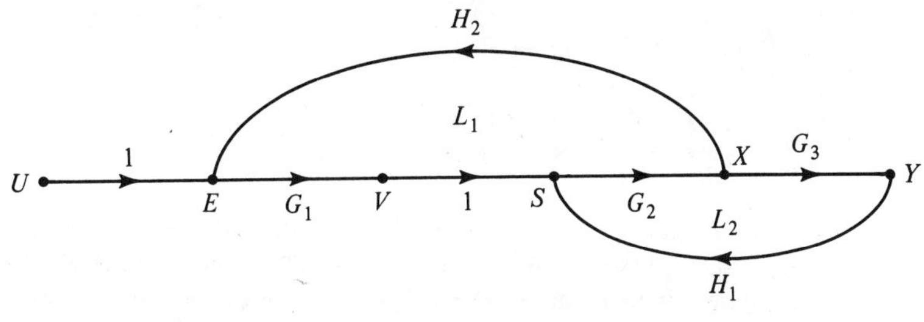

Figure 8.17 A flow graph equivalent to the system in Fig. 8.4.

touching, the determinant $\Delta = 1 - G_1G_2H_2 - G_2G_3H_1$. The cofactor of the determinant along path 1 is evaluated by removing loops 1 and 2, since these loops touch path 1. Therefore, setting $L_1 = L_2 = 0$, $\Delta(1) = 1$. Using Mason's rule, the transfer function is

$$T = \frac{P_1\,\Delta(1)}{\Delta} = \frac{G_1G_2G_3}{1 - G_1G_2H_3 - G_2G_3H_1}$$

This agrees with the answer obtained in Example 5. (See Fig. 8.5(c).)

With practice, the reader should be able to apply Mason's rule to the block diagram without having to draw a corresponding flow graph. For instance, the results in Example 11 can be deduced directly from Fig. 8.4 using Mason's rule, without having to draw Fig. 8.17.

Signal flow graphs and the gain formula have numerous applications besides biology [5], such as the analysis of feedback control systems, analog computer diagrams, electronic circuits, statistical systems, mechanical systems, and so on [6].

EXERCISES

1. Draw signal flow graphs equivalent to the block diagrams in Exercise 6 of Section 8.4.

2. Find the transfer functions of the block diagrams or systems described in the following exercises of Section 8.4 by using Mason's rule:
(a) Exercise 7
(b) Exercise 8
(c) Exercise 9
(d) Exercise 10

3. **(a)** Find the signal flow graph representing the system

$$2x_1 + x_2 = 4$$

$$x_1 + 3x_2 = 14$$

where x_1 and x_2 are the dependent variables and 4 and 14 are the inputs.
(b) Determine the value of each variable by using the gain formula and verify your answer by a direct solution.

4. The following notation is used for a multiple-loop model of an urban ecological system: P = number of people in the city, M = modernization, C = migration into the city, S = sanitation facilities, D = number of diseases, B = bacteria/area, and G = amount of garbage/area. The following causal loops are postulated:

(1) $P \to G \to B \to D \to P$
(2) $P \to M \to C \to P$
(3) $P \to M \to S \to D \to P$
(4) $P \to M \to S \to B \to D \to P$

(a) Draw a signal flow graph for the four given relations, indicating whether you believe each gain transmission is positive or negative. For example, the causal link $P \to G$ is positive, since as the number of people increase, so does the amount of garbage per unit area.
(b) State which of the four loops are positive or negative feedback loops.

5. Pollution from Ohio, Michigan, New York, and Canada has greatly increased nutrients for blue-green algae, resulting in an increase in this algae population and a corresponding decrease in the original plankton population. These algae clog filters, release obnoxious odors, and drive away swimmers. Formulate a signal flow graph model for the pollution process that includes the population along the lake, industrial development, and other variables you consider important [4,15].

6. A model of the human body for vibration studies considers it as a spring–mass–damper system as shown in the figure below. The distance between mass M_i and the ground is denoted by x_i. The model agrees well with experimental results [10].

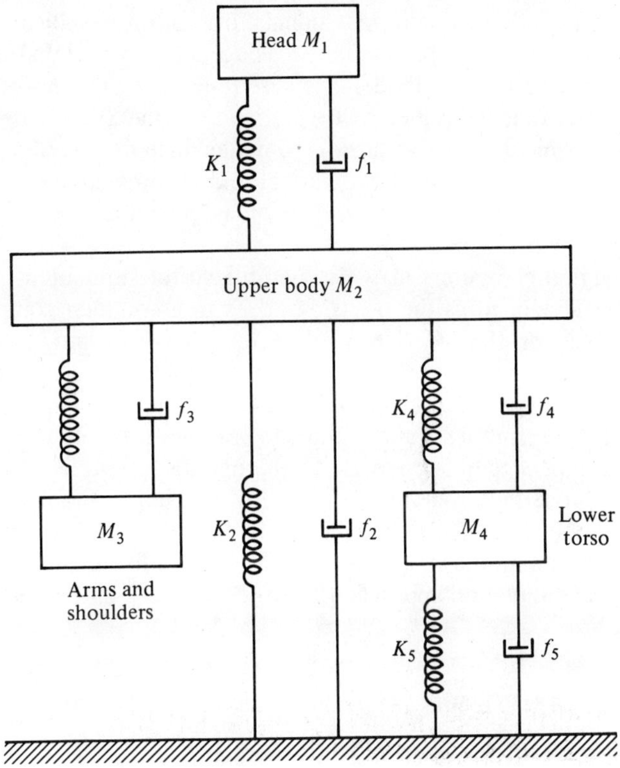

(a) Derive the equations that describe the dynamics of the system.
(b) Find the Laplace transform of the equations obtained in (a).
(c) Obtain a signal flow graph representation for the system.
(d) Determine the transfer function $X_4(s)/X_3(s)$.

8.6 LINEARITY AND NONLINEARITY

The study of linear systems is made relatively easy by the transfer function method. Using an impulse input, the transfer function of a system or its components can be obtained. Thus a complex system may often be reduced to a simpler equivalent system. Furthermore, the response of the system to a new, untested stimulus can be predicted.

Unfortunately, many real-life problems are nonlinear. The simplest to analyze are pseudo-linear systems. These systems, though nonlinear, behave as linear over a certain range of signals that are of interest.

For some systems, the departure from linearity is so severe that the assumption of linearity is not only valueless but also deceptive. The transfer function technique cannot be used. Such systems must be analyzed by more sophisticated mathematical methods. Some of these methods will be introduced in Volume II, Chapter 4. However, even such methods are inapplicable for complex systems.

Fortunately, the behavior of complicated nonlinear systems can be investigated by utilizing a computer. A computer simulation uses a model to study the system under the actual conditions and input signals to which it is subject.

A system may be simulated using analog or digital computers. An analog computer establishes a model of a system by using the analogy between the voltage of an electronic amplifier and a variable of the system being modeled [8]. The available functions on an analog computer include integration, multiplication by a constant, multiplication of two variables, and summation of several variables. These functions are often sufficient to develop a simulation model of a system (see Section 2.1).

Digital computers may also be utilized for simulation. A computer simulation may be developed as the occasion arises in a common computer language such as FORTRAN or BASIC. However, a knowledge of programming is required. An alternative approach is to use a special general-purpose simulation program for continuous time systems, such as CSMP, DYNAMO [11,12], or MIMIC [1]. A knowledge of programming is not required to use such simulation programs, but some effort must be expended to learn to use such programs. CSMP is an acronym for Continuous System Modeling Program and is an IBM product. A graphic feature is also available with CSMP. CSMP provides up to 42 functions such as integration and random-number generation and also provides many nonlinear functions. Some examples of biological simulation appear in [3], [5], [13], [14], and in many issues of the journal *Simulation*.

REFERENCES

1. CDC Reference Manual, Pub. No. 44610400.
2. Evans, C. L. *Principles of Human Physiology*. Churchill Livingstone, New York, 1949.
3. Grodins, F. S. *Control Theory and Biological Systems*. Columbia University Press, New York, 1963.
4. Kormondy, E. J. *Concepts of Ecology*, Prentice-Hall, Englewood Cliffs, NJ, 1969.
5. Lewis, E. R. *Network Models in Population Biology*. Springer-Verlag, Berlin, 1977.
6. Lorens, C. S. *Flowgraphs*. McGraw-Hill, New York, 1964.
7. Mason, S. J., and Zimmerman, H. J. *Electronic Circuits, Signals and Systems*. Wiley, New York, 1960.
8. Milhorn, H. T. *Applications of Control Theory to Physiological Systems*. W. B. Saunders, Philadelphia, 1966.
9. O'Kelly, L. I., Falk, L. J., and Flint, D. Water regulation in the rat: 1. Gastrointestinal exchange rates of water and sodium chloride in thirsty animals, *J. Comp. Physio. Psychol.* *51*, 16–21, 1958.

10. Phillips, C. Newsletter. *Simulation 6*, 6, 1966.
11. Pugh, A. L. *DYNAMO User's Manual*. MIT Press, Cambridge, MA, 1976.
12. Shaffer, W. A. *Mini-DYNAMO User's Guide*. Pugh-Roberts Associates, Cambridge, MA, 1978.
13. Toates, F. T. *Control Theory in Biology and Experimental Psychology*. Hutchison Education, London, 1975.
14. Toates, F. M., and Oately, K. Computer simulation of thirst and water balance. *Med. Bio. Eng. 8*, 71–87, 1970.
15. Turk, A., Turk, J., and Wittes, J. T. *Ecology, Pollution and Environment*. W. B. Saunders, Philadelphia, 1972.

9

Response of Linear Systems

9.1 INTRODUCTION

In this chapter the structure of the transfer function will be examined in order to determine its relation to the dynamical behavior of the corresponding system. Specifically, it will be shown that the location of the poles and zeroes of the system transfer function has a profound effect on how the system transmits different types of inputs. Thus a linear system can be synthesized to pass certain classes of inputs while rejecting others. It will also be shown that the poles of the system transfer function have a direct influence on the system's speed-of-response characteristics.

Transfer functions will also be used to study the performance of feedback control systems. Their sensitivity, accuracy in terms of the steady-state error, transient performance, response to noise, and static gain will be investigated.

9.2 DECOMPOSITION OF THE OUTPUT

The output of a continuous system will be decomposed into the input signal mode and the system mode. The analogous result for discrete systems will be left as an exercise.

First, the Laplace transform of the output and its decomposition will be found for a certain class of inputs occurring in many practical problems. The transforms of these inputs are the ratio of polynomials; that is,

$$U(s) = \frac{N(s)}{(s - q_1)(s - q_2) \cdots (s - q_k)} \tag{1}$$

where $N(s)$ is a polynomial in s of degree less than k. The response of the system with the transfer function

$$H(s) = \frac{M(s)}{(s - p_1)(s - p_2) \cdots (s - p_n)} \tag{2}$$

to the given input, where M is a polynomial in s of degree less than n, is

$$Y(s) = H(s)U(s) = \left[\frac{M(s)}{(s-p_1)(s-p_2)\cdots(s-p_n)}\right]\left[\frac{N(s)}{(s-q_1)\cdots(s-q_k)}\right]$$

To determine $y(t)$, a partial fraction expansion of $Y(s)$ must be made. For expository ease, it will be assumed that all the poles are distinct. The more general situation of multiple poles may be treated in a completely analogous manner. The partial fraction expansion of $Y(s)$ is of the form

$$Y(s) = \underbrace{\left(\frac{a_1}{s-q_1} + \frac{a_2}{s-q_2} + \cdots + \frac{a_k}{s-q_k}\right)}_{\text{input signal mode}} + \underbrace{\left(\frac{b_1}{s-p_1} + \frac{b_2}{s-p_2} + \cdots + \frac{b_n}{s-p_n}\right)}_{\text{system mode}} \tag{3}$$

where

$$a_i = \lim_{s\to q_i}(s-q_i)U(s)H(s)$$
$$b_i = \lim_{s\to p_i}(s-p_i)U(s)H(s)$$

The Laplace transform of the response has been decomposed into two parts called the *input signal mode* and the *system mode*.

The response corresponding to (3) is

$$y(t) = \underbrace{(a_1e^{q_1t} + \cdots + a_ke^{q_kt})}_{\text{input signal mode}} + \underbrace{(b_1e^{p_1t} + \cdots + b_ne^{p_nt})}_{\text{system mode}} \tag{4}$$

Note that the response is determined not only by the modes of the input signal [via the poles of $U(s)$] but also by the input signal exciting the natural modes of the system [via the poles of $H(s)$]. This result will be applied to solve two problems in system design.

Suppose that a system is desired in which the output signal resembles the input signal after a short time. It is clear, from (4), that this requires a system mode that rapidly decays to zero. Hence, in a fast-responding system all of the system transfer function poles, p_i, have real parts sufficiently smaller than zero. On the other hand, in a sluggish system the system mode decays to zero very slowly (i.e., at least one of the p_i has its real part close to zero). Analogous results can be formulated for discrete systems using the conditions that the magnitude of any pole of the system transfer function be less than one.

The poles and zeroes of the system transfer function can be selected so as to readily transmit or reject certain input signal modes. The study of the selective transmission of input signals is called *filter theory*. The following examples illustrate the fundamental ideas underlying this theory. Further details on the theory of filters can be found in [2].

Example 1

Suppose a system is desired that will not transmit the input signal $u(t) = e^{at}$ or, equivalently, $U(s) = 1/(s - a)$. The response of a linear system with transfer function $H(s)$ to this input is

$$Y(s) = \frac{H(s)}{(s - a)}$$

which, after making a partial fraction expansion, gives

$$Y(s) = \frac{A}{s - a} + [\text{terms due to poles of } H(s)]$$

The goal of rejecting the given input signal can be accomplished by forcing the coefficient

$$A = \lim_{s \to a} (s - a)Y(s) = \lim_{s \to a} H(s) = H(a)$$

to be small in magnitude. This may be accomplished by locating a zero of the system transfer function in a vicinity of $s = a$.

In general, input signals whose poles lie in a close vicinity of the zeroes of the system transfer function will tend to be rejected by the system. In a completely analogous manner, input signals whose poles lie in a close vicinity of the poles of the transfer function will tend to be transmitted by the system. This follows since, for example, the coefficient $A = H(a)$ becomes extremely large if a pole of H lies close to $z = a$.

In summary, *the poles of the system transfer function should be located about those values for which the natural modes of the input signal are to be transmitted. The zeroes of the system transfer function should be located about those values for which the natural modes of the input signal are to be suppressed.* However, the design of a properly functioning system is much more complex than this pole–zero placement would indicate. For example, it is not obvious in many applications how to determine which modes are to be suppressed and which transmitted. Furthermore, signals whose poles lie close to the suppressed signals will also tend to be suppressed.

Example 2

Filters are often required to suppress 60-Hz sinusoidal pickup in electrocardiogram recordings. The input signal to be suppressed is

$$u(t) = \sin \omega t$$

where $\omega = 120\pi$ rad/sec. The Laplace transform of u is

$$U(s) = \frac{\omega}{s^2 + \omega^2}$$

A filter will suppress this input signal if it has its zeroes located in the vicinity of the poles of $U{:}s = \pm j\omega$. A possible transfer function for the filter would be

$$H(s) = \frac{(s - j\omega)(s + j\omega)}{s^2 + as + b} = \frac{s^2 + \omega^2}{s^2 + as + b}$$

where the parameters a and b are selected to satisfy other dynamical considerations.

EXERCISES

1. Find the analog of (4) for a discrete system.
2. Verify that the same principle for transmitting or suppressing modes of the input signal remain valid for a discrete system.
3. Find a possible transfer function that will transmit sin ωt but will suppress $u = 1$.
4. Find a possible system transfer function that will suppress

$$u(t) = \frac{e^{-at} - e^{-bt}}{b - a}$$

5. Find a possible system transfer function that will suppress the input described in Exercise 4 but will transmit $u(t) = t^2$.

9.3 TRANSIENT AND STEADY-STATE RESPONSE

It will be assumed in this section that the linear time-invariant systems that will be studied are *stable* (c.f. Definition 1 and Theorem 1 in Chapter 10). For a continuous system, this means that the real parts of all the poles of the system transfer function are less than zero. For a discrete system, stability means that all the poles of the system transfer function lie within the unit circle in the complex plane.

From (4) and its analog for discrete systems, it follows that in a stable system the part of the response caused by the input exciting the natural modes of the system will always decay to zero. In many practical situations the part of the response caused by the input signal never decays to zero. The inputs studied in this section will be assumed to be of this form. With this in mind, the response signals given by (4) will be renamed as follows:

$$y(t) = y_{ss}(t) + y_t(t) \tag{5}$$

where

$$y_{ss}(t) = a_1 e^{q_1 t} + \cdots + a_k e^{q_k t} \tag{6}$$

$$y_t(t) = b_1 e^{p_1 t} + \cdots + b_n e^{p_n t} \tag{7}$$

The part denoted by y_{ss} is called the *steady-state response* because it is always present (never tends to zero); y_t is called the *transient response* because it eventually disappears (decays to zero). The rate at which the transient response decays to zero gives a measure of the system's speed of response. For example, if it is desired that the transient-response terms decay to zero rapidly, then the real parts of all the poles of the system transfer function must be much less than zero.

The transient-response terms generated by transfer functions with poles having real parts near the imaginary axis decay at a much slower rate than those whose real parts are farther from the imaginary axis. In many practical systems, the transfer function is characterized by having one or two poles whose real parts are very close to the imaginary axis, while the remaining poles have real parts that are relatively remote from the imaginary axis. Those poles whose real parts are close to the imaginary axis are called the system's *dominant poles,* because they tend to dominate the characteristics of the resultant transient response. For large t, a good approximation to the transient response is the response due to the system's dominant poles. For example, in (7), assume that the Re $p_i < 0$ for all i and p_1 and p_2 are complex conjugates, with

$$\text{Re } p_1 = \text{Re } p_2 >> \text{Re } p_i \qquad (i = 3, 4, \ldots, n)$$

Then (7) can be rewritten as

$$y_t(t) = e^{u_1 t}(b_1 e^{jv_1 t} + b_2 e^{jv_2 t} + \sum_{i=3}^{n} b_i e^{(u_i - u_1)t} e^{jv_i t}) \tag{8}$$

where $u_i = \text{Re } p_i$ and $v_i = \text{Im } p_i$. Since $u_i - u_1 << 0$ $(i = 3, \ldots, n)$,

$$\left| \sum_{i=3}^{n} b_i e^{(u_i - u_1)t} e^{av_i t} \right| \le \sum_{i=3}^{n} |b_i| \, e^{(u_i - u_1)t}$$

which approaches zero as $t \to \infty$. Hence $y_t(t)$ is approximated by the two terms corresponding to the poles p_1 and p_2 for large values of t. Therefore the transient response is approximately given by

$$Ae^{u_1 t} \cos(v_1 t + \alpha)$$

where A and α are constants.

More generally, the transient response can be decomposed into $y_d(t)$ and $y_{nd}(t)$, the dominant and nondominant parts, in a manner similar to (8). As in the example above, $y_{nd}(t)$ approaches zero more rapidly than its counterpart $y_d(t)$. This observation is summarized below to emphasize its importance.

Dominant Poles' Response Property. The response time of a linear time-invariant continuous system is directly dependent on the locations of the system transfer function's dominant poles. The dominant poles of the transfer function must be located according to the particular response-time requirement for a given application. A fast-responding system necessitates dominant poles whose real parts are much less than zero.

To analyze a system one frequently applies special test inputs and observes whether the system has desirable dynamics. A popular test input is the unit step function. In general, it is difficult to obtain an easily interpretable analytic formula for the response of a higher-order linear system to a unit step input. However, an approximation to its behavior can be obtained by using the notion of dominant poles. The use of dominant poles reduces the study of higher-order systems to a second-order system. Hence the response of such systems will be investigated first.

Consider the second-order system with the transfer function

$$H(s) = \frac{b^2}{s^2 + 2abs + b^2} \tag{9}$$

Here a and b are real nonnegative numbers. The normalizing coefficient b^2 in the numerator of H is included so that the system's unit step response is itself a unit step for large t. This may be verified by the use of the final value theorem.

The poles of the system are obtained by setting the denominator polynomial equal to zero:

$$s^2 + 2abs + b^2 = 0$$

This yields the poles

$$s = -ab \pm b\sqrt{a^2 - 1} \tag{10}$$

A number of interesting possibilities arise, depending on the value of the nonnegative parameter a as shown in Table 9.1. These possibilities will be considered separately below (c.f. Example 14 in Chapter 4).

TABLE 9.1 POSSIBLE POLE CHARACTERISTICS FOR THE STABLE SYSTEM(9)

a	Pole characteristics	System description
$0 < a < 1$	Two complex conjugate poles	Underdamped
$a = 1$	Two equal real poles	Critically damped
$a > 1$	Two unequal real poles	Overdamped

A. Underdamped Unit Step Reponse ($0 < a < 1$)

If $0 < a < 1$, then the roots given by (10) are

$$s_1 = -ab + ib\sqrt{1 - a^2} \tag{11}$$

and its complex conjugate $\bar{s}_1$. The response to the unit step function $U(s) = 1/s$ is

$$Y(s) = \frac{b^2}{s(s^2 + 2abs + b^2)} = \frac{A_0}{s} + \frac{A_1}{s - s_1} + \frac{A_2}{s - \bar{s}_1} \tag{12}$$

The constants

$$A_0 = \lim_{s\to 0} sY(s) = 1 \tag{13}$$

$$A_1 = \lim_{s\to s_1}(s - s_1)Y(s) = \lim_{s\to s_1} \frac{(s - s_1)b^2}{s(s - s_1)(s - \bar{s}_1)} = \frac{b^2}{s_1(s_1 - \bar{s}_1)} \tag{14}$$

$$= \frac{1}{(-a + i\sqrt{1 - a^2})(2i\sqrt{1 - a^2})} = \frac{1}{-2(1 - a^2 + ia\sqrt{1 - a^2})}$$

$$A_2 = \lim(s - s_1)Y(s) = A_1 \tag{15}$$

It is convenient to rewrite (14) in polar form:

$$A_1 = |A_1|e^{i\phi} \qquad \phi = \pi - \phi_1 \tag{16}$$

where

$$|A_1| = \frac{1}{2\sqrt{1 - a^2}}$$

$$\phi_1 = \tan^{-1}\left(\frac{a\sqrt{1 - a^2}}{1 - a^2}\right) \tag{17}$$

Taking the inverse transform of the partial expansion of Y given in (12) yields

$$\begin{aligned} y(t) &= 1 + A_1 e^{s_1 t} + A_1 e^{\bar{s}_1 t} \\ &= 1 + |A_1| e^{-abt}\left(e^{i(b\sqrt{1-a^2}+\phi)t} + e^{-i(b\sqrt{1-a^2}+\phi)t}\right) \end{aligned}$$

by using (16) and (11). From De Moivre's Theorem and (17),

$$y(t) = 1 - \frac{e^{-abt}}{\sqrt{1-a}}[\cos(b\sqrt{1-a^2}t - \phi_1)] \tag{18}$$

The solution (18) approaches 1 in an oscillatory manner as $t \to \infty$ (c.f. Fig. 4.4).

B. Critically Damped (a = 1)

If $a = 1$, then the response to the unit step function is

$$Y(s) = \frac{b^2}{s(s+b)} = \frac{A_0}{s} + \frac{A_1}{s+b} + \frac{A_2}{(s+b)^2} \tag{19}$$

$$A_0 = \lim_{s \to 0} sY(s) = 1 \tag{20}$$

$$A_2 = \lim_{s \to -b} (s+b)^2 Y(s) = -b \tag{21}$$

By substituting the values of A_0 and A_2 from (20) and (21) into (19) and setting $s = 1$, the value

$$A_1 = -1 \tag{22}$$

can be determined. Hence the inverse transform of (19) is

$$y(t) = 1 - (1 + bt)e^{-bt}$$

Note that the output signal is always less than 1 and approaches 1 as $t \to \infty$.

C. Overdamped (a > 1)

If $a > 1$, then, as indicated by (10), there are two real, negative, unequal roots. The response to the unit step function is

$$y(t) = 1 - \frac{1}{2}\left(1 + \frac{a\sqrt{a^2-1}}{a^2-1}\right)e^{s_1 t} + \frac{1}{2}\left(-1 + \frac{a\sqrt{a^2-1}}{a^2-1}\right)e^{s_2 t}$$

where

$$s_1 = -ab + b\sqrt{a^2-1} \qquad s_2 = -ab - b\sqrt{a^2-1}$$

Once again the output is always less than 1, but rises more rapidly to 1, as $t \to \infty$, than the output given in B.

The approximate response of a higher-order linear system to a unit step function can be found from the previous results. To simplify the following analysis it will be assumed that H has simple poles. From (4), the response to the unit step input is

$$y(t) = a_0 + b_1 e^{p_1 t} + \cdots + b_n e^{p_n t} \tag{23}$$

where

$$a_0 = H(0)$$

Since the system is stable, the steady-state unit step response,

$$y_\infty = \lim_{t\to\infty} y(t) = H(0)$$

is itself a step function of amplitude $H(0)$. The speed with which this steady-state value is approached is dependent on how quickly the transient response, given by (7), goes to zero. This depends on the magnitude of the system transfer function's poles. If the system possesses clearly dominant poles, the transient response may be approximated by much simpler expressions. Two frequently occurring situations will be considered.

One Dominant Real Pole. Let p_i be the dominant pole, where $|p_1| << |p_i|$ for $i = 2, 3, \ldots, n$. A good approximation to the transient response for large values of t is

$$y_t(t) \approx b_1 e^{p_1 t}$$

Hence a good approximation to the unit step response for sufficiently large values of t is

$$y(t) \approx H(0) + b_1 e^{p_1 t}$$

This response is overdamped. The approximation may be poor for small values of t because the terms $b_i e^{p_i t}$ for $i = 2, 3, \ldots, n$ may not have decayed sufficiently close to zero for such t.

Two Dominant Complex Conjugate Poles. Suppose that p_1 and p_2 are complex conjugate poles with $|p_1| = |p_2| << |p_i|$ for $i = 3, 4, \ldots, n$. The unit step response can be approximated by

$$y(t) \approx H(0) + b_1 e^{p_1 t} + \bar{b}_1 e^{\bar{p}_1 t}$$

for values of t such that

$$|b_1 e^{p_1 t}| >> |b_i e^{p_i t}| \qquad (i = 3, 4, \ldots, n)$$

From result 1 for a second-order system, the response to the unit step input will be an underdamped step response.

In conclusion, if the system transfer function has dominant poles, then the system's unit step response will be

1. Overdamped if the dominant pole is real;
2. Underdamped if the dominant poles are complex conjugates.

If a system has dominant poles, their location may be obtained experimentally by applying a unit step to the system's input and recording the resultant response. The relevant parameters are found by curve-fitting.

Theoretically, given the transfer function, the dominant poles can be found by selecting them from the totality of poles obtained by factoring the denominator polynomial. Practically, this approach is difficult if the denominator polynomial is of degree greater than 2. Factorization is equivalent to finding all the zeroes of the denominator polynomial, and this requires techniques of numerical analysis. Some of these techniques will be discussed in the next chapter.

Since the responses for higher-order systems are usually difficult to obtain, a few properties of the output are selected to distinguish between such systems. These characteristics of the output are usually defined for a unit step input and are illustrated in Fig. 9.1. They are defined as follows:

rise time, T_r = the first time the output reaches e_{ss}, the steady-state output

peak time, T_p = the time when the largest output (first peak), M_p, is observed

percent overshoot, PO = $(M_p - e_{ss})100\%/e_{ss}$

interval rise time, IRT = time interval required for the output to change from $0.1e_{ss}$ to $0.9e_{ss}$

settling time, T_s = the time required for the system to settle between $e_{ss}(1 + \delta)$ and $e_{ss}(1 - \delta)$

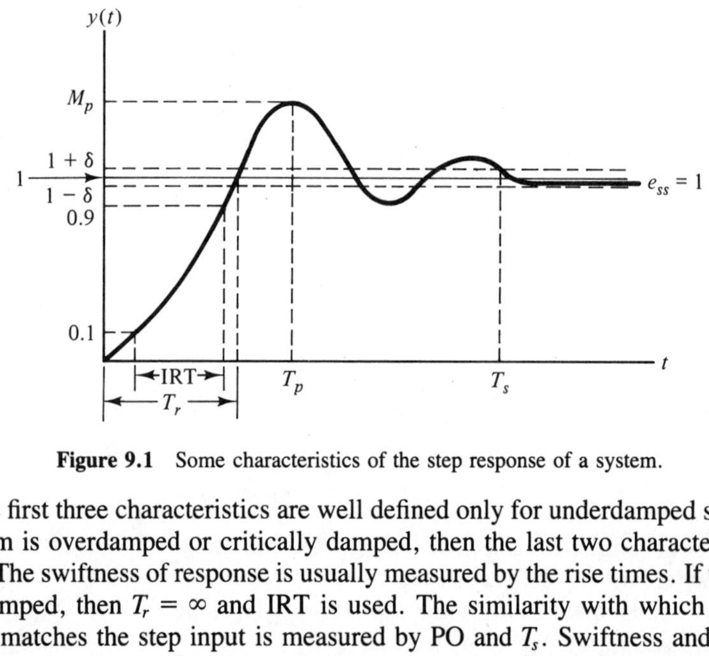

Figure 9.1 Some characteristics of the step response of a system.

The first three characteristics are well defined only for underdamped systems. If the system is overdamped or critically damped, then the last two characteristics can be used. The swiftness of response is usually measured by the rise times. If the system is overdamped, then $T_r = \infty$ and IRT is used. The similarity with which the actual response matches the step input is measured by PO and T_s. Swiftness and closeness of response are often contradictory requirements and a compromise must be made.

Another approach is used in modern control theory. It is assumed that the systems engineer can specify quantitatively the required system's performance in terms of a *performance index*. Some examples of a performance index are $\int_0^\infty e^2(t)\, dt$ or $\int_0^\infty t|e(t)|\, dt$, where $e(t)$ is the error at time t. If the output is to follow the input, then $e(t) = y(t) - u(t)$. Having chosen an appropriate performance index, the system that minimizes it is called *optimal*. In modern control theory, systems are frequently designed so that they are optimal with respect to a performance index [8]. This approach is used in Volume II, Chapter 5.

EXERCISES

1. If the system transfer function is

$$H(s) = \frac{6}{(s+1)(s+2)(s+3)}$$

decompose the output into its steady-state and transient response for a unit step input.

2. Answer Exercise 1 for the discrete linear system whose transfer function is

$$H(z) = \frac{z^3}{(2z-1)(4z+1)(3z-1)}$$

3. What are the dominant pole or poles for the systems whose transfer functions are

(a) $H(s) = 3.001/(s + 0.001)(s + 2)(s + 3)$

(b) $H(s) = 1/(s + 0.0002 - i)(s + 0.0002 + i)(s + 4)$

4. Find an approximation to the output for a unit step input applied to the systems described in Exercise 3 that is valid for large values of t.

5. The transfer function of a heart pump and pacemaker is

$$G(s) = \frac{K}{s(s/10 + 1)}$$

(c.f. Exercise 9 of Section 9.4). Suppose that the settling time to a step disturbance should be less than 1 sec and the PO should be less than 15%. Find a suitable range of K.

6. For the second-order system with the transfer function

$$H(s) = \frac{\omega^2}{s^2 + 2\zeta\omega s + \omega^2}$$

and $0 < \delta < 1$ (underdamped), show that

(a) $T_s \approx 4/\zeta\omega$ when $\delta = 2\%$

(b) $T_p = \pi/\omega\sqrt{1-\zeta^2}$

(c) $M_p = 1 + \exp(-\pi\zeta/\sqrt{1-\zeta^2})$

(d) $\text{PO} = 100\exp(-\pi\zeta/\sqrt{1-\zeta^2})$

7. The open-loop transfer function of a unity negative system is $G(s) = K/s(s+2)$. The system response to a step input is specified as follows: $T_p = 1$ sec; PO = 5%. Determine whether both specifications can be met simultaneously; if not, determine a compromise value for K so that both specifications are reduced the same percentage.

9.4 CHARACTERISTICS OF FEEDBACK CONTROL SYSTEMS

Open- and closed-loop systems will be introduced in this section. It will be shown that a closed-loop negative feedback system has the following advantages over an open-loop system:

1. The sensitivity of the system to variations in the parameters of the process is decreased.
2. The transient response can be adjusted and controlled more easily.
3. The rejection of disturbances and noise is improved.
4. The steady-state error is reduced.

The systems discussed in this section consist of one loop or can be reduced to one loop as shown in Chapter 8 for stationary linear systems with multiple loops. They can be analyzed mathematically to give clear and simple results. Biological systems are more complex and generally consist of multiple loops (see Section 2.3). These more complicated systems are rarely amenable to the techniques used for one-loop systems. This fact may cause the reader to doubt the relevance of such techniques. However, science progresses by a series of approximations to reality. At present, we must be satisfied with a qualitative description of some biological control systems accompanied by a mathematical model that provides a good first approximation.

A. Open- and Closed-Loop Systems

An open-loop system is shown in Fig. 9.2. In a closed-loop system, the input, u, is modified in some way by using some facts about the output, y. Figure 9.3 shows a closed-loop system having linear components that may be characterized by transfer functions. As usual, the Laplace transforms of the input and output are denoted by $U(s)$ and $Y(s)$, respectively. $G(s)$ is the *forward loop* transfer function or the *process* transfer function. $H(s)$ is the *feedback loop* transfer function.

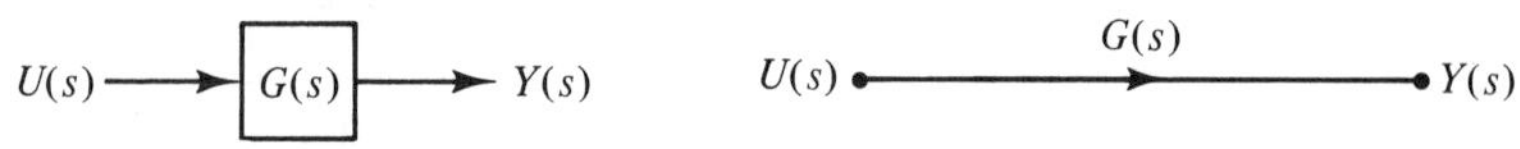

(a) Transfer-function block diagram (b) Corresponding signal-flow graph

Figure 9.2 An open-loop system.

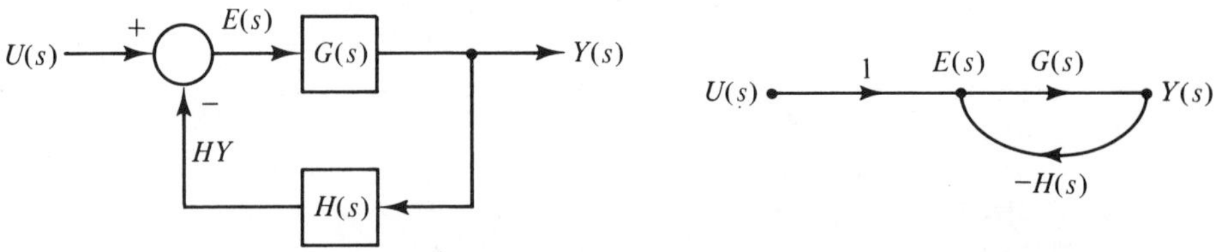

(a) Transfer-function block diagram (b) Corresponding signal-flow graph

Figure 9.3 A negative-feedback closed-loop system.

The significant difference between the open- and the closed-loop systems is the generation and the utilization of the *error signal,*

$$E(s) = U(s) - H(s)Y(s) \tag{24}$$

The term *negative feedback* system is used because of the minus sign in (24). Such a system, when operating correctly, reduces the error to a minimum value. In a *positive feedback* system the feedback output, $H(s)Y(s)$, is added to the input $U(s)$. Only negative feedback systems will be discussed in detail; the corresponding results for positive feedback systems are left as an exercise for the reader (see Exercise 1).

The output of the open-loop system is

$$Y(s) = G(s)U(s) \tag{25}$$

The output of the closed-loop system is

$$Y(s) = G(s)E(s)$$

In order to express $Y(s)$ in terms of the system's transfer functions, $E(s)$ is replaced by its value from (24). This gives

$$Y(s) = G(s)U(s) - H(s)Y(s)$$

Solving for $Y(s)$ yields

$$Y(s) = \frac{G(s)}{1 + G(s)H(s)}U(s) \tag{26}$$

The actuating error signal can also be expressed in terms of the system's transfer functions by substituting the value of $Y(s)$ from (26) into (24). The resulting expression simplifies to

$$E(s) = \frac{1}{1 + G(s)H(s)}U(s) \tag{27}$$

The error is reduced when $1 + G(s)H(s)$ is greater than one over the range of s under consideration.

Example 3

Suppose that the system process is an integrator and the feedback pathway has a gain of 1 $[H(s) = 1]$. Find the output for the open-loop and closed-loop systems in Figs. 9.2 and 9.3 for a unit impulse input.

Here $G(s) = 1/s$. Hence, from (25), the output of the open-loop system is $Y(s) = 1/s$, since $U(s) = 1$. The inverse Laplace transform is a unit step function.

The output for the closed-loop system is obtained from (26) by substituting $G(s) = 1/s$; $H(s) = U(s) = 1$. The output simplifies to $Y(s) = 1/(s + 1)$, and so the inverse Laplace transform is $y(t) = \exp(-t)$.

Further examples of open and closed loops are provided by models of protein synthesis.

Example 4 (Linear Model of Protein Synthesis)

A simple model of protein synthesis will be developed and analyzed. The following notation will be used: $u(t)$, $n(t)$, and $y(t)$ will denote the concentration of messenger RNA (mRNA), the enzyme, and the protein, respectively. Assume that the rate of enzyme and protein synthesis is directly proportional to the mRNA and enzyme concentration, respectively. The enzyme and protein are removed from the reaction (by decomposition or transport) at a rate proportional to their concentration. The mathematical translation of these assumptions leads to the system

$$\frac{dn}{dt} = a_1 u - a_2 n$$

$$\frac{dy}{dt} = a_2 n - a_3 y$$

where a_1, a_2, and a_3 are the distinct, positive proportionality constants. Initially, $n(0) = n_0$ and $y(0) = y_0$ are given.

The Laplace transform of the given system is

$$sN - n_0 = a_1 U - a_2 N$$

$$sY - y_0 = a_2 N - a_3 Y$$

Eliminating N and solving the system for Y gives

$$Y = GU + \frac{Gn_0}{a_1} + \frac{y_0}{s + a_3}$$

where

$$G = \frac{a_1 a_2}{(s + a_2)(s + a_3)}$$

The transfer function $G = Y/U$ when the initial conditions are zero. Hence, the protein synthesis system is an open-loop system.

The above model is a gross simplification of reality, because translation of mRNA actually involves an enormous number of separate, nonlinear reaction steps. Furthermore, the rate of protein synthesis was unaffected by its concentration. Clearly this is not economical, since if the protein were provided from some other source, the synthesis of the protein and the enzyme would continue at the same rate. It is known that biochemical processes are usually regulated by negative feedback, so that an increase in the protein concentration would decrease its net rate of synthesis. The next example is still a linear model, but it regulates protein synthesis by negative feedback. The linear model and a nonlinear model are analyzed by Rapp [6].

Example 5 (Linear Model of Protein Synthesis with Negative Feedback)

Negative feedback will be introduced into the model in Example 4 as follows. The output y is compared to a reference output y_r and their difference is used to modify the first system equation to

$$\frac{dn}{dt} = a_1[u - k(y - y_r)] - a_2 n$$

where $k > 0$ is a constant. The second equation for the rate of protein synthesis is unchanged. If $y > y_r$, then the new rate of change of the enzyme concentration, dn/dt, is decreased, while if $y < y_r$, dn/dt is increased. Thus n will decrease (increase) if $y > y_r (y < y_r)$. It follows from the equation for dy/dt that this quantity will decrease (increase) if $y > y_r (y < y_r)$. Therefore the modified system exhibits negative feedback. This will be verified by the Laplace transform method.

The Laplace transform of the new system equations are

$$sN = a_1[U - k(Y - Y_r)] - a_2 N$$

$$sY = a_2 N - a_3 Y$$

using zero initial conditions. Solving for Y yields

$$Y = \frac{G}{1 + kG}(U + kY_r)$$

where

$$G = \frac{a_1 a_2}{(s + a_2)(s + a_3)}$$

is the system transfer function appearing in Example 4. The equation for the output Y in Laplace transform space corresponds to the negative-feedback closed loop depicted in Fig. 9.3 under the following correspondences. The output Y is compared to a reference output Y_r with $Y - Y_r$ fed back through the feedback transfer function $H(s) = k$.

In Example 5 the desired concentration of protein, y_r, at any time t is specified by nature. The actual concentration of synthesized protein, y, should be close to the reference value. A control system is said to be *accurate* if the output closely tracks the reference setting. A useful measure of accuracy is the *offset,* which is defined as the difference between the output and the reference signal. In Example 5 the offset would be $y(t) - y_r(t)$. For greatest accuracy the offset should be zero. The Laplace transform of the offset is

$$Y - Y_r = \frac{GU - Y_r}{k(G + 1/k)}$$

which seems to suggest that $Y - Y_r$ approaches zero as $k \to \infty$. Thus it appears that the offset could be decreased by increasing the gain k of the feedback loop. Frequently, this approach is successful for improving control accuracy. However, in some situations, increasing k causes $G(s) + 1/k$ to have at least one zero with a positive real part. The corresponding time function $y(t) - y_r(t)$ would then contain a term of the form e^{at} with $a > 0$ and so approach infinity as $t \to \infty$. This is an example of a general phenomenon called the control–stability conflict, in which the accuracy of control is related to the stability so that if the accuracy is improved the system becomes unstable.

The stability of the system in Example 5 can be determined by examining the roots of the characteristic equation $G(s) + 1/k = 0$. These roots can easily be found because of the simple form of $G(s)$. However, even in this simple example it is not easy to determine if the system is stable for all values of the parameters a_1, a_2, and a_3. The general stability analysis of this system will be deferred until Chapter 10, where more powerful stability criteria are developed [see Example 9(b)]. These methods can be applied even if the roots of the characteristic equation cannot be found directly.

B. Sensitivity of Systems to Parameter Variations

To study the effect of parameter variations, consider a change ΔG in the process transfer function so that the new process is $G + \Delta G$. The change in the transform of the output for the open-loop system is

$$\Delta Y = \Delta G U \tag{28}$$

For the closed-loop system,

$$Y + \Delta Y = \frac{G + \Delta G}{1 + (G + \Delta G)H} U$$

and so the change in the output is

$$\Delta Y = \frac{\Delta G}{(1 + GH + \Delta GH)(1 + GH)} U \tag{29}$$

Often, $GH >> \Delta GH$, and then

$$\Delta Y \approx \frac{\Delta G}{(1 + GH)^2} U \tag{30}$$

Comparing (28) and (30), the change in output of the closed-loop system is reduced by the factor $(1 + GH)^2$, which is usually much greater than 1 over the range of interest. The requirement $GH >> 1$ may cause the system to oscillate or even be unstable.

Noting that the change in output is proportional to the change in the system transfer function, T, the following definition is sensible.

Definition 1. The *system sensitivity, S*, is defined as the ratio of the percentage change in the system transfer function to the percentage change in the process transfer function. The mathematical definition is

$$S = \frac{\Delta T(s)/T(s)}{\Delta G(s)/G(s)} \tag{31}$$

In the limit, (31) can be written in various ways,

$$S = \frac{\partial T/T}{\partial G/G} = \frac{\partial \ln T}{\partial \ln G} = \frac{G}{T}\frac{\partial T/\partial s}{\partial G/\partial s} = \left(\frac{\partial T}{\partial G}\right)\frac{G}{T} \tag{32}$$

which are useful in applications.

For the open-loop system $T_0 = G$ and so $S_0 = 1$. For the closed-loop system $T_c = G/(1 + GH)$ from (26). Therefore, for small increments,

$$S_c = \left(\frac{\partial T}{\partial G}\right)\frac{G}{T} = \frac{1}{(1 + GH)^2} \cdot \frac{G}{G/(1 + GH)} = \frac{1}{1 + GH}$$

Hence, for sufficiently large $|GH|$, $S_c < S_0$.

The sensitivity of the feedback to changes in the feedback element is

$$S_H^T = \frac{\partial T}{\partial H} \cdot \frac{H}{T} = \frac{-GH}{1 + GH}$$

If $|GH|$ is large, then $|S_H^T| \approx 1$ and the output $Y \approx U/H$. Thus changes in H affect the output response directly. Hence it is important to use feedback components that will not vary too much with environmental changes or that can be maintained constant.

Example 6

Consider an open-loop system with gain $G(s) = 100$. If the gain is decreased to 50, the percentage change in the output to a unit step input is 50%. This result is a consequence of (25).

Now consider a closed-loop system with feedback-loop transfer function $H = 1$. If the gain of the process function undergoes the same changes as the open-loop system described in the previous paragraph, then the percentage change in the output is less than 1%. This result follows from (26) and illustrates the advantage of a negative feedback system.

C. Control of the Transient Response of a System

The transient response is one of the most important characteristics of a control system. If an open-loop control system does not have a satisfactory response, then there is no choice but to replace the process function with a suitable one. By contrast, a closed-loop system can often be made to yield the desired response by adjusting the feedback loop parameters.

Example 7

The response of the open-loop system in Fig. 9.1, with $G(s) = 1/s$, to the unit step input $U(s) = 1/s$ is $Y(s) = 1/s^2$ according to (25). Taking the inverse Laplace transform of $Y(s)$ yields $y(t) = t$. The output deviates further and further from the input as time increases.

Suppose it is desired that the output approach the input as time increases while still using the same process function. The desired type of performance can be obtained by simply introducing a negative feedback loop with gain $H(s) = 1$. By (26),

$$Y(s) = \frac{1/s}{1 + 1/s} \times \frac{1}{s} = \frac{1}{s(s + 1)}$$

Decomposing the output into partial fractions gives $Y(s) = 1/s - 1/(s + 1)$, whose inverse Laplace transform is $y(t) = 1 - \exp(-t)$. Therefore, as t increases, $y(t)$ approaches 1 as desired.

A generalization of this example appears in Exercise 3.

D. Disturbances in Negative Feedback Systems

A disturbance can occur at several places in a system. For example, consider a disturbance, D, in the process transfer function as shown in the signal flow graph in Fig. 9.4.

By (25), the output of the open-loop system in Fig. 9.4(a) is

$$Y = G_1G_2U - DG_2$$

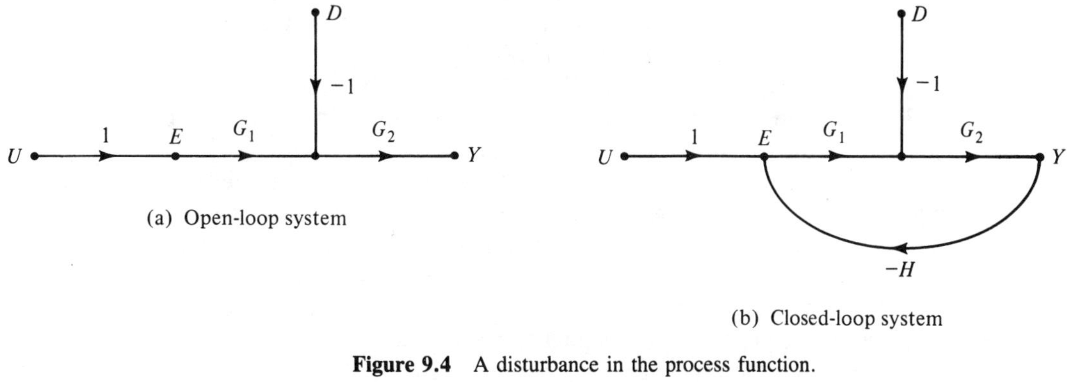

Figure 9.4 A disturbance in the process function.

The output, Y, can be decomposed into two parts: $Y_0 = G_1G_2U$, the undisturbed output when $D = 0$, and $Y_D = -DG_2$, the disturbed output. The disturbed output cannot be eliminated if $G_2 \neq 0$ is fixed.

By utilizing the signal-flow graph gain formula or directly (see Exercise 2), the disturbed output for the closed-loop system is

$$Y_D = \frac{-G_2}{1 + G_1G_2H}D \approx -\frac{D}{G_1H} \qquad (\text{if } |G_1G_2H| >> 1) \tag{33}$$

Therefore, if G_1 is sufficiently large, the effect of the disturbance can be reduced by closed-loop feedback.

A noise signal is frequently generated by the measurement sensor as depicted in Fig. 9.5. Its effect is to produce the disturbed output

$$Y_D = \frac{-G_1G_2H_2N}{1 + G_1G_2H_1H_2} \approx -\frac{N}{H_1} \qquad (\text{if } |G_1G_2H_1H_2| >> 1) \tag{34}$$

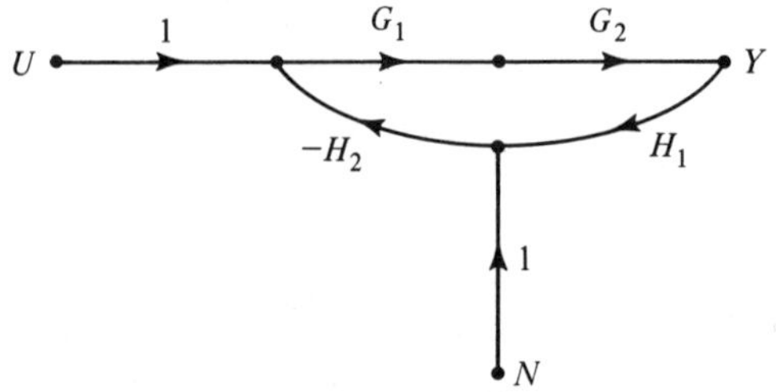

Figure 9.5 Closed-loop system with measurement noise.

The disturbance can be minimized by maximizing $H_1(s)$, which is equivalent to maximizing the signal-to-noise ratio of the measurement sensor. Hence the feedback element $H_1(s)$ should be well designed and operate with minimum noise, drift, and parameter variation. This is the same as the requirement determined from the sensitivity function, which showed that $S_H^T \approx 1$, where $H = H_1H_2$.

The system in Fig. 9.5 illustrates the fact that the proper choice of a feedback element can reduce the sensitivity of the system to parameter variations and the effects of disturbances. Fortunately, this simultaneous reduction can be achieved in other negative feedback systems. For example, the disturbed output and the sensitivity with respect to G_2 for the system in Fig. 9.3(b) are $Y_D = -G_2N/(1 + G_1G_2H)$; $S_{G_2}^T = 1/(1 + G_1G_2H)$. If $|G_1G_2H| >> 1$, then $Y_D \approx -N/G_1H$ and $S_{G_2}^T \approx 1/G_1G_2H$. Both of these quantities can be decreased simultaneously by increasing $|G_1(s)|$. Another example appears in Exercise 5.

Often noise is present in the input to a control system. For example, the input might be $u(t) + n(t)$, where $u(t)$ is the desired system response and $n(t)$ is the noise. In general, the feedback control will simply process the noise as well as the input signal and will not be able to improve the signal-to-noise ratio that is present in the input. However, if the frequency spectrum of the noise and input signals are of a different character, the output signal-to-noise ratio can often be maximized by designing a closed-loop transfer with the proper frequency response [2].

EXERCISES

1. In a positive feedback system the minus signs in Fig. 9.2 are replaced by plus signs. Show that $Y = GU/(1 - GH)$ and $E = U/(1 - GH)$.

2. For the system shown in Fig. 9.3(b), verify (33) by showing that $Y = (EG_1 - D)G_2$ and $E = U - HY$.

3. **(a)** The error in the open-loop system shown in Fig. 9.1 is

$$E_0(s) = U(s) - Y(s) = [1 - G(s)]U(s)$$

By using the final value theorem, show that the steady-state error, $e(\infty) = \lim_{t\to\infty} e(t)$, for a unit step input is $e_0(\infty) = 1 - G(0)$.

(b) Repeat part (a) for the closed-loop system in Fig. 9.2 and show that $e_c(\infty) = 1/[1 + G(0)H(0)]$.

(c) The value of $G(s)$ when $s = 0$ is often called the dc gain and is normally greater than 1. Show that if $H(s) = 1$ and $|G(0)| > \sqrt{2}$, then $|e_0(\infty)| > |e_c(\infty)|$.

(d) The open-loop system may be calibrated so that $G(0) = 1$. What is wrong with trying to reduce e_0 to zero by this method?

4. The following terminology will be used in this problem concerning the system in Fig. 9.2. The inputs $u(t) = A$, vt, and $at^2/2$ are called step, ramp, and parabolic, respectively. The positional, velocity, and acceleration error constants are defined by $k_0 = \lim_{s\to 0} GH(s)$, $k_v = \lim_{s\to 0} sGH(s)$, and $k_a = \lim_{s\to 0} s^2 GH(s)$. The product GH, the open-loop gain, is of the form

$$GH = \frac{k(s + 1/\tau_1)(s + 1/\tau_2)\cdots(s + 1/\tau_m)}{s^n(s + 1/\sigma_1)(s + 1/\sigma_2)\cdots(s + 1/\sigma_r)}$$

and the system is said to be of type n. Verify the entries below, where $e = e(\infty)$ denotes the steady-state error.

	Step input		Ramp input		Parabolic input	
Type	k_0	e_0	k_v	e_v	k_a	e_a
0	Constant	$A/(1 + k_0)$	0	∞	0	∞
1	∞	0	Constant	v/k_v	0	∞
2	∞	0	∞	0	Constant	a/k_a

5. Show that the effects of the disturbed output in the diagrammed system is $Y/D = 1/(1 + GH)$ = the sensitivity S_G^T.

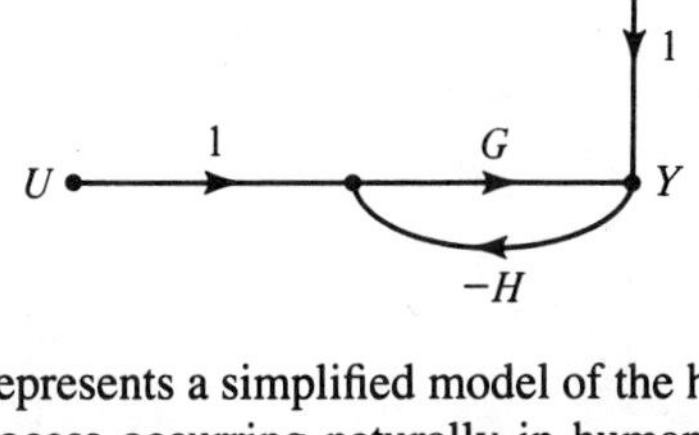

6. The following diagram represents a simplified model of the human speech system. Speech control is a feedback process occurring naturally in humans; when the system becomes unstable, a person stutters [1]. The vocal tract is a resonant cavity and is approximated by $G_2(s)$, $\omega = 2\pi(500)$, $\zeta = 0.032$, and normally, $K_2 = 0.01$. The central nervous system is approximated by the synaptic delay of the nerves.

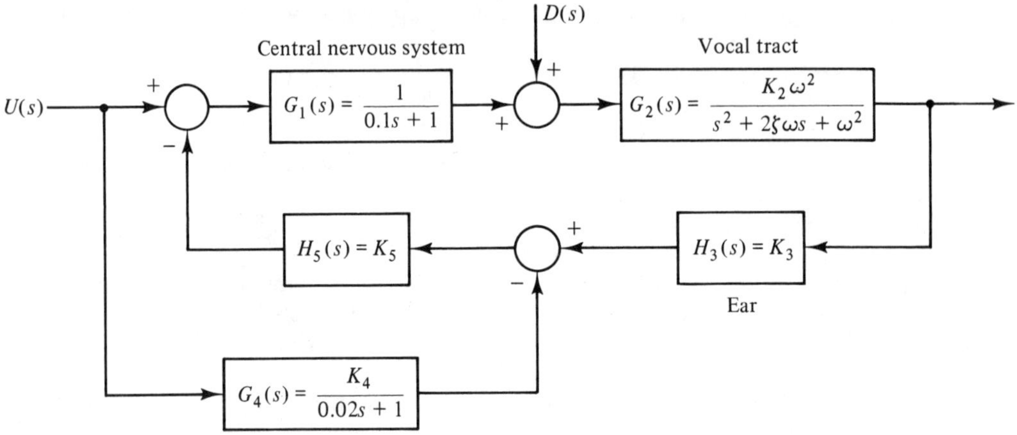

(a) Determine the effect of the disturbance on the output. The disturbance represents an extraneous noise signal generated in the central nervous system.

(b) Determine the sensitivity of the system to the parameters K_2 and K_4 in order to investigate their effect on stuttering.

7. For the following feedback with $K_1 = K_2 = 100$, find the closed-loop transfer function and the sensitivity of the system with respect to the parameter K_1.

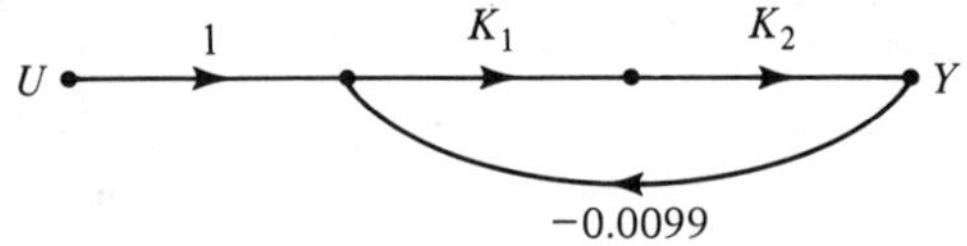

8. Repeat the calculations in Exercise 7 for the following system:

9. A closed-loop system that includes a pacemaker and the measurement of the heart rate is shown below:

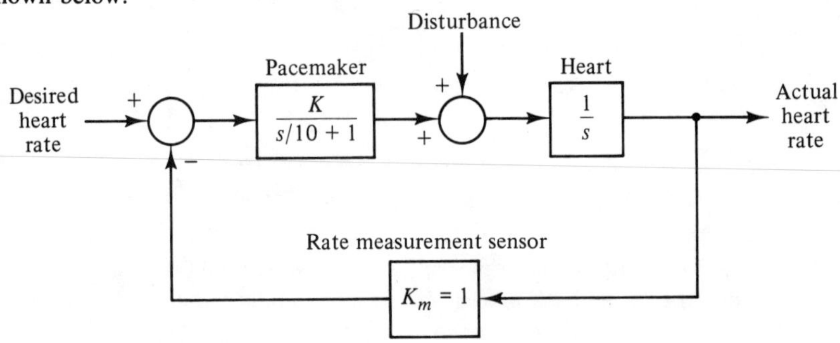

(a) Find the system transfer function when $D = 0$.
(b) If $K = 10$, find the sensitivity of the system to small changes in K.
(c) Evaluate the sensitivity at DC. (This means that $s = 0$.)

10. Show that the system in Example 5 is stable for $a_1 = 0.5$, $a_2 = 1$, $a_3 = 2$, and all values of k. [*Hint:* Use the quadratic formula to find the roots of $G(s) + 1/k = 0$.]

9.5 STATIC GAIN

The term *static* refers to the characteristics that a system possesses in the steady state. Consider the system in Fig. 9.2. The final value theorem justifies defining the steady-state factor, G_0, of G by $G_0 = \lim_{s\to 0} G(s)$ and the steady-state factor of H by $H_0 = \lim_{s\to 0} H(s)$.

The *gain* of a system at any time is defined to be the output divided by the input. Hence

$$\text{system gain} = \frac{Y}{U} = \frac{G}{1 + GH} \tag{35}$$

This suggests the following definition for the gain in the steady state.

Definition 2

1. The *steady-state* (or *static*) *system gain* $= G_0/(1 + G_0H_0)$.
2. The *open-loop gain* $= G_0H_0$.

In physiological systems the open-loop gain is often called the *homeostatic index*.

There are three methods for calculating the open-loop gain.

Method 1

1. Open the loop as shown below (any position will do).

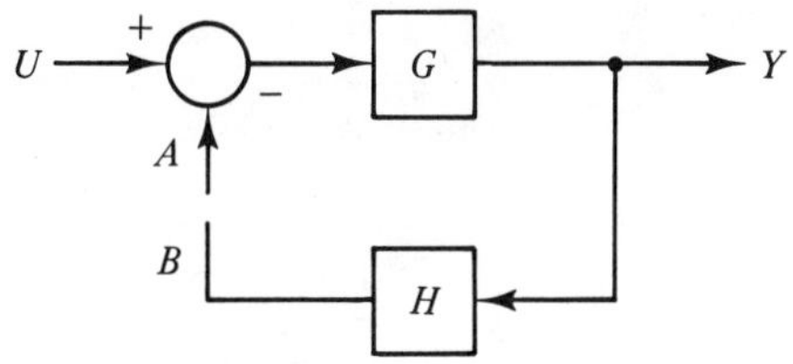

2. Send in a signal at A and measure the steady-state output at B.
3. Repeat step 2 with a different signal. Then it can be shown that (see Exercise 1)

$$\text{open-loop gain} = \frac{\Delta(\text{output at } B)}{\Delta(\text{input at } A)} \tag{36}$$

The arterial pressure regulating system will be considered in the following example to illustrate the application of method 1.

Example 8 (Gain of Arterial Pressure Regulating System)

Suppose that the open-loop gain of the carotid sinus loop of the arterial pressure regulating system is desired [4]. The other pressoreceptors are denervated so that the carotid sinus loop is the only one remaining. Then a method must be determined for opening this loop and changing the value of the input while simultaneously measuring the output. One method for doing this is illustrated in Fig. 9.6. The vessels entering and leaving the sinus region are ligated so that changes in the body arterial pressure cannot affect the sinus pressure. Hence the loop is open. The pressure in the sinus region is only a function of the height (h) of the container above the sinus (hydrostatic pressure) as shown in Fig. 9.5. The pressure of the sinus (input at A) can be changed by altering the height of the container, and the change in the body pressure (output at B) can be measured. The open-loop gain can then be calculated from (36).

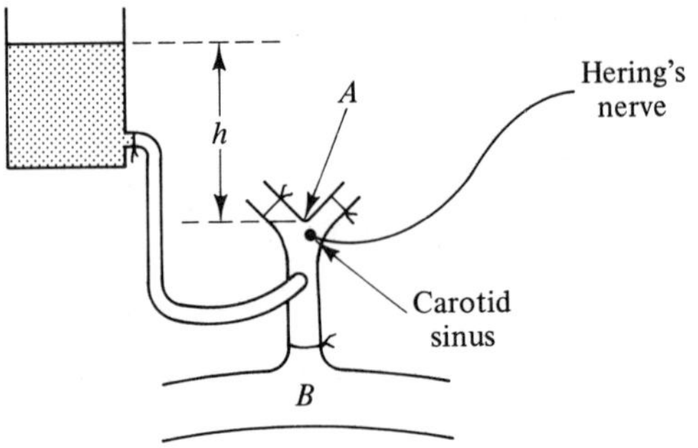

Figure 9.6 Open-loop state of the carotid sinus loop.

Method 2

1. Disturb the output in the open loop with the same process function as the closed loop under study. Allow the disturbed system to reach a steady state. Hence, determine $(\Delta Y)_{\text{open}}$, the disturbed steady-state output minus the undisturbed steady-state output.
2. Introduce a disturbance in the closed-loop system as shown below. Determine $(\Delta Y)_{\text{closed}}$, the disturbed steady-state output minus the undisturbed steady state output. Then the

$$\text{open-loop gain} = \frac{(\Delta Y)_{\text{open}}}{(\Delta Y)_{\text{closesd}}} - 1 \tag{37}$$

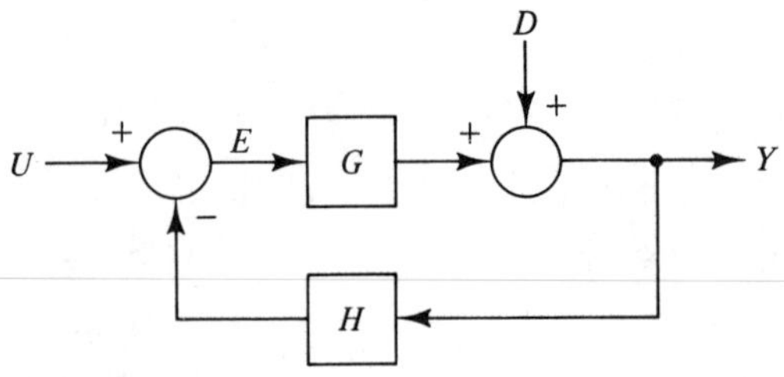

The proof of (37) is a consequence of the following facts. The output of the system illustrated in method 2, step 2, is

$$\frac{UG}{1 + GH} + \frac{D}{1 + GH}$$

Subtracting the undisturbed output given in (26) gives

$$(\Delta Y)_{\text{closed}} = \frac{D}{1 + GH}$$

Since $(\Delta Y)_{\text{open}} = D$, Eq. (37) reduces to the open-loop gain appearing in Definition 2.

The following definition is useful as a measurement of how well the control system performs its job in reducing the steady-state error.

Definition 3

$$\text{minification} = \frac{1}{1 + G_0 H_0} = \frac{(\Delta Y)_{\text{closed}}}{(\Delta Y)_{\text{open}}}$$

For example, it follows from (27) that if the minification $= \frac{1}{3}$ with $(\Delta Y)_{\text{open}} = 1$, the system has reduced the steady-state error of the output to only one-third of what it would have been if no regulation at all had occurred.

Method 2 will be applied to calculate the open-loop gain of a system that regulates the blood concentration of a substance. It is assumed that the steady-state error resulting from a disturbance can be measured. Thus the open-loop gain of the system cannot be too high, since the higher the open-loop gain, the smaller the steady-state error. In practical terms, the resulting steady-state error must be large enough to be measured accurately. The change in output in the open-loop state is obtained by dividing the quantity of the administered substance by the blood volume or estimating it from the peak of the transient change. The next example will illustrate this procedure.

Example 9 (Blood Glucose Regulating System)

Consider the blood glucose regulation system. Suppose that 1500 mg of glucose are suddenly injected intravenously into a person. If the person has a blood volume of 5 liters, the change in the blood glucose concentration in the open-loop state would be 1500 mg/5000 ml = 30 mg/ml. The blood glucose concentration was measured as a function of time by periodically sampling the person's blood. The blood glucose concentration (mg/100 ml) was plotted as a function of time. From the resulting graph it was determined that the change in the blood glucose concentration in the closed-loop state was 5 mg/100 ml. Therefore

$$\text{open-loop gain} = \frac{(\Delta Y)_{\text{open}}}{(\Delta Y)_{\text{closed}}} - 1 = \frac{30}{5} - 1 = 5$$

The minification is 6.

The blood glucose concentration after an injection will eventually return to its normal level due to passive effects. Thus the calculations of open-loop gain must be restricted to the short-term active effects.

The definition of open-loop gain can be extended to nonlinear feedback loops.

To motivate this extension, consider the feedback system in Fig. 9.2. The following relations hold:

$$Y = f(E) = EG$$

$$E = G(Y) = U - HY$$

Note that G is the slope of one feedback and $-H$ is the slope of the other. Hence

$$\text{open-loop gain} = -(\text{product cf slopes}) \tag{38}$$

Since the slope of the equation $y = f(x)$ is $df(x)/dx$ and the slope of the equation $x = g(y)$ is $dg(y)/dy$, relation (38) leads to the following method for finding open-loop gain of stable systems.

Method 3. The open-loop gain of a multivariable feedback loop at any particular steady state can be calculated from the functional equations $y = f(x)$ and $x = g(y)$ relating any two of the variables x and y in the loop as follows:

$$\text{open-loop gain} = -\left(\frac{df(x)}{dx}\right)\left(\frac{dg(y)}{dy}\right) \tag{39}$$

The following example illustrates the use of method 3.

Example 10

Consider the following equations relating the variables in a feedback loop:

$$u = 4y$$

$$z = 2u$$

$$x = \frac{1}{z}$$

$$y = x^2$$

Eliminating u and z from the first three equations gives the following feedback relations:

$$y = x^2 \tag{40}$$

$$x = \frac{1}{8y} \tag{41}$$

By (39) the open-loop gain is $-(2x)(-1/8y^2)$. Solving (40) and (41) simultaneously yields $x = \frac{1}{2}$ and $y = \frac{1}{8}$. Hence the open-loop gain is $-(1)(-8) = 8$.

Method 3 is applicable only when the two feedback equations are known. It may be necessary to obtain one or both of the feedback equations empirically. Often it is possible to derive one of the feedback equations theoretically. The following example will illustrate this possibility.

The body temperature, T_b, control system will be considered. Let T_h be the temperature of the temperature-sensitive region of the hypothalamus. By heating the hypothalamus locally so as to increase T_h by a known amount and measuring the resulting steady-state change in T_b, one of the feedback equations can be obtained empirically:

$$T_b = f(T_h)$$

Next, the temperature of the body could be increased and the resulting temperature of the hypothalamus region measured in order to obtain the other equation. However, this is not necessary since the temperatures would be the same; hence

$$T_h = T_b$$

The open-loop gain could be calculated from these two equations.

The next example is an actual application of method 3 to a physiological system.

Example 11 (Open-Loop Gain of the Respiratory System)

The respiratory tract is composed of the conducting airway (nose, mouth, pharynx, larynx, trachea, bronchi, and bronchioles) and the alveoli. Rapid exchange of oxygen and carbon dioxide between gas and blood occurs only in the alveoli and not in the airway.

The conducting airway, at the end of a normal expiration, is filled with alveolar gas. This alveolar gas is drawn back into the alveoli during the early part of the next inspiration and does not raise the alveolar pressure of O_2 or CO_2. Only the fresh air that goes beyond the conducting airway into the alveoli raises the alveolar pressure of O_2 and CO_2. Alveolar volume, V_A, thus refers to the volume of fresh air entering the alveoli and not to the volume of gas that enters.

The respiratory response to carbon dioxide will be considered. During the inhalation of higher-than-normal carbon dioxide concentration, the volume of air entering the alveoli per minute, $\dot{V}_A$, increases. This is an attempt by the respiratory system to regulate the blood carbon dioxide level. If alveolar ventilation, $\dot{V}_A$, and alveolar carbon dioxide partial pressure, $P_{CO_2,A}$, are chosen as the two variables of interest in the loop, one of the feedback equations can be determined theoretically. The following simplifying assumptions are used:

1. The inspired air is already at body temperature and saturated with water vapor.
2. The respiratory quotient (the rate of CO_2 output to the rate of O_2 intake) is 1. This implies that the volume of inspired air equals the volume of expired air.
3. Gas flow through the lungs is assumed to be continuous.
4. Diffusion equilibrium exists between the alveolar air and the pulmonary venous blood.
5. There are no arteriovenous shunts across the lungs.
6. The total volume of CO_2 produced will be measured at standard pressure, 760 mmHg.

The alveoli are lumped into one compartment. The following notation will be used:

$\dot{Q}_{CO_2,I}$ = moles of CO_2 inspired into the alveoli per minute
$\dot{Q}_{CO_2,E}$ = moles of CO_2 exhaled per minute
$\dot{Q}_{CO_2}$ = moles of CO_2 produced by the body per minute
$P_{CO_2,I}$ = partial pressure of CO_2 in inspired air
$P_{CO_2,A}$ = partial pressure of CO_2 in alveoli or expired air
P_B = barometric pressure

In the steady state, the rate at which CO_2 enters the alveoli must equal the rate at which it leaves. Thus

$$\dot{Q}_{CO_2,I} + \dot{Q}_{CO_2} = \dot{Q}_{CO_2,E} \tag{42}$$

The ideal gas law states that

$$P_{gas} V_{gas} = RTQ_{gas}$$

Differentiating V and Q in the ideal gas law with respect to time gives

$$P_{gas} \dot{V}_{gas} = RT\dot{Q}_{gas} \tag{43}$$

Multiplying both sides of (42) by RT and using the appropriate version of (43) gives

$$\dot{V}_{CO_2,I} P_{CO_2,I} + \dot{V}_{CO_2} P_B = P_{CO_2,A} \dot{V}_{CO_2,E} \tag{44}$$

Since, by assumption 2, $\dot{V}_A = \dot{V}_{CO_2,I} = \dot{V}_{CO_2,E}$, Eq. (44) can be written in the form

$$P_{CO_2,A} = P_B C_I + P_B \dot{V}_{CO_2}/\dot{V}_A \tag{45}$$

where, by Dalton's law of partial pressure, $P_{CO_2,I}/P_B = C_I$, the concentration of CO_2 in the inspired air.

The second feedback equation for $\dot{V}_A$ must be determined empirically. It is obtained by observing how alveolar ventilation varies as $P_{CO_2,A}$ is altered by inhalation of gas mixtures containing different concentrations of carbon dioxide. According to Gray [3], this relation can be approximated by

$$\dot{V}_A = 2P_{CO_2,A} - 75 \tag{46}$$

The open-loop gain is the negative product of the first derivatives of the functions defined by (45) and (46) with respect to $\dot{V}_A$ and $P_{CO_2,A}$, respectively. Hence

$$\text{open-loop gain} = \frac{2\dot{V}_{CO_2} P_B}{\dot{V}_A^2} \tag{47}$$

The numerical value of the open-loop for a normal adult will be found. By assumption, $P_B = 760$ mmHg. In the normal state, $C_I = 0$, $\dot{V}_A$ is approximately 5 liter/min, and $P_{CO_2,A} = 40$ mmHg. Substituting these values in (45) yields $\dot{V}_{CO_2} = 0.263$ liter/min. Therefore, from (47), the normal open-loop gain is $2(0.263)(760)/(5)^2 = 16$.

Further details on calculating the open-loop gain for nonlinear systems appears in [7]. This book also contains many interesting, nontrivial examples of physiological feedback mechanisms.

A note of caution is necessary at this point. In many engineering systems and abstract models, it is easy to modify inputs and break feedback loops. However, such procedures are difficult, if not impossible, in intact organisms, where the feedback loops may be inaccessible. Furthermore, even if such manipulations were possible, the disturbances could perturb the system so much that it would become atypical or, more seriously, it may even cease to function. Nevertheless, there are certain operational tricks for overcoming these difficulties. Machin [5] has described some methods for breaking feedback loops in biological situations. Sometimes a subsystem can be isolated from the animal and the required experiments performed on the preparation. Another approach is to synthesize an abstract model of the organism that mimics the behavior of the organism under ordinary conditions. This model can then be used to predict the outcome of dangerous or painful experiments.

EXERCISES

1. Prove (36).
2. Suppose that the peak change in blood glucose concentration in Example 7 was 28 mg/100 ml. Find the open-loop gain and compare the answer with the result in Example 7.
3. The open-loop gain in a negative feedback system is 8. What is the steady-state gain if $H_0 = 1$?
4. Suppose that 0.30 g of glucose is injected into an animal having a blood volume of 1.5 liters. The blood sugar level increases from 70 mg/100 ml to a new steady-state value of 75 mg/100 ml. Find the open-loop gain and the minification of the system.
5. Answer Exercise 4 if the peak was 90 mg/100 ml and the blood volume was not known.
6. Choose two new variables in Example 8 and show that the open-loop gain calculated from method 3 is the same.
7. Can the open-loop gain be found for the system without the disturbance described in Exercise 9 in Section 9.4?
8. The following equations form a feedback loop describing thyroid hormone metabolism:

$$C = \frac{\dot{Q}}{\dot{V}_{\text{cl}}}$$

$$\dot{V}_{\text{cl}} = kC$$

   where C is the concentration of hormonal iodine in the plasma, $\dot{Q}$ is the rate of supply of thyroid hormone, $\dot{V}_{\text{cl}}$ is the rate of plasma clearance of hormonal iodine, and k is a constant. Find numerical values for the open-loop gain and minification.
9. A feedback loop describing the pupillary reaction to light is defined by the following equations:

$$\dot{Q}_e = kBA_p$$

$$A_p = \frac{160\dot{Q}_e^{-0.4} + 94.8}{4\dot{Q}_e^{-0.4} + 15.8}$$

   in which $\dot{Q}_e$ is the rate at which light enters the eye, B is the brightness of the light, A_p is the area of the pupil, and k is a constant. Determine the open-loop gain and the minification.

REFERENCES

1. Butler, B. R. The stuttering problem considered from an automatic control point of view. *Proc. San Diego Symp. Biomed. Eng.*, pp. 47–62, 1964.
2. Cadzow, J. A. *Discrete Time Systems*. Prentice-Hall, Englewood Cliffs, NJ, 1973.
3. Gray, J. S. *Pulmonary Ventilation and its Physiological Regulation*. Charles C Thomas, Springfield, IL, 1950.
4. Guyton, A. C. *Textbook of Medical Physiology,* 6th ed., W. B. Saunders, Philadelphia, 1981.

5. Machin, K. E. Homeostasis and feedback mechanisms. *Symp. Soc. Exp. Biol. 18,* 421–446, 1964.
6. Rapp, P. E. Biological applications of control theory. In *Continuum Models in Molecular and Cellular Biology,* I. A. Segel (ed.). Cambridge University Press, New York, 1984, chap. 2.
7. Riggs, D. S. *Control Theory and Physiological Feedback Mechanisms.* Williams & Wilkins, Baltimore, 1970.
8. Schultz, W. C., and Rideout, Y. C. Control system performance measures: Past, present, and future. *IRE Trans. Automatic Control,* February 1961, pp. 22–35.

10

Stability Criteria for Linear Stationary Systems

10.1 INTRODUCTION

The feedback systems of a healthy body provide homeostasis, coordinated physiological reactions that maintain most of the normal steady states in the body. Thus the output of one of these feedback systems should not oscillate wildly when the system is subject to disturbances in its input or anywhere in its feedback loop. In other words, the system should be stable. To discuss this concept rigorously, the following definition suffices for linear stationary systems.

Definition 1. A stationary linear system is *stable* if its outputs are bounded for all bounded inputs.

From the above definition a criterion for stability can be formulated in terms of the roots of the denominator of the system transfer function. Consider a linear stationary system with transfer function $T(s)$, input $u(t)$, and output $y(t)$. The relation between the input and the output in terms of Laplace transforms is

$$Y(s) = T(s)U(s)$$

Assume that

$$T(s) = \frac{N(s)}{D(s)}$$

is a rational function of s. According to the theory of the Laplace transform, developed in Chapter 7, the output y can be characterized in terms of the three possible types of roots of $D(s) = 0$. The equation $D(s) = 0$ is called the *characteristic equation* of the system.

1. *Roots with positive real parts*. Such roots correspond to terms with positive real exponents of e. Such terms increase in magnitude without bound as time increases and so the output is unbounded. Therefore the system is not stable.

2. *Roots with zero real parts*. There are two different situations to consider. First, suppose that the imaginary parts of the roots are not zero. Such roots occur in conjugate pairs and if repeated $n > 1$ times correspond to terms of the form

$$(a_0 + a_1 t + \cdots + a_{n-1} t^{n-1})(c_1 \sin \theta t + c_2 \cos \theta t)$$

As time increases, such terms increase without bound and indicate an unstable system. If $n = 1$, the bounded input $c_1 \sin \theta t + c_2 \cos \theta t$ will cause the denominator of $T(s)U(s)$ to have a conjugate pair of roots repeated twice. Therefore, by the previous discussion, the output will be unbounded and the system will be unstable.

Next, suppose that the denominator has an identically zero root repeated $n > 1$ times. This root corresponds to the terms

$$a_0 + a_1 t + \cdots + a_{n-1} t^{n-1}$$

which clearly give an unbounded output. For a single zero root, a step input will cause the denominator of $T(s)U(s)$ to have a double zero root, and so the system will be unstable.

3. *Roots with negative real parts*. These roots correspond to output terms with negative real exponents of e. Such terms always decrease toward zero as time increases, even when (as may be true with multiple roots) some integral power of time also appears in the term as a coefficient of e. They are therefore compatible with a stable system.

The above discussion leads to the following theorem.

Theorem 1. A linear time-invariant system whose system transfer function is rational is stable if, and only if, the zeroes of its characteristic equation (that is, the roots of the denominator of its transfer function) has negative real parts.

Based on Theorem 1, the following two approaches are used for determining stability.

1. *Calculating the exact roots of $D(s) = 0$*. The classical approach of finding the roots directly will be outlined in the next section. In engineering work another approach, the Root-locus method [6,7], is used frequently. However, in biology this method is not used very often and so it will not be discussed.
2. *Determination of the region of system parameters guaranteeing that the roots of $D(s) = 0$ have negative real parts*. The first approach can be tedious and lead to extra work, especially if only bounds on certain system parameters for stability are required. Three different methods (the Routh-Hurwitz criterion, the Nyquist criterion, and the Bode diagram) for carrying out the second approach will be discussed in this chapter.

10.2 DIRECT DETERMINATION OF THE ZEROES OF A POLYNOMIAL

The direct determination of the roots of $D(s) = 0$ has increased in popularity due to the advent of digital computers. First, methods of finding the real roots of a polynomial and then methods of finding complex roots will be discussed.

A. Real Roots

There are four steps in locating the roots of a low-degree polynomial that cannot be factored easily:

1. Reduce the given polynomial to simple polynomial with the same roots.
2. Approximate the position of each root in the simple polynomial.
3. Improve the accuracy of any approximation found in step 2.
4. Find the multiple roots.

These four steps will be discussed in turn.

Since the above four steps depend on *Euclid's algorithm* [5]* for finding the greatest common divisor (g.c.d.) for two polynomials, this procedure will be outlined first.

Euclid's Algorithm

1. Divide the lower-degree polynomial into the higher-degree polynomial, obtaining a remainder r_1.
2. Divide the remainder, r_1, into the lower-degree polynomial, obtaining the remainder r_2.
3. Divide r_2 into r_1 to get the remainder r_3.
4. Repeat step 3, dividing r_n into r_{n-1} to get the remainder r_{n+1} for $n = 3, 4, \ldots$ until a remainder is zero. The preceding divisor is the g.c.d. multiplied by some numerical constant.

Note that the g.c.d. can be found as follows. First, divide the two polynomials by their highest common numerical factor k. Then apply Euclid's algorithm to the resulting polynomials, obtaining the g.c.d. multiplied by a numerical constant, say, cg, where c is a constant. Next factor out all the numerical factors c_1 from the nonconstant term g, yielding $g = g_1c_1$. Replacing the resulting constant, c_1c, by k gives the actual g.c.d., g_1k, of the original polynomials.

Example 1

Consider the polynomials

$$p(x) = x^3(x + 1)^2(x - 1)^2 = x^7 - 2x^5 + x^3 \tag{1}$$

* Also known as the highest common factor (HCF) algorithm.

and

$$q(x) = x^2(x + 1)^2(x + 2) = x^5 + 4x^4 + 5x^3 + 2x^2 \tag{2}$$

From their factored form their g.c.d. is $x^2(x + 1)^2$.

The g.c.d. will now be found by applying Euclid's algorithm to the unfactored form of these polynomials. Dividing p by q yields

$$\frac{p}{q} = x^2 - 4x + 9 + \frac{(-18x^4 - 36x^3 - 18x^2)}{q}$$

Hence

$$r_1 = -18x^4 - 36x^3 - 18x^2$$

Dividing r_1 into q gives a remainder of zero. Hence, by Euclid's algorithm, r_1 is the g.c.d. multiplied by a numerical constant. Since the highest common numerical factor of p and q is 1, the g.c.d. of p and q is $x^4 + 2x^3 + x$

Note that fractions could have been avoided in the division of q by r_1 by first dividing r_1 by -18 and using the new polynomial as the divisor. In general, any of the r_i in Euclid's algorithm can be multiplied by a numerical constant in order to avoid fractions in division. This only alters the final numerical constant multiplying the g.c.d. Fortunately, the exact value of this constant is not important.

The four steps for finding the real roots of a polynomial will now be discussed.

1. Reducing the given polynomial to a simple polynomial with the same roots. To carry out this procedure, find the g.c.d. of the given polynomial and its derivative by applying Euclid's algorithm. Then divide the given polynomial by this g.c.d. to get the simple polynomial with the same roots.

Example 2

Consider the polynomial

$$p(x) = x^3(x + 1)^2(x - 1)^2 = x^7 - 2x^5 + x^3$$

Then

$$p'(x) = 7x^6 - 10x^4 + 3x^2$$

From Euclid's algorithm, the g.c.d. of p and p' is $x^4 - x^2$. Dividing p by this polynomial gives

$$q(x) = x^3 - x = x(x - 1)(x + 1)$$

a simple polynomial with the same roots as p.

2. Approximating the roots of a simple polynomial. The following two methods are helpful in finding the approximate position of each root.

Drawing the Graph of the Polynomial. This method is more suitable for lower- than for higher-degree polynomials. The approximate zeroes of the polynomial are the x coordinates of the points where the graph of the polynomial cuts or touches the x axis. These points of intersection are determined graphically or by linear interpolation as in Example 4. Some information about the multiplicity of a root can be found by using the fact that if a zero is repeated k times, then it is repeated $k - 1$ times in the derivative of the polynomial.

The next method is useful for higher-degree polynomials. It gives the precise number of roots in a segment of the real axis.

Sturm's Method [14]. Sturm's method requires the formation of a sequence of polynomials called the *Sturm sequence*. The Sturm sequence, Σ_i, is constructed as follows:

Σ_1 = the given polynomial

Σ_2 = the derivative of the given polynomial and for $i \geq 3$

Σ_i = minus the remainders when Euclid's algorithm is applied to Σ_1 and Σ_2 with the following provisos:

(a) If the last remainder is zero, the preceding remainder is the g.c.d. of Σ_1 and Σ_2. Neither the g.c.d. nor zero should be included in the Sturm sequence.

(b) If Σ_1 and Σ_2 have no g.c.d. involving the variable, the last remainder will be a constant. This constant should be included as a member of the Sturm sequence.

For ease in calculations, the numerical fractions that appear in a remainder can be eliminated by multiplying by a suitable positive integer.

After the Sturm sequence has been constructed, the next theorem shows how to locate the roots of the given polynomial in an interval of the real axis.

Sturm's Theorem [14]. The number of nonrepeated, real roots lying between any two real numbers c_1 and $c_2 (c_1 < c_2)$ is the number of sign changes in the Sturm sequence with $x = c_1$ minus the number of sign changes in the Sturm sequence with $x = c_2$.

Example 3

Consider the polynomial

$$p(x) = x^3 - x$$

Since $p(\frac{1}{2}) = -\frac{3}{8} < 0$, while $p(2) = 6 > 0$, there is at least one real root between $\frac{1}{2}$ and 2. In order to find the exact number of such roots without factoring the polynomial, Sturm's theorem will be applied.

The Sturm sequence is constructed as follows. Let $\Sigma_1(x) = p(x)$ and $\Sigma_2(x) = p'(x) = 3x^2 - 1$. Dividing Σ_2 into Σ_1 leaves the remainder $-2x/3$. For ease in calculation, let $\Sigma_3(x) = 2x$. When Σ_3 is divided into Σ_2, the remainder is -1 and so $\Sigma_4(x) = 1$. In summary, the Sturm sequence is

$$\Sigma_1 = x^3 - x$$

$$\Sigma_2 = 3x^2 - 1$$

$$\Sigma_3 = 2x$$

$$\Sigma_4 = 1$$

Let $c_1 = \frac{3}{2}$ and c_2 be very large. The numerical value of the Sturm sequence corresponding to c_1 is

$$\Sigma_1(\tfrac{3}{2}) = \tfrac{15}{8} \qquad \Sigma_2(\tfrac{3}{2}) = \tfrac{23}{4} \qquad \Sigma_3(\tfrac{3}{2}) = 3 \qquad \Sigma_4(\tfrac{3}{2}) = 1 \tag{3}$$

while the Sturm sequence corresponding to c_2 is approximated by

$$\Sigma_1(c_2) \doteq c_2^3 \qquad \Sigma_2(c_2) \doteq 3c_2^2 \qquad \Sigma_3(c_2) \doteq 2c_2 \qquad \Sigma_4(c_2) = 1 \tag{4}$$

There are no sign changes in either sequence (3) or (4). Therefore, by Sturm's theorem, there are no real roots larger than $\frac{3}{2}$. The Sturm sequence corresponding to $c_1 = \frac{1}{2}$ is

$$\Sigma_1(\tfrac{1}{2}) = -\tfrac{3}{8} \qquad \Sigma_2(\tfrac{1}{2}) = -\tfrac{1}{4} \qquad \Sigma_3(\tfrac{1}{2}) = 1 \qquad \Sigma_4(\tfrac{1}{2}) = 1 \tag{5}$$

Since there is one sign change in (5) and no sign changes in (3), Sturm's theorem states that there is exactly one real root between $\frac{1}{2}$ and $\frac{3}{2}$. By reducing the size of this interval—for example, by bisecting it—the root can be located in a smaller interval. Better approximations can be obtained by repeating this bisecting procedure.

The work involved in constructing a Sturm sequence or in applying Euclid's algorithm can be reduced by using synthetic division. In this division process the variable and its powers need not be written down. A detailed description of this process appears in [21]. (See Exercise 2.)

3. Improving the accuracy of an approximate root of a simple polynomial. Let x_0 be an approximation to a root of a simple polynomial equation $p(x) = 0$. The successive approximations x_n for $n = 1, 2, 3, \ldots$ are defined recursively by the Newton-Raphson formula [18]:

$$x_{n+1} = x_n - \frac{p(x_n)}{p'(x_n)} \tag{6}$$

It can be shown that if x_0 is sufficiently close to the zero it is approximating, then the sequence x_n converges to this zero. The assumption that p is simple is required in order that x_n does not approximate a zero of p', which would result in the denominator in (6) becoming very large.

Frequently, it is less work numerically to find $p(x_n)$ as the remainder on dividing $p(x)$ by $x - x_n$ using synthetic division [21]. If this remainder is not zero, the quotient can be divided by $x - x_n$ to get $p'(x_n)$.

The next example illustrates the use of formula (6).

Example 4

Consider the polynomial

$$p(x) = x^3 + x^2 - 2x - 2$$

Since $p(1) = -2$ and $p(2) = 6$, the equation $p(x) = 0$ has a root between 1 and 2. The two points $(1, -2)$ and $(2, 6)$ can be joined by a line and the x intercept of this line can be used as a first approximation to the root. This process of linear interpolation (see Exercise 6) gives $x_0 = 1.25$. The approximation will be improved by using (6), which in this situation is

$$x_{n+1} = x_n - \frac{x_n^3 + x_n^2 - 2x_n - 2}{3x_n^2 + 2x_n - 2} \tag{7}$$

Setting $n = 1$ in (7) gives

$$x_1 = 1.25 + \frac{0.984375}{5.1875} = 1.25 + 0.189753 = 1.44$$

Setting $n = 2$ in (7) gives

$$x_2 = 1.44 - \frac{0.179584}{7.1008} = 1.4147$$

However,

$$p(x) = (x - \sqrt{2})(x + \sqrt{2})(x + 1)$$

and so the sought for root is $\sqrt{2} = 1.4142$. Note that the second iteration x_2 differs from the correct result by only 0.0005. If more accuracy is desired, x_3, x_4, and so on can be calculated from (6). When the true solution is not known, the iteration process is continued until two successive iterates agree to the desired number of decimal places.

4. Finding multiple roots. Each real root r of the simple polynomial corresponds to the factor $x - r$. Divide the original polynomial by $x - r$ and test the quotient polynomial to see if r is still a zero. Repeat the above division procedure until r is no longer a zero. The number of repetitions gives the multiplicity of the root.

The above steps for finding the real roots of a polynomial require a great deal of mechanical work by the investigator, especially for high-degree polynomials. A program based on the above four steps could be written for a digital computer. However, there is another numerical approach, known as Lehmer's method, which is not sensitive to multiple roots and can also find complex roots. Lehmer's method is a procedure for searching the complex plane in such a way that a zero is isolated in a sequence of disks of decreasing radii. The approximate zero is divided out of the given polynomial and the process is restarted using the quotient polynomial. A program in ALGOL based on this approach appears in [20], but it has an error that is corrected in a later volume [20].

In some situations, the stability of a system may be the primary concern. According to Theorem 1, stability is equivalent to determining whether the zeroes of the denominator of the system transfer function have negative real parts. The exact location of the zeroes is not required. If all the roots are real, then the number of positive and negative roots could be found by applying Sturm's theorem. However, this approach can involve a greal deal of work. The following simpler approach, known as Descartes' rule, can sometimes be applied.

Descartes' Rule [5]. The polynomial must be written in canonical form with each term placed to the left of every term of lower degree and the coefficient of the highest degree term must be 1. Then the number of positive real roots is either equal to the number of sign changes in the coefficients of the given polynomial, or less than it by an even integer.

Note that if there are two or more sign changes, the exact number of real roots may not be able to be determined by this method. The position of the roots is not determined by this method.

The number of negative real roots can be obtained by replacing x by $-x$ in the given polynomial, writing the resulting polynomial in canonical form, and applying Descartes' rule. The number of positive roots of the transformed polynomial is equal to the number of negative roots of the original polynomial.

Example 5

Descartes' rule will be applied to the polynomial

$$x^4 + 2x^3 - x^2 - 2x$$

to determine the number of positive and negative real roots. The signs of the coefficients from left to right are $+$, $+$, $-$, $-$. Hence there is one sign change and so one positive root by Descartes' rule. Replacing x by $-x$ in the given polynomial transforms it into

$$x^4 - 2x^3 - x^2 + 2x$$

This polynomial is positive for large values of x and negative for $x = 1.5$. Therefore it has at least one positive root. There are two sign changes in the transformed polynomial and so it has two positive real roots. Thus the original polynomial has two negative roots. The original polynomial also has a zero root.

The above results can be verified from the factored form of the given polynomial:

$$x(x - 1)(x + 1)(x + 2)$$

B. Finding Complex Roots

Once all of the real roots have been found, the given polynomial can be divided by the corresponding factors, giving a polynomial that has only complex roots. It is of even degree, since complex roots occur in conjugate pairs. Repeated roots can be removed from it by applying Euclid's algorithm as discussed in Section A. Therefore, only simple polynomials with nonrepeated complex roots occurring in conjugate pairs will be discussed in this section. If the degree of this simple polynomial is 2, the roots can be found directly.

If the degree of the simple polynomial is greater than 2, a numerical method for finding a quadratic factor will be described below. An alternative approach, useful when extracting a quadratic factor is tedious, is Lehmer's method, which was discussed in Section A.

Bairstow's method [15,18] of finding a quadratic factor exhibits quadratic convergence if the initial estimate of the quadratic factor is quite close to the target. However, until a reasonably accurate estimate has been obtained, it is not too efficient. Therefore, two methods of estimating a quadratic factor, which do not require an accurate, initial estimate, will be discussed before describing Bairstow's method.

The first method is a simple, trial-and-error search. The simple polynomial is divided by a trial quadratic factor and the remainder inspected. Then, using one's judgment, the trial remainder is corrected and the division is repeated. The object is to reduce the coefficients of the remainders to small values. By observing the effects of successive corrections on these coefficients, an educated guess at the subsequent corrections can be made. Once the coefficients are sufficiently small, Bairstow's method can be applied.

A more systematic, iterative approach that does not require an accurate initial estimate is Friedman's method [8]. Convergence in this method becomes slower as the magnitudes of the complex root pairs become comparable; in fact, Friedman's method may not converge when the magnitudes are nearly equal.

Friedman's Method [8]. To find an estimate of a quadratic factor of the polynomial

$$p(x) = x^n + a_1x^{n-1} + \cdots + a_n$$

from the initial estimate

$$g(x) = x^2 - mx - p$$

proceed as follows:

1. Arrange both polynomials from left to right in order of descending powers of x. Divide p by g to obtain the quotient

$$q(x) = q_0x^{n-2} + q_1x^{n-1} + \cdots + q_{n-2}$$

whose coefficients are given by

$$q_0 = 1$$

$$q_1 = mq_0 + a_1$$

$$q_j = mq_{j-1} + pq_{j-2} + a_j \qquad (j = 2, 3, \ldots, n-2)$$

2. Divide each term in the quotient q by q_{n-2}. Arrange the result in order of ascending powers, obtaining the polynomial

$$r(x) = 1 + r_1x + r_2x^2 + \cdots + r_{n-2}x^{n-2}$$

where

$$r_j = \frac{q_{n-2-j}}{q_{n-2}} \qquad (j = 0, 1, \ldots, n-2)$$

3. Arrange $p(x)$ in order of ascending powers. Divide it by $r(x)$ to obtain the new quotient,

$$s(x) = s_0 + s_1x + s_2x^2$$

where

$$s_0 = a_n \qquad s_1 = a_{n-1} - r_1a_n \qquad s_2 = a_{n-2} - r_2a_n - r_1s_1$$

4. Divide each term of s by s_2 and arrange the result in order of descending powers to get the next candidate quadratic factor,

$$g(x) = x^2 - mx - p$$

where

$$-m = \frac{s_1}{s_2} \qquad -p = \frac{s_1}{s_2}$$

5. Go to step 1 and repeat until $g(x)$ satisfies your accuracy criterion.

Example 6

Friedman's method will be used to find a quadratic factor of

$$p(x) = x^4 - 2x^3 + 3x^2 - 2x + 2 \tag{8}$$

The initial quadratic factor, selected arbitrarily, is

$$g(x) = x^2 - x + 1$$

and so $m = 1$ and $p = -1$. Applying the steps in Friedman's method gives

$$\begin{array}{llll} & q_0 = 1 & r_1 = -1 & s_0 = 2 \\ a_1 = -2 & q_1 = -1 & r_2 = 1 & s_1 = 0 \\ a_2 = 3 & q_2 = 1 & & s_2 = 1 \\ a_3 = -2 & & & \\ a_4 = 2 & & & \end{array}$$

from which the corrected coefficients of g are $m = 0$ and $p = -2$ and

$$\begin{array}{llll} & q_0 = 1 & r_1 = -2 & s_0 = 2 \\ a_1 = -2 & q_1 = -2 & r_2 = 1 & s_1 = 2 \\ a_2 = 3 & q_2 = 1 & & s_2 = 5 \\ a_3 = -2 & & & \\ a_4 = 2 & & & \end{array}$$

from which $m = -0.4$ and $p = -0.4$ and

$$\begin{array}{llll} & q_0 = 1 & r_1 = -0.6741573 & s_0 = 2 \\ a_1 = -2 & q_1 = -2.4 & r_2 = 0.2808989 & s_1 = -0.6516854 \\ a_2 = 3 & q_2 = 3.56 & & s_2 = 1.9988637 \\ a_3 = -2 & & & \\ a_4 = 2 & & & \end{array}$$

from which $m = 0.3260279$ and $p = -1.0005685$ and

$$\begin{array}{llll} & q_0 = 1 & r_1 = -1.151549 & s_0 = 2 \\ a_1 = -2 & q_1 = -1.6739721 & r_2 = 0.687914 & s_1 = 0.303098 \\ a_2 = 3 & q_2 = 1.4536699 & & s_2 = 1.9732042 \\ a_3 = -2 & & & \\ a_4 = 2 & & & \end{array}$$

from which $m = -0.153607$ and $p = -1.0135798$. The convergence is not fast, and so a program was written for a digital computer. After 10 iterations $p = -1.000004$ and $m = 0.00198894$. After 15 iterations $p = -1.00000001$ and $m = 0.0000619$. This last result gives a quadratic factor close to the exact factor $x^2 + 1$. Since

$$p(x) = (x^2 + 1)(x^2 - 2x + 2)$$

one might think that using the initial estimate $x^2 - x + 1$ in Friedman's method would yield an approximation to $x^2 - 2x + 2$. However, Friedman's method converges to the factor $x^2 + 1$ even when the initial estimate $x^2 - 2x + 1.95$ is used (see Exercise 7).

After a reasonably accurate estimate of a quadratic factor has been obtained, Bairstow's method can be used to sharpen it.

Bairstow's Method. To iteratively estimate, with arbitrary accuracy, a quadratic factor of the polynomial

$$p(x) = x^n + a_1x^{n-1} + \cdot\cdot\cdot + a_n$$

from the reasonably accurate initial estimate

$$g(x) = x^2 - mx - p$$

proceed as follows.

1. Step 1 is the same as step 1 of Friedman's method except that two additional constants are required:

$$q_{n-1} = mq_{n-2} + pq_{n-3} + a_{n-1}$$
$$q_n = mq_{n-1} + pq_{n-2} + a_n$$

   The constant q_{n-1} appears in the remainder $q_{n-1}x + q_{nn}$ obtained when $p(x)$ is divided by $g(x)$. The constant q_n is related to the constants in the remainder by the relation

$$q_n = q_{nn} + mq_{n-1}$$

2. Divide the first quotient $q(x)$ by $g(x)$, obtaining a second quotient,

$$r(x) = r_2x^{n-4} + r_3x^{n-5} + \cdot\cdot\cdot + r_{n-2}$$

   and the remainder $r_{n-1}x + r_{nn}$. The coefficients in these polynomials are obtained numerically from the sequence

$$r_2 = 1$$
$$r_3 = mr_2 + q_1$$
$$r_j = mr_{j-1} + pr_{j-2} + q_{j-2} \qquad (j = 4, 5, \ldots, n)$$

   The constant r_n is related to the constants in the remainder:

$$r_n = r_{nn} + mr_{n-1}$$

3. Calculate

$$R = pr_{n-1} + mr_n \qquad D = (r_n)^2 - Rr_{n-1}$$

4. The estimated errors in m and p are

$$e_m = \frac{q_{n-1}r_n - q_n r_{n-1}}{D}$$

$$e_p = \frac{r_n q_n - R q_{n-1}}{D}$$

5. Correct the estimates of m and p by subtracting the corresponding estimated errors from each. Then, using the corrected values of m and p, go back to step 1 and repeat the cycle until the desired accuracy is obtained.

Example 7

The quadratic factor $x^2 - 2x + 1$ of the polynomial

$$p(x) = x^4 - 2x^3 + 3x^2 - 2x + 2$$

will be estimated by Bairstow's method from the initial estimate

$$g(x) = x^2 - 2.1x + 2.1$$

Here $m = 2.1$ and $p = -2.1$ and

	$q_0 = 1$	$r_2 = 1$	$R = 3.003$	$e_m = 0.0926$
$a_1 = -2$	$q_1 = 0.1$	$r_3 = 2.2$	$D = 6.5703$	$e_p = -0.0977901$
$a_2 = 3$	$q_2 = 1.11$	$r_4 = 3.63$		
$a_3 = -2$	$q_3 = 0.121$			
$a_4 = 2$	$q_4 = -0.0769$			

The corrected values are $m = 2.0074$ and $p = -2.0022$. Repeating the steps in Bairstow's method with these corrected values gives

$q_0 = 1$	$r_2 = 1$	$R = 2.0985028$	$e_m = 0.007382$
$q_1 = 0.0074$	$r_3 = 2.0148$	$D = 5.1047434$	$e_p = -0.0022657$
$q_2 = 1.0126548$	$r_4 = 3.0549643$		
$q_3 = 0.017987$			
$q_4 = 0.0085697$			

Correcting again, $m = 2.000018$ and $p = -1.9999343$, which are very close to the coefficients of the actual quadratic factor.

EXERCISES

1. **(a)** Prove the *remainder theorem,* which states that if a polynomial $p(x)$ is divided by $x - a$ until a constant remainder r is obtained, then $f(a) = r$.
   (b) Prove the *factor theorem,* which states that if $p(a) = r = 0$, then $x - a$ is a factor of $p(x)$.
   (c) Find the remainder when $x^3 + 6x^2 - 5x + 4$ is divided by $x - 2$.

(d) Without dividing $x^3 + 2x^2 - 5x - 6$ by $x + 1$, show that it is a factor of the given polynomial.

(e) How do you use the factor theorem to determine if $2x - 3$ is a factor of $2x^3 - 3x^2 + 8x - 12$?

2. The polynomial $p(x) = p_n x^n + p_{n-1}x^{n-1} + \cdots + p_1 x + p_0$ can be divided by $x - a$ by *synthetic division,* which uses the following array:

$$\begin{array}{cccccc|l} p_n & p_{n-1} & p_{n-2} & \cdots & p_1 & p_0 & \underline{a} \\ & aq_{n-1} & aq_{n-2} & \cdots & aq_1 & aq_0 & \\ \hline q_{n-1} & q_{n-2} & q_{n-3} & \cdots & q_0 & r & \end{array}$$

where

$$q_{n-1} = p_n$$

$$q_{n-k-1} = aq_{n-k} + p_{n-k} \qquad (k = 1, 2, \ldots, n-1)$$

$$r = p_0 + aq_0$$

The polynomial must be written so that its powers are in descending order and zeroes are placed in for missing powers. Note that the quotient is

$$q(x) = q_{n-1}x^{n-1} + q_{n-2}x^{n-2} + \cdots + q_0$$

and the remainder is r.

Verify that the setup for synthetic division of $3x^4 - 5x + 6$ by $x - 4$ is

$$\begin{array}{ccccc|l} 3 & 0 & 0 & -5 & 6 & \underline{4} \\ & 12 & 48 & 192 & 748 & \\ \hline 3 & 12 & 48 & 187 & 754 & \end{array}$$

Check the answer by long division.

3. Use the following result:

if the coefficient of the highest power in a polynomial is 1, then any integral roots are factors of the constant term,

to find the integral roots of $x^3 + x^2 - 14x - 24 = 0$.

4. (a) Use the theorem in Exercise 3 to deduce the following: Any rational roots of a polynomial equation

$$p(x) = a_n x^n + \cdots + a_0 = 0$$

must be integral factors of a_0 divided by integral factors of a_n.

(b) Use the result in (a) to find the 16 possible candidates for the rational roots of

$$4x^3 - 3x^2 - 25x - 6 = 0$$

(c) Find the roots of the equation in (b). (*Hint:* Try the roots by synthetic division. When a root is discovered, the corresponding quotient factor appears in the array for synthetic division. It will be a quadratic and so the remaining roots can be found from the quadratic formula.)

5. Prove the following results, which are sometimes helpful in eliminating some of the roots obtained by 4(a).

(a) The number of positive roots of a polynomial equation $p(x) = 0$ cannot exceed the number of sign changes in $p(x)$ in going from one term to the next in $p(x)$. The number of negative roots cannot exceed the number of sign changes in $p(-x)$.
(b) If the bottom row in a synthetic division array contains all positive numbers when a root is being tried, then there are no roots larger than the one tried.
(c) If the signs alternate in the bottom row of a synthetic division array, then there are no roots less than the one tried.

6. Suppose that only one root of a continuous function $f(x)$ lies between x_1 and x_2, where $x_1 < x_2$. Assume that $f(x_1) = y_1 > 0$ and $f(x_2) = y_2 < 0$. Approximating the curve $y = f(x)$ by a straight line joining (x_1, y_1) and (x_2, y_2) and the root by the x intercept of the line is called *linear interpolation*. The closer x_1 is to x_2, the better the approximation.
(a) Show that the root lying between x_1 and x_2 is approximated by

$$x = \frac{x_1 y_2 - y_1 x_2}{y_2 - y_1}$$

(b) Prove that if $f(x) > 0$ (< 0), the required root can be approximated as in (a) by joining x and x_2 (x_1).
(c) Apply steps (a) and (b) to approximate the irrational root of $x^3 + 2x^2 + 8x - 2 = 0$, which lies between 0 and 1. Use synthetic division to evaluate $f(x)$, where x is an approximate root obtained from step (a).

7. Using the initial estimate $x^2 - 2x + 1.95$, show that the application of Friedman's method to the polynomial defined in (8) gives the factor $x^2 + 1$.

8. Find a quadratic factor of the polynomial $x^4 + x^2 + 2x + 6$ using Friedman's method with the initial estimate $x^2 + x + 1$.

9. Find the quadratic factor of the polynomial appearing in Exercise 8 using the initial estimate $x^2 + 2.1x + 2.1$.

10. (a) Do you think Definition 1 is a suitable definition of stability for discrete time systems?
(b) Formulate the analog of Theorem 1 for such systems.

10.3 ROUTH-HURWITZ STABILITY CRITERION

The Routh-Hurwitz stability criterion is an algebraic procedure for determining whether a polynomial has any zeroes in the right half-plane. It involves examining the signs and the magnitude of the coefficients of the characteristic equation without actually having to determine its roots. Although this method overcomes the difficulties associated with finding the roots of a polynomial, it still does not indicate the relative degree of stability or instability. A measure of stability will be developed when the Nyquist criterion is studied in the next section.

Hurwitz [9] and Routh [16] independently determined necessary and sufficient conditions for stability. Although their criteria differ somewhat, both furnish the same information. A useful form of their approach is described below without proof [2,14].

For stability, the roots of the characteristic equation

$$a_n s^n + a_{n-1} s^{n-1} + \cdots + a_1 s + a_0 = 0 \tag{9}$$

must be either

1. Real and negative, implying factors of Eq. (9) of the form

$$s + a \qquad (a > 0), \text{ or}$$

2. Complex with negative real parts, involving conjugate factors of the form [since the coefficients of Equation (9) are real]

$$(s + a + jb)(s + a - jb) = s^2 + 2as + a^2 + b^2 \qquad (a > 0)$$

The characteristic polynomial is the product of these two types of factors, all having positive coefficients. Hence all the coefficients must be present, and they all must be positive (the coefficient of the leading term s^n is unity). By multiplying the resulting equation by a constant, the roots are not changed but a slightly more general polynomial of the form appearing on the right side of Eq. (9) is obtained. Again, all terms must be present and all must be of the same sign. Therefore, necessary conditions for stability are the following:

1. All the coefficients $a_n, a_{n-1}, \ldots, a_0$ must be present, and
2. They must all have the same sign.

If (a) or (b) is violated, then the corresponding system is unstable. If (a) and (b) are satisfied, further investigation is required to determine if the system is stable.

The Routh-Hurwitz criterion gives a necessary and sufficient condition for stability. This criterion requires the construction of the following array of coefficients and values dependent on the coefficients:

$$\begin{array}{llllll}
\text{row } 1 & a_n & a_{n-2} & a_{n-4} & a_{n-6} & \cdots \\
\text{row } 2 & a_{n-1} & a_{n-3} & a_{n-5} & a_{n-7} & \cdots \\
\text{row } 3 & b_1 & b_2 & b_3 & b_4 & \cdots \\
\text{row } 4 & c_1 & c_2 & c_3 & c_4 & \cdots \\
\vdots & \vdots & \cdot & \cdot & \cdot & \cdots \\
\text{row } (n+1) & \cdot & \cdot & \cdot & \cdot & \cdots
\end{array}$$

Row 1 contains the coefficients of s^n, s^{n-2}, s^{n-4}, . . . in the characteristic polynomial. It is assumed that a_n is positive. Row 2 contains the coefficients of s^{n-1}, s^{n-3}, s^{n-5}, . . . in the characteristic polynomial. Zeroes can be inserted at the ends of rows if needed.

The remaining coefficients in the Routhian array are generated by the following rules:

$$b_1 = -\frac{1}{a_{n-1}}\begin{vmatrix} a_n & a_{n-2} \\ a_{n-1} & a_{n-3} \end{vmatrix} \qquad b_2 = -\frac{1}{a_{n-1}}\begin{vmatrix} a_n & a_{n-4} \\ a_{n-1} & a_{n-5} \end{vmatrix} \qquad b_3 = -\frac{1}{a_{n-1}}\begin{vmatrix} a_n & a_{n-6} \\ a_{n-1} & a_{n-7} \end{vmatrix}$$

$$c_1 = -\frac{1}{b_1}\begin{vmatrix} a_{n-1} & a_{n-3} \\ b_1 & b_2 \end{vmatrix} \qquad c_2 = -\frac{1}{b_1}\begin{vmatrix} a_{n-1} & a_{n-5} \\ b_1 & b_3 \end{vmatrix} \qquad c_3 = -\frac{1}{b_1}\begin{vmatrix} a_{n-1} & a_{n-7} \\ b_1 & b_4 \end{vmatrix}$$

and so on. The 2×2 arrays of numbers enclosed by parallel lines that appear in the formulas for the b's and c's are called *determinants* (see Volume II, Section 1.6). The numerical value of such a determinant is evaluated as follows:

$$\begin{vmatrix} a & b \\ c & d \end{vmatrix} = ad - bc$$

Thus, for example,

$$b_1 = \frac{a_{n-1}a_{n-2} - a_n a_{n-3}}{a_{n-1}}$$

There will be $n + 1$ rows, provided no element in the first column is zero. If there are zeroes in the first column, then the system is not stable. Methods for completing the array in these circumstances appear in [2], but will not be discussed here. (See Exercises 5 and 6.)

Once the Routh array has been constructed, the following result can be used to determine the stability of the system.

Routh-Hurwitz Stability Criterion

1. The roots of the characteristic equation all have negative real parts if, and only if, the elements of the first column of the Routh array are positive.
2. The number of roots with positive real parts is equal to the number of sign changes of the elements in the first column.

Corollary. A system is stable if, and only if, the elements of the first column of its associated Routh array are positive.

The corollary is a direct consequence of part 1 of the Routh-Hurwitz criterion by Theorem 1.

Example 8

The Routh-Hurwitz criterion will be illustrated with two polynomials.

(a) The Routh array corresponding to the characteristic polynomial

$$s^3 + 4s^2 + 100s + 500$$

is

$$\begin{array}{rr} 1 & 100 \\ 4 & 500 \\ -25 & 0 \\ 500 & 0 \end{array}$$

The system is not stable. There are two sign changes in the first column, and so there are two roots with positive real parts.

(b) The characteristic polynomial

$$s^5 + s^4 + 4s^3 + 4s^2 + 2s + 1$$

has the following Routh array:

$$\begin{array}{ccc} 1 & 4 & 2 \\ 1 & 4 & 1 \\ 0 & 1 & 0 \end{array}$$

Since there is a zero in the last row, the array cannot be completed by the method described above. By the Routh-Hurwitz criterion, the system is unstable and the complete array is not required.

In Example 8 no fraction appeared in the Routh arrays. If a fraction is encountered, the entire row containing the fraction can be multiplied by an appropriate constant, which changes the fraction into an integer without affecting the conclusion.

The next example shows how the Routh-Hurwitz criterion can be used to find a range of values for a system parameter for stability.

Example 9

(a) A third-order system has the characteristic polynomial

$$s^3 + 2s^2 + 4s + K$$

where K is an adjustable loop gain. The Routh array is

$$\begin{array}{cc} 1 & 4 \\ 2 & K \\ \dfrac{8-K}{2} & 0 \\ K & 0 \end{array}$$

Hence, for stability $K > 0$ and $(8 - K)/2 > 0$, that is, $0 < K < 8$.

(b) *Protein synthesis model*. The stability of the protein synthesis model, described in Example 4 of Chapter 9, involves the characteristc polynomial

$$s^2 + (a_2 + a_3)s + ka_1a_2 + a_2a_3 = 0$$

where the constants a_i are all positive. Its Routh array is

$$\begin{array}{cc} 1 & ka_1a_2 + a_2a_3 \\ a_2 + a_3 & 0 \\ ka_1a_2 + a_2a_3 & 0 \end{array}$$

The system will be stable if $ka_1a_2 + a_2a_3 > 0$ or $k > -a_3/a_1$.

Frequently, some indication of the relative stability of a stable system is required. For example, it may be desirable that all the system modes decay at least as fast as $e^{-\sigma t}$. In other words, the real parts of the roots of the characteristic equation should be less than $-\sigma$, where σ is some positive real number. In this situation the system is said to have a *stability margin* σ. If the characteristic equation $C(s) = 0$ has roots $s_1, s_2, \ldots, s_n$, then the equation $C(s - \sigma) = 0$ has roots $s_1 + \sigma$, $s_2 + \sigma, \ldots, s_n + \sigma$. Applying the Routh-Hurwitz criterion to $C(s - \sigma) = 0$ will determine if $s_i + \sigma$ $(i = 1, 2, \ldots, n)$ will all have negative real parts or if $\text{Re } s_i < -\sigma$ for all i.

Example 10

Determine whether the system with the associated characteristic polynomial

$$C(s) = s^3 + 7s^2 + 17s + 15$$

has stability margin 1.

$$\begin{aligned} C(s-1) &= (s-1)^3 + 7(s-1)^2 + 17(s-1) + 15 \\ &= s^3 + 4s^2 + 6s + 4 \end{aligned}$$

The Routh array for $C(s-1)$ is

$$\begin{matrix} 1 & 6 & 0 \\ 4 & 4 & 0 \\ 5 & 0 & \\ 4 & 0 & \end{matrix}$$

Hence all the roots of $C(s-1)$ have negative real parts, and so the stability margin is 1.

The Routh-Hurwitz criterion and any other test of stability based on detecting the roots of a polynomial equation in the right half-complex plane can be applied to determine the stability of a discrete system if a transformation of its characteristic equation $C(z) = 0$ can be found that satisfies the following conditions for the transform variable s:

1. The exterior of the unit circle in the z plane is transformed into the right half of the s plane.
2. The interior of the unit circle in the z plane is transformed into the left half of the s plane.
3. The boundary of the unit circle in the z plane is transformed into the imaginary axis of the s plane.

The effect of such a transformation is illustrated in Fig. 10.1.

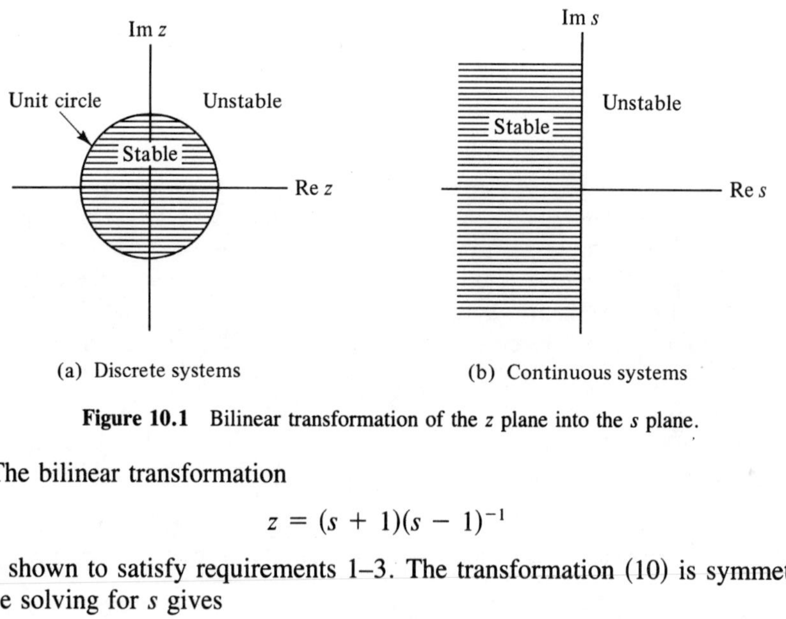

Figure 10.1 Bilinear transformation of the z plane into the s plane.

The bilinear transformation

$$z = (s+1)(s-1)^{-1} \tag{10}$$

can be shown to satisfy requirements 1–3. The transformation (10) is symmetrical, because solving for s gives

$$s = (z+1)(z-1)^{-1} \tag{11}$$

As an example of the transformation, consider the point $z = \frac{1}{2}$. From (11), this point is transformed into $s = (\frac{1}{2}+1)(\frac{1}{2}-1)^{-1} = -3$. Similarly, it can be shown that the points $z = 1 + i, i$ are transformed into the points $s = 1 - 2i, -i$, respectively. These three examples agree with the results described in 2, 1, and 3, respectively.

The effect of the bilinear transformation (11) is to transform the characteristic equation $C(z) = 0$ into the equation $\overline{C}(s) = 0$. There is a one-to-one correspondence between any root z_0 of $C(z) = 0$ and the root s_0 of $\overline{C}(s) = 0$ defined by (10) and (11). Moreover, $|z_0| < 1$ if, and only if, Re $s_0 < 0$. Hence the discrete system is stable if, and only if, the roots of $\overline{C}(s) = 0$ have negative real parts. The Routh-Hurwitz criterion can be used to test this last fact. The next example illustrates the above procedure.

Example 11

The discrete system with the characteristic equation

$$C(z) = 6z^2 - 5z + 1 = 0$$

is stable, since the roots of $C(z) = 0$ are $\frac{1}{2}$ and $\frac{1}{3}$.

The stability of this discrete system can be verified from the Routh-Hurwitz criterion by first applying the transformation (11) to $C(z)$, which gives

$$C(z) = \frac{6(s+1)^2}{(s-1)^2} - \frac{5(s+1)}{(s-1)} + 1$$

$$\overline{C}(s) = 2s^2 + 10s + 12$$

The Routh array for $\overline{C}(s)$ is

$$\begin{matrix} 2 & 12 \\ 10 & 0 \\ 12 & \end{matrix}$$

and so the discrete system is stable.

EXERCISES

1. Utilizing the Routh-Hurwitz criterion, determine the stability of the following polynomials. If possible, determine the number of roots in the right half-plane.
   (a) $s^3 + 3s^2 + s + 20$
   (b) $s^4 + 3s^2 + 7s + 10$
   (c) $s^3 - 3s^2 + s + 20$
   (d) $s^5 + 5s^4 + 11.5s^3 + 17s^2 + 14.5s + 5$
2. If possible, find values of K so that the systems corresponding to the following characteristic equations are stable.
   (a) $s^3 + 2s^2 + K = 0$
   (b) $s^3 + 2s^2 + s + K = 0$
3. Determine the stability of the discrete system with the characteristic polynomial

   $$C(z) = z^3 - 0.20z^2 - 0.25z + 0.05$$

   by applying the bilinear transformation (11) and using the Routh-Hurwitz criterion.
4. A system has the characteristic polynomial

   $$C(s) = s^3 + 9s^2 + 26s + 24$$

   Can the system stability margin be 2?

5. If the first term in any row is zero, the Routh array can be continued by replacing this zero by a small positive number ϵ. After completing the array in the usual manner, ϵ is allowed to approach zero in order to determine the sign changes in the first column. Apply this result to investigate the stability of the system whose characteristic equation is

$$s^5 + s^4 + 4s^3 + 4s^2 + 2s + 1 = 0$$

6. If there is a zero in the first column of the Routh array and all other elements of the row containing the zero are also zero, then the array can be completed by utilizing the auxiliary polynomial whose coefficients are the row immediately preceding the zero row. If the characteristic polynomial is of degree n and the coefficients of the auxiliary polynomial appear in the rth row, then the auxiliary polynomial is of degree $n - r + 1$ and only powers that differ from the degree by multiples of 2 can appear. The auxiliary polynomial is differentiated and the resulting coefficients are used to complete the array. Use the described procedure on the following characteristic equation,

$$s^3 + 10s^2 + 16s + 160 = 0$$

and show that the auxiliary polynomial is $10s^2 + 160$. Verify that the characteristic equation has no zeroes with positive real parts but there are a pair of imaginary roots. (*Hint*: The auxiliary polynomial is always a factor of the characteristic polynomial.)

7. For the simple stuttering system described in Exercise 6 of Section 9.4 with $K_2 = 0.01$, $K_3 = 0.1$, and $K_4 = 1$, stuttering occurs when the system is unstable. Determine the value of the gain K_5 when stuttering occurs.

8. Determine the range of values of K for the pacemaker system described in Exercise 9 of Section 9.4 to be stable in the absence of disturbances.

9. A system enabling a person paralyzed below the neck to drive himself about in a motorized wheelchair is shown in the following diagram. The system utilizes velocity sensors mounted at 90° intervals in a headgear. Typical values for the time constants are $\tau_1 = 0.5$, $\tau_3 = 1$ sec, and $\tau_4 = 0.33$ sec. Determine the limiting gain $K = K_1 K_2 K_3 K_4$ for a stable system.

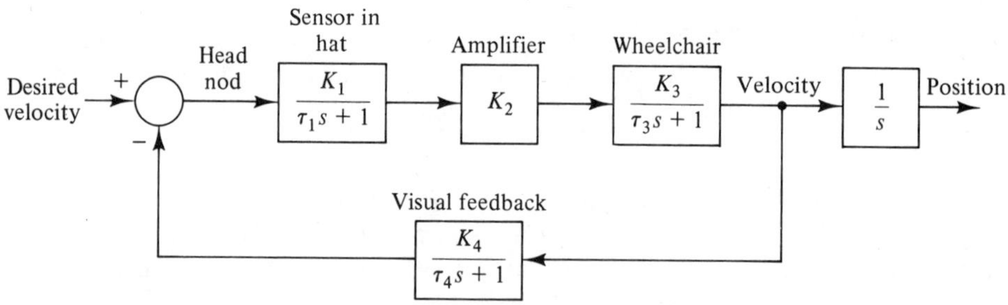

10.4 FREQUENCY RESPONSE METHOD; NYQUIST CRITERION

A. Frequency Response Method

The previous methods of stability analysis depended on a detailed knowledge of the system dynamics. If the transfer function of the system is not known, it is often possible to obtain it experimentally by exciting the system with a range of sinusoidal inputs of fixed amplitude but varying frequency and observing the resulting changes in amplitude and phase shift at the output. This information is known as the *system frequency response*. Since the system dynamics of many biological systems are not known completely, the frequency response technique is useful in the investigation of such systems.

If an input is applied to a system at rest, steady-state oscillatory conditions cannot apply until the transient part of the solution has nearly vanished. The transient part of the response is the homogeneous solution of the linear differential equation describing the system and can be difficult to obtain explicitly. The steady-state frequency response can be obtained in a particularly simple way from the system transfer function. The sinusoidal output is considered a steady state in that its amplitude and relative phase do not vary in time.

The frequency response method is more general than it first appears. It can be used to calculate the output of complex repeated waveform inputs from the system response to sine and cosine waves. First, consider a waveform that is a linear combination of sines and cosines. Its output, by the principle of linear superposition, is just the same linear combination of the known outputs corresponding to each component input. Next, the output for a complex repeated waveform can be found from the first result by using Fourier's theorem, which expresses such waveforms as a sum of sine and cosine waves of various amplitudes, comprising, in general, the fundamental and all of its harmonics. The frequency response method has been extended to some nonlinear systems by means of describing functions.

The disadvantage of the frequency response method for analysis and design is the indirect link between the frequency response and the time domain. In practice, the frequency characteristic is adjusted by using various design criteria that will normally result in a satisfactory transient response.

A polar plot of the frequency response of an open-loop system is called a *Nyquist diagram*. The stability of the corresponding closed-loop system is determined by visual inspection of the Nyquist diagram and depends on the position of the curve relative to the point $(-1, 0)$. An important feature of this method is that it employs the open-loop transfer function, which is easier to obtain and handle than the closed-loop transfer function.

The basic result underlying the frequency response method is that if a sinusoidal wave is input into a linear system, then the steady-state output is also a sinusoidal wave of the same frequency but generally of different amplitude and phase relative to the input. A similar result holds for a cosine wave. The assertion for the sine wave will be investigated and the corresponding result for a cosine wave will be left as an exercise.

If the system has the transfer function $T(s)$, then the transform of the output is $Y(s) = T(s)U(s)$. For a sinusoidal input of unit amplitude $u(t) = \sin \omega t$, $U(s) = \omega/(s^2 + \omega^2)$ and $Y(s) = T(s)\omega/(s^2 + \omega^2)$. If $T(s)$ has poles at $s = p_1$, $p_2, \ldots, p_n$, $Y(s)$ can be expanded in partial fractions as

$$Y(s) = \sum_{k=1}^{n} \frac{A_k}{s - p_k} + \frac{B_1}{s + i\omega} + \frac{B_2}{s - i\omega} \tag{12}$$

where $i^2 = -1$ and

$$A_k = \left[\frac{(s - p_k)\omega T(s)}{s^2 + \omega^2}\right]_{s=p_k}$$

$$B_1 = \left[\frac{(s + i\omega)\omega T(s)}{s^2 + \omega^2}\right]_{s=-i\omega} = -\frac{1}{2i}T(-i\omega)$$

$$B_2 = \left[\frac{(s - i\omega)\omega T(s)}{s^2 + \omega^2}\right]_{s=i\omega} = \frac{1}{2i}T(i\omega)$$

Taking the inverse Laplace transform of (12) gives

$$y(t) = \underbrace{\sum_{k=1}^{n} A_k e^{p_k t}}_{\text{transient terms}} + \underbrace{B_1 e^{-i\omega t} + B_2 e^{i\omega t}}_{\text{steady-state terms}} \tag{13}$$

If the system is stable, then all the poles have negative real parts, so that as t increases, the transient terms decrease and become negligible relative to the steady-state terms in (13).

The steady-state terms can be written as

$$\frac{1}{2i}[T(i\omega)e^{i\omega t} - T(-i\omega)e^{-i\omega t}] = \text{Im}[T(i\omega)e^{i\omega t}] \tag{14}$$

Since $T(i\omega) = |T(i\omega)|e^{i\phi}$, in polar form, where

$$\phi = \tan^{-1}\left(\frac{\text{Im } T(i\omega)}{\text{Re } T(i\omega)}\right)$$

Eq. (14) can be written as

$$\text{Im}[|T(i\omega)|\, e^{i(\omega t+\phi)}] = |T(i\omega)| \sin(\omega t + \phi) \tag{15}$$

by using Euler's identity,

$$e^{ix} = \cos x + i \sin x$$

Equation (15) shows that the steady-state output is also sinusoidal and that the input amplitude is multiplied by $|T(i\omega)|$. The frequency is unchanged; however, there is a phase shift ϕ relative to the input.

The *frequency diagram* is the graph of $T(i\omega)$ in the complex plane as the frequency varies from $-\infty$ to ∞. For systems whose transfer functions are the ratios of two polynomials with real coefficients, it follows that $T(i\omega)$ is the complex conjugate of $T(-i\omega)$. It is sufficient to plot $T(i\omega)$ for $0 \leq \omega < \infty$, since the remainder of the graph is symmetrical about the real axis to this plot.

A frequency-domain stability criterion was developed by H. Nyquist [12] in 1932 for a negative-feedback closed-loop system, depicted in Fig. 9.3. The application of this criterion requires a polar plot of the open-loop transfer function. This plot is usually referred to as a *Nyquist diagram*. The open-loop transfer function is the ratio of the transform of the feedback signal HY to the transform of the error signal E. It follows from Eqs. (26) and (27) in Chapter 9 that the required ratio is

$$\frac{H(s)Y(s)}{E(s)} = G(s)H(s)$$

Hence the *open-loop transfer function* is GH. The Nyquist criterion will now be investigated.

In order to develop the stability criterion for a more general control system, its characteristic equation, $F(s) = 0$, where

$$F(s) = 1 + R(s)$$

will be considered. Note that for a single loop, $R(s) = G(s)H(s)$, while for a multiloop system, in terms of signal flow graphs,

$$R(s) = -1 + \Delta(s)$$

where $\Delta(s)$ is the graph determinant, defined by (11) in Chapter 8. For a system to be stable, all of the zeroes of its characteristic equation must lie in the left-half s plane, that is, to the left of the imaginary axis. To obtain information about the location of the zeroes of the characteristic equation from the polar plot of R requires the following result in the theory of complex variables, known as Cauchy's argument principle [13].

Let γ be a closed contour in the s plane enclosing P poles and Z zeroes of $F(s)$ but not passing through any poles or zeroes of $F(s)$. Let F be analytic within and on the contour except at the P poles. As s traces out γ in the positive direction, $F(s)$ traces out a closed contour Γ in the $F(s)$ plane, which encircles the origin $N_0 = Z - P$ times in the positive direction.

By convention, a contour is traversed in the positive direction when it is traversed counterclockwise. The rigorous definition of an analytic function can be found in [13]. All rational functions are analytic except at the poles of the denominator polynomial.

To apply Cauchy's theorem to F requires a closed contour in the s plane avoiding the poles and zeroes of F. Recall that the frequency response required s to trace out the imaginary axis. Hence the imaginary axis will be chosen to be part of the contour. The remainder of the contour is a semicircular path of large radius r (where $r \to \infty$) in the right half of the s plane. The semicircle is assumed large enough to enclose all of the poles and zeroes of $F(s)$ that may lie in the right half-plane. The combination of the above two contours defines the Nyquist path as shown in Fig. 10.2. The small semicircular indentations are used to avoid the zeroes and poles of $R(s)$ that lie on the imaginary axis.

It is convenient to map γ using R instead of $F = R + 1$, since R is typically available in factored form. Now $F(s)$ lies to the right of 0 if and only if $R(s)$ lies to the right of -1, and $F(s)$ lies to the left of 0 if and only if $R(s)$ lies to the left of -1. Hence the number of encirclements of the origin in the $F(s)$ plane equals the number of encirclements (N_{-1}) of the $-1 + i0$ point in the $R(s)$ plane. Hence the number of zeroes of F,

$$Z = N_{-1} + P \tag{16}$$

Note the P = the number of poles of F = the number of poles of R. Clearly, from (16), if the number of poles of $R(s)$ in the right-hand s-plane is zero ($P = 0$), then $N_{-1} = 0$ for a stable system since $Z = 0$. If $P = 0$, Z must be zero for a stable system and so $N_{-1} = -P$; that is, there are P clockwise encirclements of the *critical point* $(-1, 0)$. The results are summarized below. A rigorous proof of the Nyquist criterion can be found in [11].

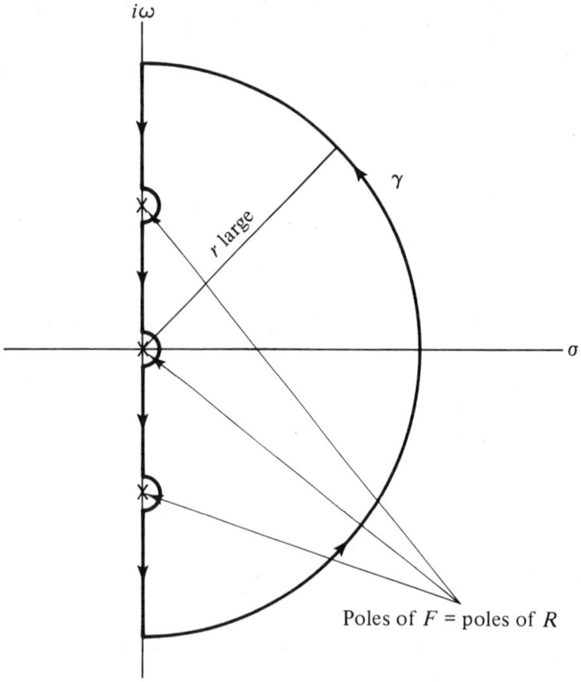

Figure 10.2 Nyquist path in the complex s plane.

B. The Nyquist Criterion

To determine the stability of a linear multiloop feedback system with the characteristic function $F = 1 + R$, consider the map $R(\gamma) = \Gamma$ in the $R(s)$ plane of the Nyquist contour γ traversed in a positive sense.

The system is stable if there are no zeroes of F on the imaginary axis and if:

1. Γ does not encircle the point $(-1, 0)$ when the open-loop transfer function R has no poles in the right half-plane.
2. Γ encircles the point $(-1, 0)$ P times in the negative direction when the open-loop transfer function R has P poles in the right half-plane.

Otherwise the system is unstable.

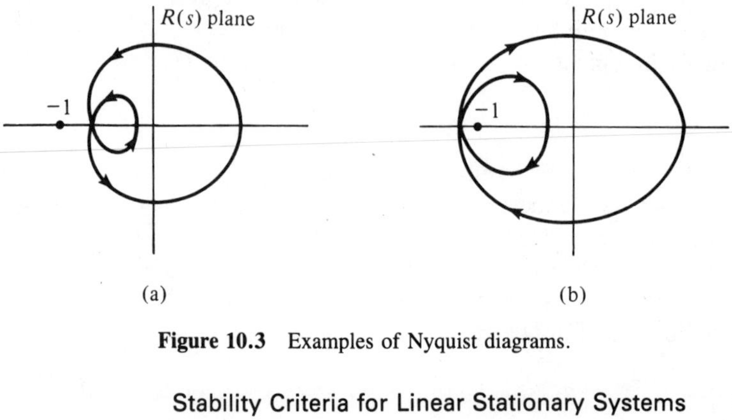

Figure 10.3 Examples of Nyquist diagrams.

Some examples of Nyquist diagrams appear in Fig. 10.3. If $P = 0$, the system corresponding to Fig. 10.3(a) is stable while the system corresponding to Fig. 10.3(b) is unstable since Γ encircles the point $(-1, 0)$ twice. However, if the system corresponding to Fig. 10.3(b) has two poles in the right half-plane, then it is stable.

The use of the Nyquist criterion will be illustrated by several examples.

Example 12

A single-loop feedback has the open-loop transfer function

$$GH(s) = \frac{K}{s(s+1)}$$

Here $R(s) = GH(s)$. The Nyquist path γ in the s plane is shown in Fig. 10.4(a). The

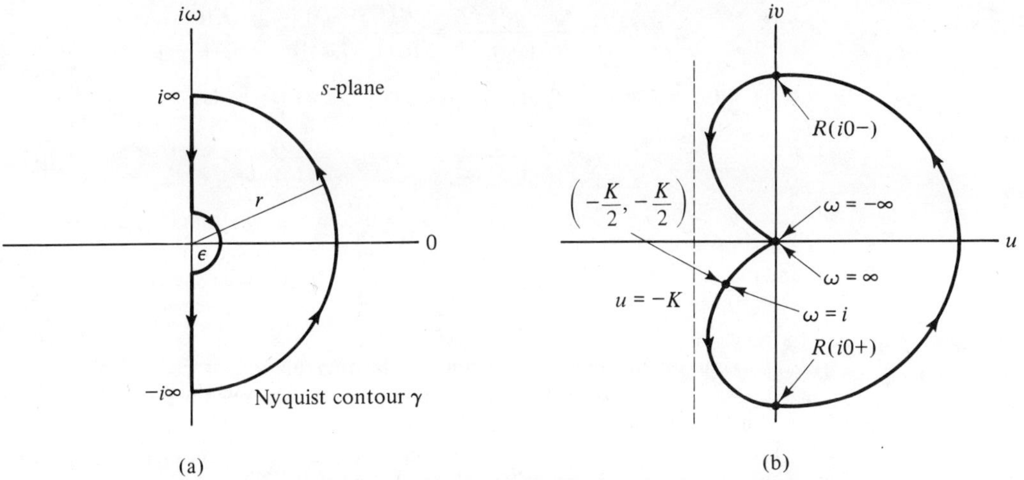

Figure 10.4 Nyquist path and its image for $R = K/s(s+1)$.

semicircle of radius ϵ is a consequence of the condition that the Nyquist path does not pass through a pole of R at $s = 0$.

A sketch of the contour $\Gamma = R(\gamma)$ appears in Fig. 10.4(b). The details of the mapping follow. The notations $i\infty$ and $-i\infty$ refer to very large positive and negative frequencies, while $i0-$ and $i0+$ refer to values of the frequency just less than zero and just greater than zero, respectively.

(a) *The image of the large semicircle of radius r.* A point s on this semicircle can be represented in polar form as $s = re^{i\phi}$. Since s is large, $s + 1 \approx s$, and so

$$\lim_{r\to\infty} GH(s) = \lim_{r\to\infty} \frac{Ke^{-2i\phi}}{r^2} = 0$$

Hence the large semicircle is mapped into the origin in the R plane. When $\omega = \infty$, $\phi = \pi/2$ and so the Γ contour approaches the origin at an angle of $-2(\pi/2) = -180°$. Similarly, when $\omega = -\infty$, the contour approaches the u axis at an angle of $180°$.

In most physical systems the degree of the numerator of R is at most equal to the degree of the denominator. Hence the magnitude of $R(s)$ when r approaches infinity is always zero or a constant. The main problem is mapping the portion of the γ contour on the $i\omega$ axis (the frequency response).

(b) *The image of the small semicircle of radius* ϵ. Any point on the semicircular contour around the origin can be represented as $s = \epsilon e^{i\phi}$. Since ϵ is small, $s + 1 \approx 1$ and so

$$R(s) \approx \frac{Ke^{-i\phi}}{\epsilon}$$

As $\epsilon \to 0$, $K/\epsilon \to \infty$, so that as ϕ varies from $-90°$ to $90°$, $R(s)$ describes a semicircle of infinite radius joining the points $R(i0-) = i\infty$ and $R(i0+) = -i\infty$.

(c) *The image of the imaginary axis from* $\omega = \infty$ *to* $\omega = 0+$. Any point on the imaginary axis can be represented by $s = i\omega$ and so is mapped into the point

$$R(i\omega) = \frac{K}{i\omega(i\omega + 1)} = \frac{K(1 - i\omega)}{i\omega(1 + \omega^2)} = -\frac{K(i + \omega)}{\omega(1 + \omega^2)}$$

Separating R into its real (u) and imaginary (v) parts yields

$$u = -\frac{K}{1 + \omega^2} \qquad v = -\frac{K}{\omega(1 + \omega^2)} \tag{17}$$

Since $\omega > 0$, $u < 0$ and $v < 0$. As $\omega \to 0+$, it follows from (17) that $u \to -K$ and $v \to -\infty$. When $\omega = 1$, $u = -K/2$ and $v = -K/2$. From the above results and the fact that $|u|$ increases as ω decreases, $R(i\omega)$ can be sketched as shown in Fig. 10.4(b).

Any point on the negative portion of the imaginary axis can be represented by $s = -i\omega$ with $\omega > 0$. Such points are mapped into $R(-i\omega) = -K(-i + \omega)/\omega(i + \omega^2) = \overline{R(i\omega)}$, the complex conjugate of $R(i\omega)$. Hence the portion of the curve $R(-i\omega)$ that lies in the second quadrant is just the mirror image in the u axis of the previously sketched curve $R(i\omega)$ lying below the u axis. The complete contour appears in Fig. 10.4(b).

To investigate the stability of the given system, note that the number of poles of $R(s)$ in the right-hand s plane is zero. Therefore, for stability, $N_{-1} = 0$. Examining Fig. 10.4(b) shows that the contour Γ does not encircle the $(-1, 0)$ point irrespective of the value of the gain K. Hence the system is stable for all values of K.

The following two general conclusions may be drawn from the above example.

1. The polar plot of $R(s)$ will be symmetrical in the R plane about the u axis if the transfer function is rational.
2. The magnitude of $R(s)$ for $s = re^{i\phi}$ and $r \to \infty$ will normally approach zero or a constant.

Example 13

Suppose that a single-loop feedback system has the open-loop transfer function

$$GH = \frac{K}{s(\tau_1 s + 1)(\tau_2 s + 1)} \qquad (\tau_1 > 0,\ \tau_2 > 0)$$

To investigate the stability of this system using the Nyquist criterion, the Nyquist contour shown in Fig. 10.4(a) will be used. The polar plot of $R(s) = G(s)H(s)$ is symmetrical about the u axis, since the open-loop transfer function is rational. The origin of the s plane maps into a semicircle of infinite radius as in Example 12. Moreover, the semicircle $re^{i\phi}$ in the s plane maps into the origin in the R plane, as expected. Hence it is sufficient to plot the frequency diagram $GH(i\omega)$ for $0 < \omega < \infty$. When $s = i\infty$,

$$\begin{aligned} GH(i\omega) &= \frac{K}{i\omega(i\omega\tau_1 + 1)(i\omega\tau_2 + 1)} \\ &= \frac{-K(\tau_1 + \tau_2) - iK\omega^{-1}(1 - \omega^2\tau_1\tau_2)}{1 + \omega^2(\tau_1^2 + \tau_2^2) + \omega^4\tau_1^2\tau_2^2} \\ &= \frac{Ke^{i\phi}}{[\omega^4(\tau_1 + \tau_2)^2 + \omega^2(1 - \omega^2\tau_1\tau_2)^2]^{1/2}} \end{aligned} \tag{18}$$

where

$$\phi = -\tan^{-1}\omega\tau_1 - \tan^{-1}\omega\tau_2 - \frac{\pi}{2}$$

For $\omega = 0+$, the magnitude of the locus is infinite at an angle of $-90°$ in the GH plane. Letting $\omega \to \infty$ gives

$$\lim_{\omega\to\infty} GH(i\omega) = \lim_{\omega\to\infty}\left|\frac{c}{\omega^3}\right|(e^{\pi i/2})^{-3} = \lim_{\omega\to\infty}\left|\frac{c}{\omega^3}\right|e^{-3\pi i/2}, \qquad c = k/\tau_1\tau_2$$

Therefore $GH(i\omega)$ approaches a magnitude of zero at an angle of $-270°$. Since the locus approaches the origin at an angle of $-270°$, the locus must cross the u axis in the GH plane as shown in Fig. 10.5. Therefore it is possible to encircle the -1 point as shown in Fig. 10.5.

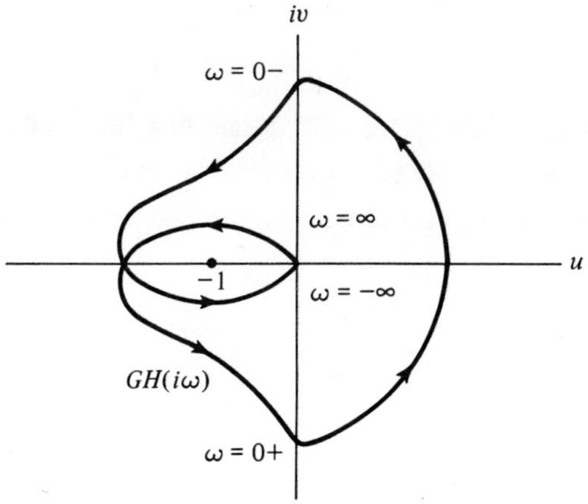

Figure 10.5 Nyquist diagram for $GH(s) = K/s(s\tau_1 + 1)(s\tau_2 + 1)$.

To determine the intersection of the GH locus with the real axis, set the imaginary part of $GH(i\omega) = u + iv$ equal to zero. Then, from (18),

$$v = \frac{-K\omega^{-1}(1 - \omega^2\tau_1\tau_2)}{1 + \omega^2(\tau_1^2 + \tau_2^2) + \omega^4\tau_1^2\tau_2^2} = 0$$

Thus $v = 0$ implies that $1 - \omega^2\tau_1\tau_2 = 0$ or $\omega = 1/\sqrt{\tau_1\tau_2}$; then

$$u = \left.\frac{-K(\tau_1 + \tau_2)}{1 + \omega^2(\tau_1^2 + \tau_2^2) + \omega^4\tau_1^2\tau_2^2}\right|_{\omega^2 = 1/\tau_1\tau_2} = \frac{-K\tau_1\tau_2}{\tau_1 + \tau_2} \tag{19}$$

The system is stable when the point of intersection lies to the right of the -1 point or $u > -1$, or

$$K < \frac{\tau_1 + \tau_2}{\tau_1\tau_2}$$

If the reverse inequality holds, the -1 point is encircled twice as shown in Fig. 10.5. Since the open-loop transfer function has no poles in the right half-plane, the system is unstable and there are two roots in the right-hand s plane.

The Nyquist contour can be used to define *relative stability* for a stable system. It is assumed that the open-loop transfer function has no poles or zeroes in the right-hand plane. The proximity of the $R(i\omega)$ contour to the $(-1, 0)$ point can be used as a measure of the leeway that a stable system has before becoming unstable. There are two commonly used definitions of relative stability based on the closeness of angle and distance that use the notation in Fig. 10.6.

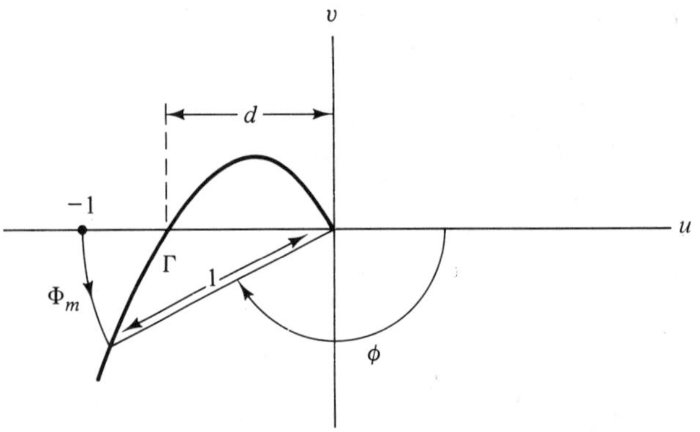

Figure 10.6 Measures of relative stability from the Nyquist locus.

Definition 2. Two measures of relative stability based on the Nyquist diagram are as follows:

1. $$G_m = \text{gain margin} = \min |R(i\omega)|^{-1} = d^{-1}$$

   when the phase angle of $R(i\omega)$ is 180°, that is, $v = 0$. In terms of logarithmic (decibel) measure,

   $$\text{db gain margin} = 20 \log\left(\frac{1}{d}\right) = -20 \log d$$

2. Φ_m = phase margin = the angle formed by the unit vector $|GH(i\omega)| = 1$ and the negative real axis as shown in Fig. 10.6.

Note that a system whose open-loop transfer function has no poles in the right-hand plane is stable if, in decibel measure, $G_m > 0$ and $\Phi_m > 0$.

Example 14

Consider the system described in Example 13. From (19), it follows that Γ passes through $-1 + i0$ when $K = (\tau_1 + \tau_2)/\tau_1\tau_2$. As K is decreased below this marginal value, the stability is increased. This is measured by the gain margin,

$$G_m = |GH(i\omega)|_{v=0}^{-1} = \frac{\tau_1 + \tau_2}{K\tau_1\tau_2}$$

For example, when $\tau_1 = \tau_2 = 1$ and $K = 0.5$, the gain margin is 4 or, in logarithmic measure, 20 log 4 = 12 db. The gain margin indicates that the system gain may be increased by a factor of 4 before the stability margin is reached.

The phase margin is more difficult to calculate algebraically in this example. It can be found more easily geometrically by using the Nyquist diagram.

The gain and phase margins may readily be calculated utilizing a computer program [4, 17].

The Nyquist stability criterion has been discussed for systems whose transfer function is rational. There are many control systems that have a time delay that affects the stability of the system. A time delay occurs in systems that transport material that requires a finite time to pass from an input or control point to an output or a measured point. The effect of the time delay on the relative stability of the control system can be determined from the Nyquist criterion. A pure time delay is represented by the transfer function $G_d(s) = e^{-sT}$, where T is the time delay. The Nyquist criterion remains valid, since the factor e^{-sT} does not introduce any additional poles or zeroes within the contour. This factor adds a phase shift to the frequency response without altering its magnitude.

EXERCISES

1. For the Nyquist diagrams of the systems sketched below, find N_{-1}, P, and Z, and state whether the system is stable.

(a)

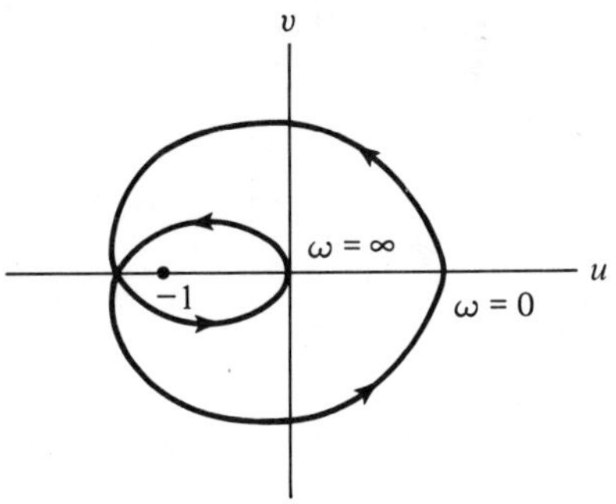

$$R(s) = \frac{K_1}{(1 + \tau_1 s)(1 + \tau_2 s)(1 + \tau_3 s)}$$

(b)

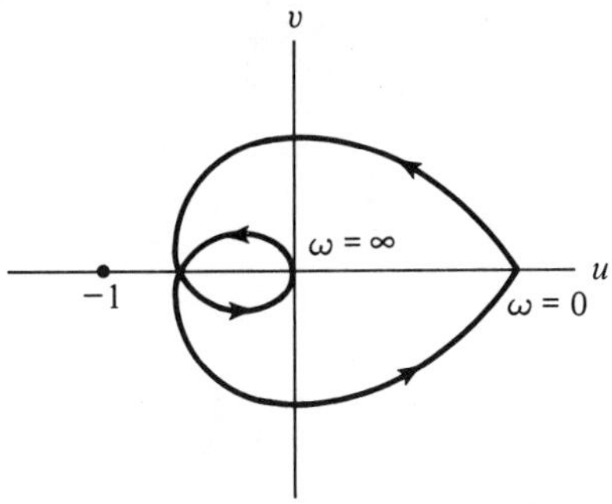

$$R(s) = \frac{K_2}{(1 + \tau_1 s)(1 + \tau_2 s)(1 + \tau_3 s)}$$

(c)

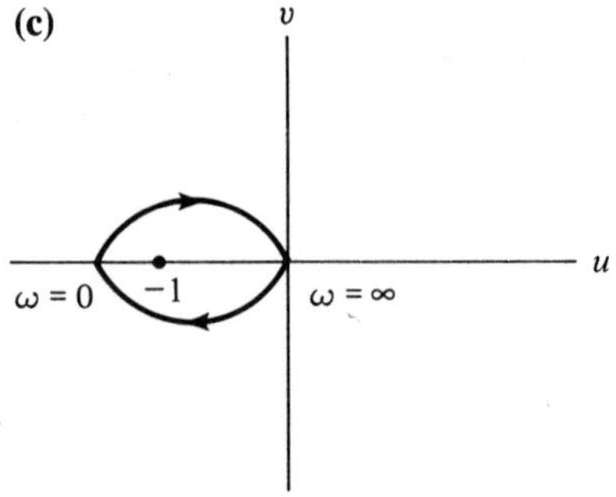

$$R(s) = \frac{K_1}{(-1 + \tau s)}$$

(d)

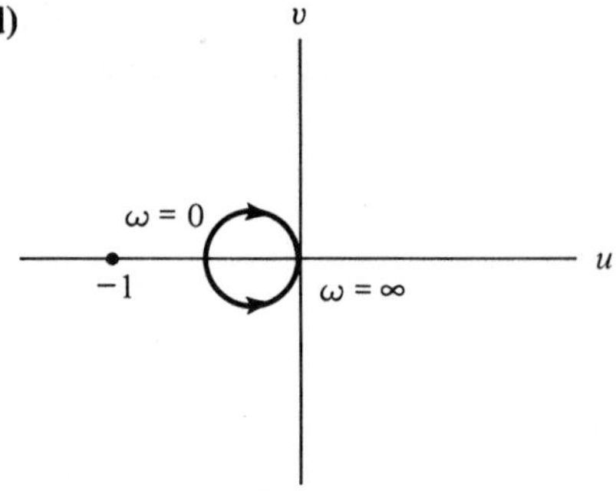

$$R(s) = \frac{K_2}{(-1 + \tau s)}$$

2. Sketch the Nyquist diagram for the following open-loop transfer functions. Find N_{-1}, P, and Z, and determine whether the system is stable.

(a) $\dfrac{1}{(1 + 0.5s)(1 + 2s)}$ **(b)** $\dfrac{1 + 0.5s}{s^2}$

(c) $\dfrac{s + 3}{s^2 + 4s + 16}$ **(d)** $\dfrac{30s(s + 8)}{s(s + 2)(s + 4)}$

3. Determine if the system illustrated below is stable.

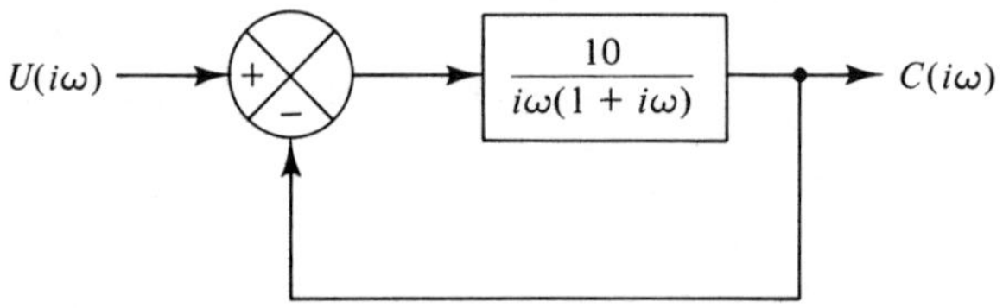

4. Use the Nyquist criterion, if possible, to determine the stability of the following system for the indicated parameter values ($K > 0$).

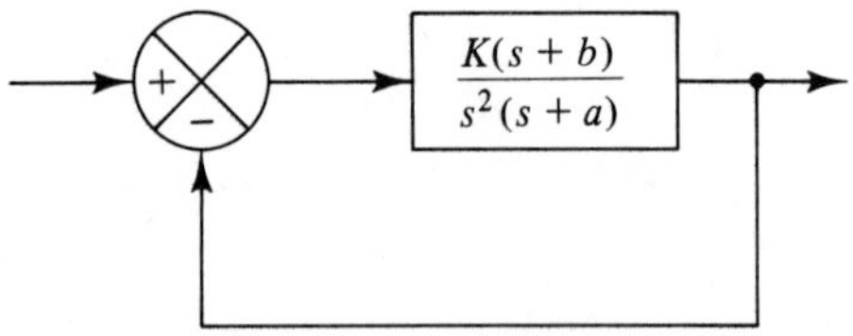

(a) $a = 2$, $b = 1$ **(b)** $a = 1$, $b = 2$
(c) $a = b$ (*Hint:* The criterion fails. Why?)

5. Determine the stability of the system having the open-loop transfer function

$$\frac{K}{s(s + 1)(s + 2)}$$

when (a) $K = 1$ and (b) $K = 2$.

6. By means of electrical pulses, a paralyzed limb can be made to contract, resulting in functional movements of the extremity [3]. The block diagram of one such process is shown below. The electronic controller $G(s)$ feeds electrical signals to the contracting muscles of the arm (agonists), which stretch the opposing muscles (antagonists). Here ϕ is the angle between the forearm and the upper arm. Typical values of the parameters are

$$J = 1 \qquad C = 20 \qquad \tau_0 = 0.1 \qquad K = 1$$

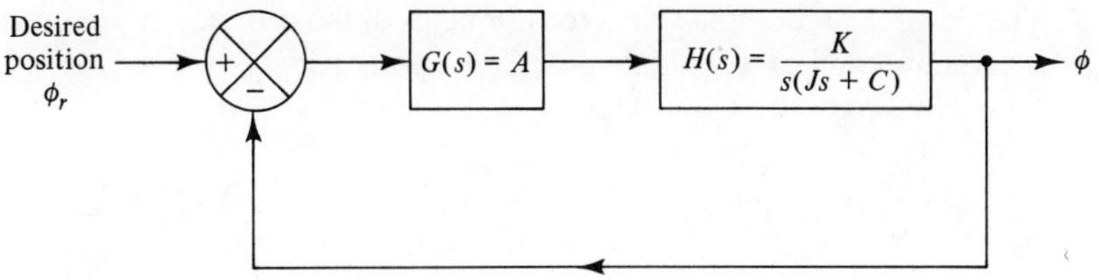

(a) Utilizing the Nyquist diagram, determine the phase margin when the electronic gain $A = 2$.
(b) Repeat (a) when $A = 100$.

10.5 THE BODE PLOT

The polar plot of the frequency response of a system may be difficult to sketch even if one has previously sketched another system whose transfer function has some of the same factors. The root of the difficulty is that there is no simple graphical interpretation of the amplitudes of the products or quotients of complex numbers in terms of the amplitudes of the individual numbers. This difficulty can be overcome by taking logarithms. It will be seen that the plot of the logarithm of the amplitude versus the frequency and the plot of the phase angle versus the frequency greatly simplify the graphical portrayal of the frequency response, since the resulting graph is the algebraic sum of the graphs of the individual factors. These logarithmic plots are called *Bode plots,* in honor of H. W. Bode, who used them extensively in his study of feedback amplifiers [1].

By convention, logarithms to the base 10 are used, and the logarithm of the magnitude is expressed in units of decibels, so that the

$$\text{logarithmic gain} = 20 \log|R(i\omega)|$$

Moreover, a linear scale of frequencies is not the most convenient, and a logarithmic scale of frequencies is also used.

Definition 3. If the frequency $\omega_2 = 10\omega_1$, then the distance between these two frequencies $(\omega_2 - \omega_1)$ is called a *decade* (dec). If $\omega_2 = 2\omega_1$, then the distance between these two frequencies is called an *octave*.

In the Bode diagram the amplitude is expressed in decibels and is plotted versus the frequency on semilogarithmic paper (or versus logarithm of frequency on ordinary paper). The phase shift $\sigma(i\omega)$ is plotted in degrees versus the frequency on semi-logarithmic paper.

The advantage of the logarithmic plot can readily be seen by considering a general open-loop transfer function such as

$$R(i\omega) = \frac{K \prod_{1}^{m} (1 + i\omega\tau_k)}{(i\omega)^n \prod_{1}^{p} (1 + i\omega\sigma_k) \prod_{1}^{q}[1 + b_k i\omega + a_k(i\omega)^2]}$$

The transfer function has m zeroes, n poles at the origin, p poles on the real axis, and q pairs of complex conjugate poles. The polar plot of such a function is a formidable task. However, the logarithmic magnitude of $R(i\omega)$ is

$$\begin{aligned}20 \log|R(i\omega)| &= 20 \log K + 20 \sum_{1}^{m} \log|1 + i\omega\tau_k| - 20 \log|(i\omega)^n| \\ &\quad - 20 \sum_{1}^{p} \log|1 + i\omega\sigma_k| - 20 \sum_{1}^{q} \log|1 + b_k i\omega + a_k(i\omega)^2|\end{aligned}$$

and the Bode diagram can be obtained by adding the plot due to each individual factor. Moreover, the phase plot

$$\phi(\omega) = \sum_{1}^{m} \tan^{-1} \omega\tau_k - n(90°) - \sum_{1}^{p} \tan^{-1} \omega\sigma_k - \sum_{1}^{q} \tan^{-1}\left(b_k\omega/(1 - a_k\omega^2)\right)$$

is the summation of the phase angles due to each individual factor of the transfer function.

Thus the four different kinds of factors that may occur in a rational transfer function are

1. K, a constant gain
2. Poles or zeroes at the origin $(i\omega)$
3. Poles or zeroes on the real axis $(i\omega\tau + 1)$
4. Complex conjugate poles (or zeroes) $[1 + bi\omega + a(i\omega)^2]$

In addition, the transfer function of a pure time delay will be considered:

5. $R(i\omega) = e^{-i\omega T}$

The logarithmic magnitude and phase angle plots will be determined for each of these five factors. These plots will then be utilized to obtain the Bode diagram of a general transfer function that can be written as an algebraic sum of factors corresponding to the types 1–5. Thé method is to obtain the curve for each factor and then add these curves graphically to obtain the Bode diagram for the given complicated transfer function. Moreover, this procedure can be simplified by using the asymptotic approximations to the curves corresponding to 1–5 and obtaining the actual curves only at specific important frequencies.

In accordance with the above discussion, the curves corresponding to 1–5 will be drawn.

1. *Constant gain K*. The logarithmic gain is

$$20 \log K = \text{a constant in db}$$

The corresponding angle is 0° or 180° depending on whether the constant is positive or negative, respectively. The Bode diagram of the phase and curves are horizontal lines.

2. *Poles or zeroes at the origin*. A pole of multiplicity n at the origin gives rise to the term $(1/i\omega)^n$. The logarithmic magnitude is

$$20 \log \left| \frac{1}{i\omega} \right|^n = -20n \log \omega \tag{20}$$

and the phase angle is

$$\phi(\omega) = -n\left(\frac{\pi}{2}\right) \tag{21}$$

Note that when the gain in db, on a linear scale, is plotted against ω, on a logarithmic scale, a straight line is obtained. The line passes through the point $\omega = 1$ and gain $= 0$ and has the slope

$$\frac{-20n \log \omega + 20n \log 10\omega}{\omega - 10\omega} = \frac{20n \log 10}{-\text{dec}} = -20n \text{ db/dec}$$

Equation (21) shows that the Bode plot of the phase angle is a straight line.

A zero of order n at the origin is treated in a similar manner. Equations similar to (20) and (21) are obtained except that they have a positive sign. The Bode diagrams for poles and zeroes are shown in Fig. 10.7.

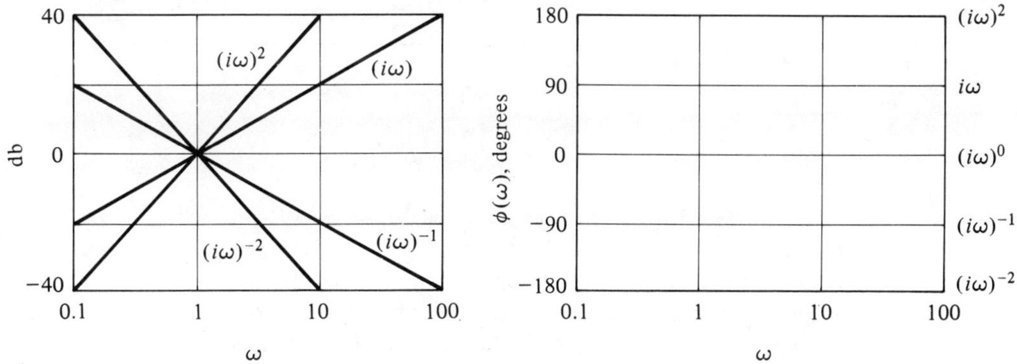

Figure 10.7 Bode diagram of the magnitude and phase of $(i\omega)^{\pm n}$.

3. *Poles or zeroes on the real axis*. The logarithmic gain of a pole factor $(1 + i\omega\tau)^{-1}$ is

$$20 \log \left| \frac{1}{1 + i\omega\tau} \right| = -20 \log |1 + i\omega\tau| = -20 \log(1 + \omega^2\tau^2)^{1/2}$$

If $\omega << 1/\tau$, then $1 + \omega^2\tau^2 \approx 1$ and the asymptotic curve is $20 \log 1 = 0$ db. The asymptotic curve for $\omega >> 1/\tau$ is $-20 \log \omega\tau$, which has a slope of -20 db/dec. The two asymptotic curves intersect when

$$20 \log 1 = 0 \text{ db} = -20 \log \omega\tau$$

or when $\omega = 1/\tau$, the *break frequency*. The actual gain when $\omega = 1/\tau$ is -3 db for this factor. The Bode diagram for the gain and its approximation is shown in Fig. 10.8(a).

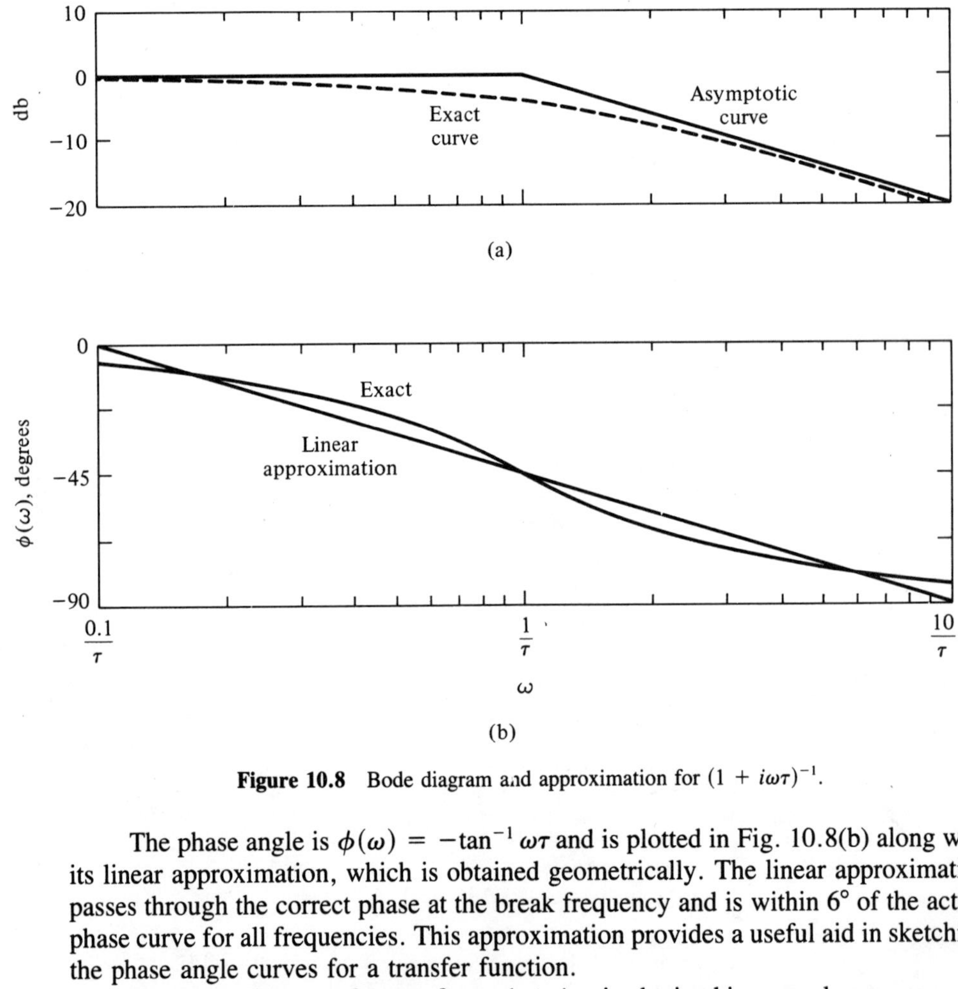

Figure 10.8 Bode diagram and approximation for $(1 + i\omega\tau)^{-1}$.

The phase angle is $\phi(\omega) = -\tan^{-1} \omega\tau$ and is plotted in Fig. 10.8(b) along with its linear approximation, which is obtained geometrically. The linear approximation passes through the correct phase at the break frequency and is within 6° of the actual phase curve for all frequencies. This approximation provides a useful aid in sketching the phase angle curves for a transfer function.

The Bode diagram of a zero factor $1 + i\omega\tau$ is obtained in an analogous manner. However, the slope is positive at 20 db/dec and the phase angle $\phi(\omega) = \tan^{-1} \omega\tau$.

4. *Complex conjugate poles or zeroes.* The quadratic factor for a pair of complex conjugate poles can be written in the normalized form as

$$G(u) = (1 + i2\zeta u - u^2)^{-1}$$

where $u = a^{1/2}\omega$ and the damping ratio $\zeta = b/2a^{1/2}$. The logarithmic magnitude is

$$20 \log|G(u)| = -10 \log[(1 - u^2)^2 + 4\zeta^2u^2]$$

and the phase angle is

$$\phi(u) = -\tan^{-1}\left(\frac{2\zeta u}{1 - u^2}\right)$$

If $u << 1$, then the magnitude is $-10 \log 1 = 0$ db and the phase angle approaches 0°. If $u >> 1$, then the gain approaches $-10 \log u^4 = -40 \log u$, which results in a curve with a slope of -40 db/dec. The phase angle, when $u >> 1$, approaches $-180°$. The magnitude asymptotes meet at the 0-db line when $u = 1$. However, the difference between the actual magnitude curve and the asymptotic approximation depends on the damping ratio and must be taken into account when $\zeta < 0.707$. The Bode diagram of a quadratic factor is shown in Fig. 10.9.

5. *Time delay factor*. A time delay factor occurs in systems that are characterized by the movement of mass that requires a finite time to pass from one point to another. The transfer function of a pure time delay factor is given by $G(i\omega) = e^{-i\omega T}$. This results in a phase shift $\phi(\omega) = -\omega T$ radians. Since $|G(i\omega)| = 1$, $\log|G(i\omega)| = 0$. Hence this factor does not affect the magnitude curve. Figure 10.10 illustrates the Bode diagram of a pure time delay factor.

The Bode diagram of a transfer function that contains several zeroes and poles is obtained by adding the plot due to each individual pole and zero. The procedure is illustrated in the next example.

Example 15

Consider the transfer function

$$G(i\omega) = \frac{5(1 + 0.1i\omega)}{i\omega(1 + 0.5i\omega)(1 + 0.6i(\omega/50) + (i\omega/50)^2)}$$

The factors are

(a) A constant gain $K = 5$
(b) A pole at the origin
(c) A pole at $\omega = 2i$
(d) A zero at $\omega = 10i$
(e) A pair of complex poles

First, the magnitude characteristic of each individual factor (a)–(e) will be plotted as shown in Fig. 10.11.

(a) The constant gain is $20 \log 5 = 14$ db.
(b) The magnitude of the pole at the origin extends from zero frequency to infinite frequencies and has a slope of -20 db/dec, intersecting the 0-db line at $\omega = 1$.
(c) The asymptotic approximation of the magnitude of the pole at $\omega = 2i$ has a slope of 20 db/dec beyond the break frequency at $\omega = 2$. The asymptotic magnitude below the break frequency is 0 db.
(d) The asymptotic magnitude for the zero at $\omega = 10i$ has a slope of 20 db/dec beyond the break frequency at $\omega = 10$.

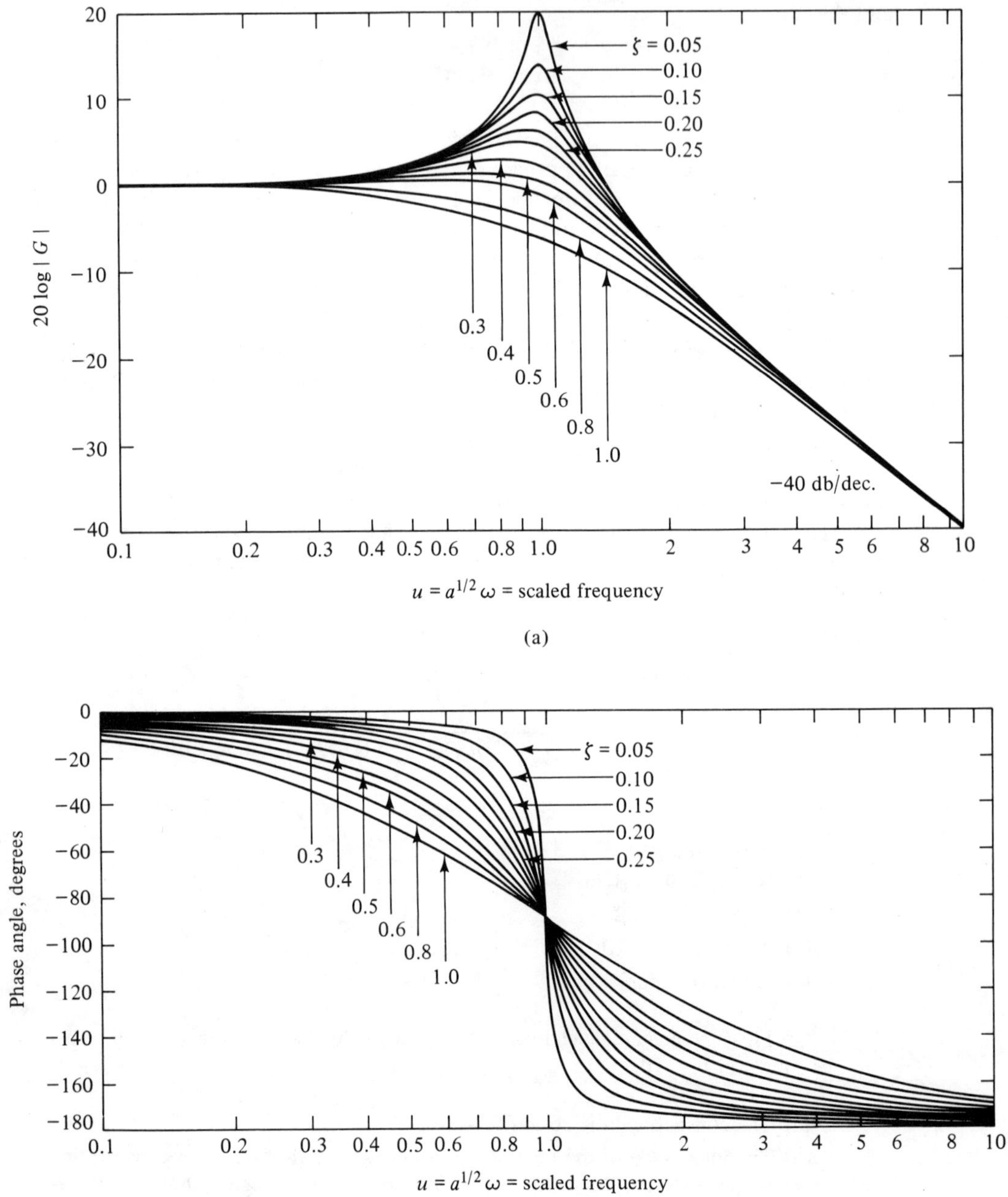

(b)

Figure 10.9 Bode diagram of the quadratic factor $G(u)$.

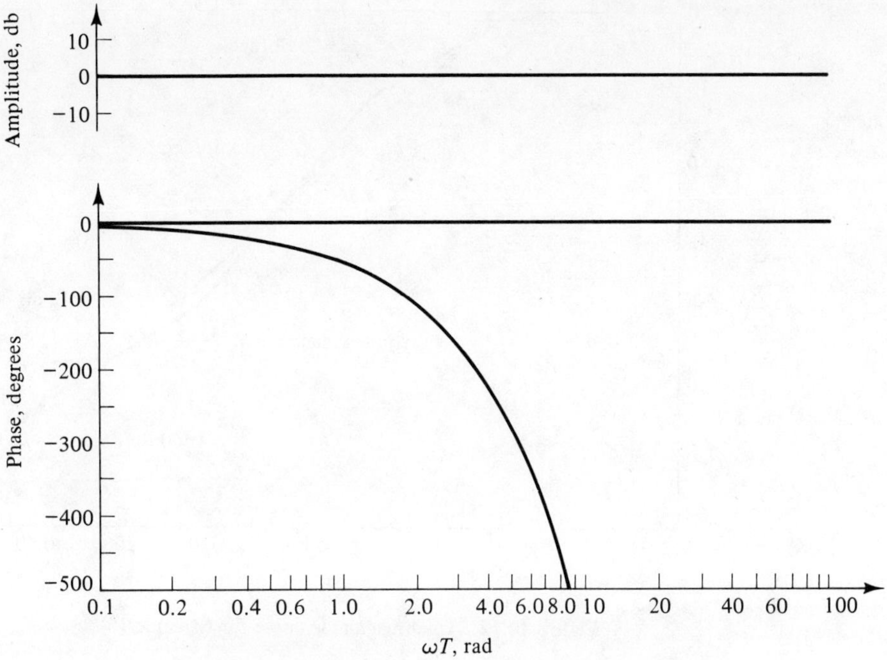

Figure 10.10 Bode diagram of a pure time delay factor.

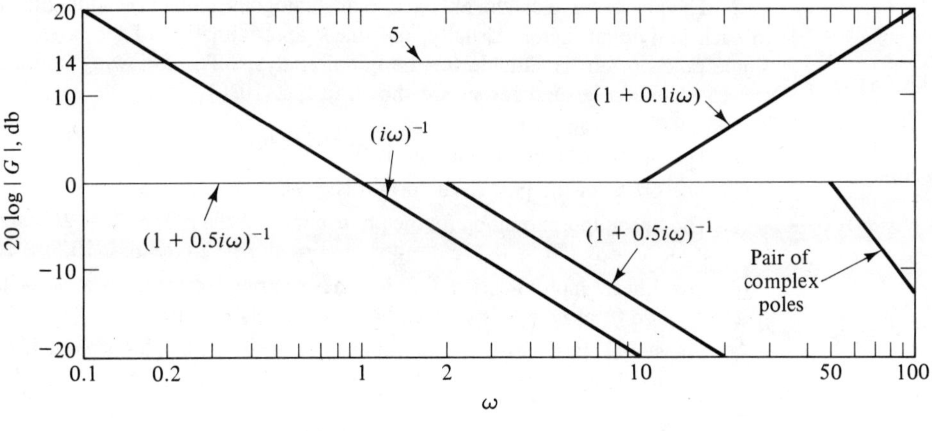

Figure 10.11 Magnitude asymptotes of poles and zeroes of Example 15.

(e) The asymptotic approximation for the pair of complex poles has a slope of -40 db/dec due to the quadratic form. The break frequency is $u = 1$ or $\omega = 50$. This approximation must be corrected to the actual magnitude, since the damping ratio $\zeta = b/2a^{1/2} = 0.3$ and the magnitude differs noticeably from the approximation, as shown in Fig. 10.12.

The total asymptotic magnitude can be plotted by adding the asymptotes due to

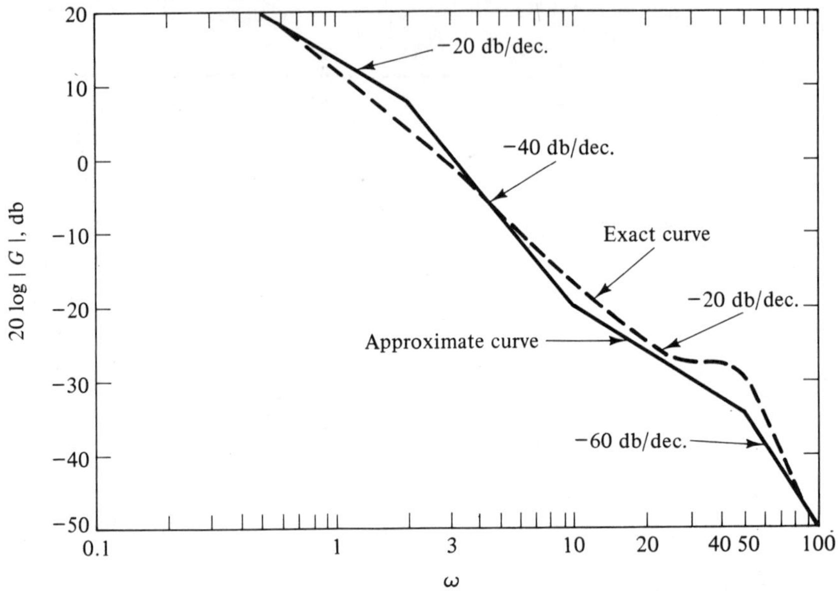

Figure 10.12 Total magnitude curve for Example 15.

each factor, as shown in Fig. 10.12 by the solid line. The exact magnitude curve for $G(i\omega)$ is indicated by the dashed line in Fig. 10.12.

The phase curve can be obtained in an analogous manner by adding the phase due to each individual factor. Usually, the linear approximation of the phase curve for a single pole or zero is suitable for the initial analysis. The following individual phase curves for the poles and zeroes are shown in Fig. 10.13.

(a) The phase of the constant gain is 0°.
(b) The phase of the pole at the origin is −9°.
(c) The linear approximation for the phase curve for the pole at $\omega = 2i$ passes through 0° at $\omega = 0.1/\tau = 0.2$ and −45° at $\omega = 1/\tau = 2$ [see Fig. 10.8(b)].
(d) The linear approximation for the phase curve for the zero at $\omega = 10i$ passes through 0° at $\omega = 1/\tau = 1$ and 45° at $\omega = 1/\tau = 10$.
(e) The actual phase characteristic for the pair of complex poles is plotted. An approximation is not used.

The total phase curve, $\phi(\omega)$, is obtained by adding the phase due to each factor as shown in Fig. 10.13.

The approximate curves for the magnitude and phase shift of a transfer function are helpful in determining important frequency ranges. Then, within the small important frequency ranges, the exact magnitude and phase shift can be evaluated by using the exact equations along with the approximations. For example, it will be seen that it is important to find the frequency for which $\phi(\omega) = -180$. The approximate phase curve in Fig. 10.13 indicates that a phase shift of −180° occurs at $\omega = 46$.

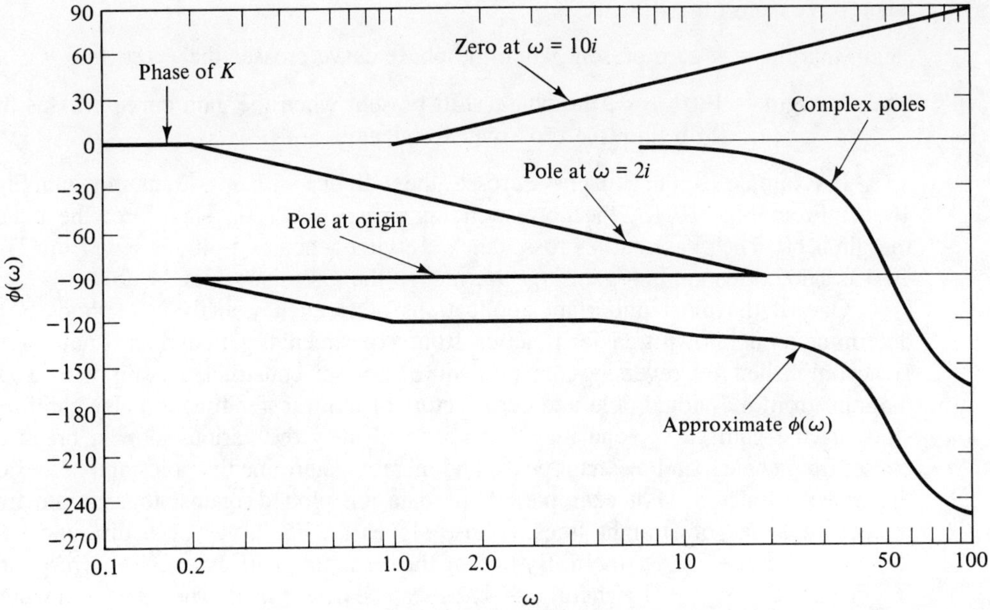

Figure 10.13 Phase curve for Example 15.

The exact equation for the phase shift in Example 15 is

$$\phi(\omega) = -90° - \tan^{-1} 0.5\omega + \tan^{-1} 0.1\omega - \tan^{-1}\frac{0.6\omega/50}{1-(\omega/50)^2} \tag{22}$$

From (22),

$$\phi(46) = -90° - \tan^{-1} 23 + \tan^{-1} 4.6 - \tan^{-1}3.59 = -174°$$

and the approximate curve has an error of 6° at $\omega = 46$. A more accurate estimate of the frequency ω such that $\phi(\omega) = -180°$ can be determined by trial and error by substituting frequencies close to 46 in (22) as follows:

$$\phi(47) = -177.9° \qquad \phi(48) = -181.6°$$

Since $\phi(48)$ is too large and $\phi(47)$ is too small, $\omega = 47.6$ is tried, giving $\phi(47.6) = -180.2°$. The above procedure can be repeated until the desired accuracy is obtained.

In the discussion of the Nyquist diagram, the degree of stability was defined in terms of phase and gain margins. This can easily be related to the Bode diagram by making use of the following two facts:

1. The unit circle of the Nyquist diagram transforms into the 0-db line of the amplitude for all frequencies.
2. The negative real axis of the Nyquist diagram transforms into the negative 180° phase line for all frequencies.

Therefore, from Fig. 10.6,

gain margin = $-$gain present when the phase curve crosses the -180 line

phase margin = $180° + \phi$, the phase shift present when the gain curve crosses the 0-db line (the crossover frequency)

In Example 15, the gain curve crosses the 0-db line with $\omega = 3$, as shown in Fig. 10.12. From Fig. 10.13, the corresponding phase is -120. Therefore, the phase margin is 60. The phase curve crosses the -180 line when $\omega = 46$, as seen from Fig. 10.13. The corresponding gain is -34, and so the gain margin is 34 db.

One of the most important applications of frequency analysis methods is to determine an unknown transfer function from experimental gain and shift data. This is accomplished by reversing the procedure used to construct a composite Bode diagram from individual pole and zero factors of the transfer function. By carefully studying the gain curve, one may attempt to identify the various slopes, break or corner frequencies, and resonant peaks and thereby determine the poles and zeroes of the transfer function. For example, if the data are plotted against logarithmic frequency, a series of straight lines, whose slopes are multiples of 6 db/octave or 20 db/dec, can be fitted to them. By finding the break frequencies, ω_i, the corresponding linear factors $\tau_i s + 1$, where $\tau_i = 1/\omega_i$, can be determined. The powers to which the factors are raised can be determined from the multiples of 6 db/octave. A constant gain can be found by noting the height of the first horizontal portion of the phase curve above the 0-db line. This procedure is illustrated in the following example.

Example 16

The data in the table below are the experimental open-loop amplitude characteristics of a certain system.

ω	377	314	251	219	157	126	100	63	44	16	8	1.2	0
db gain	−7.85	−4.3	−0.2	0.75	5.2	8.0	10.5	15.0	16.9	20.4	21.6	24.0	24.1

The data are plotted as shown in Fig. 10.14. By fitting straight lines to the data, as described above and illustrated in Fig. 10.14, the following transfer function is obtained:

$$R = \frac{15.9}{(0.049s + 1)(0.0067s + 1)}$$

Determining a model from experimental data is known as the *system identification problem*. The above procedure can be useful in practice. However, if the system is of high order or the data are noisy, it often fails. Further information on the system identification problem can be found in [7].

System identification from frequency response data is complicated by the fact that the gain plot does not uniquely determine the system. An example appears in Exercise 9. Thus the amplitude and phase curve should be used simultaneously, and the best match for both should be determined. This can often be simplified if linear asymptotes are used for both.

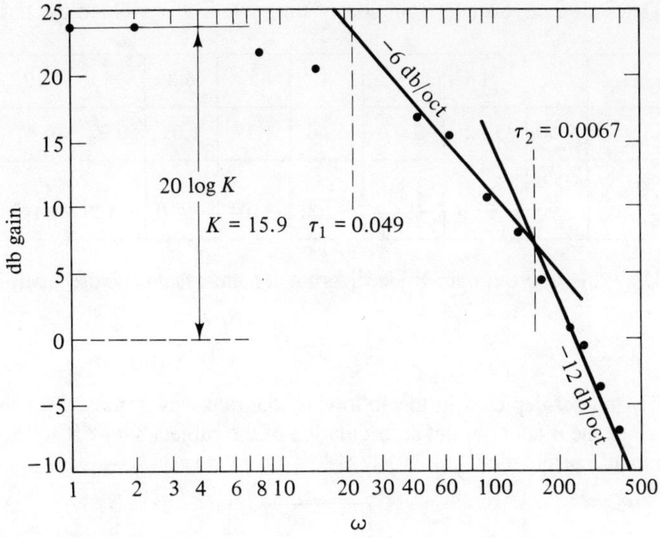

Figure 10.14 Frequency response data for Example 16.

EXERCISES

1. Draw the Bode diagrams for the transfer functions in Exercise 2 of Section 10.3.
2. A simplified model of the human respiratory control system is shown below [10]. This model provides an insight into the physiology of respiratory disorders. This system controls the effective ventilation at the lungs, $V(s)$. The percentage of CO_2, P, in the circulated blood at the chemoreceptor (measuring element) is proportional to the effective ventilation $V(s)$. The effective gain $K = 0.1$ for a normal system, and the delay time constant of the circulatory system $\tau = 10$ sec.

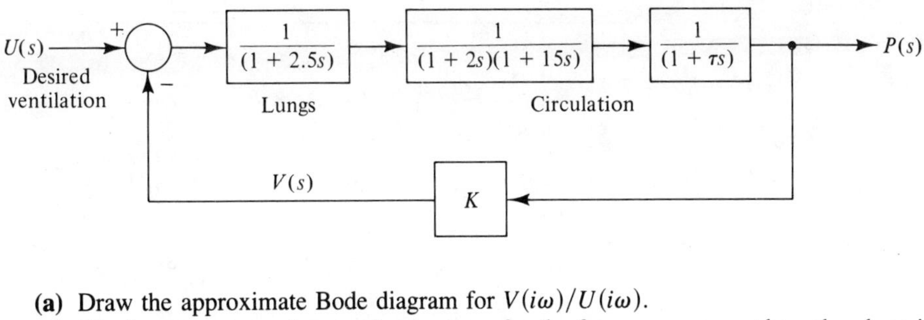

   (a) Draw the approximate Bode diagram for $V(i\omega)/U(i\omega)$.
   (b) Obtain an accurate phase and gain curve for the frequency range where the phase is -180.
3. A system with unity negative feedback has the transfer function

$$G(s) = \frac{2(s+2)}{s^2 - 1}$$

   Draw the Bode diagram.

4. Determine the transfer function of the system that yielded the data:

ω, rad/sec	1	1	2	4	5	6.3	8	10	12.5	20	31
$G(i\omega)$	0.50	5.02	2.07	1.36	1.17	1.03	0.97	0.97	0.74	0.13	0.026
phase, deg	−90	−92.4	−96.2	−100	−104	−110	−120	−143	−169	−245	−258

5. Draw the approximate Bode diagram for the system whose open-loop transfer is

$$GH = \frac{K(s+5)}{s^2(s+10)}$$

6. The model depicted in the following diagram was proposed for the "knee-jerk" test. Does the model predict an oscillation of the subject's leg? If so, at what frequency? Is the model reasonable?

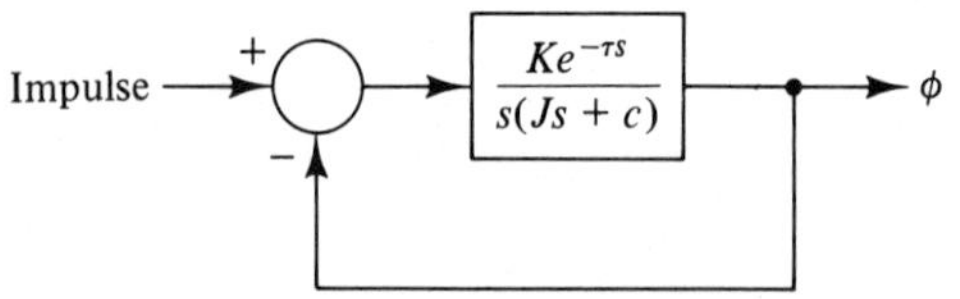

$\phi(t)$ = angle of the leg
$c/J = 2\pi/\text{sec}$
$\tau = 1/8$ sec
$K/c = 2\sqrt{2}\pi$

7. The three mutually perpendicular semicircular canals in the human ear sense rotational motion of the skull and allow for stabilization of the eye. Each canal contains the cupula, a hair-sensing transducer, which has an angular deflection related to the angular velocity of the skull motion. A human subject was instrumented and his skull was forced with a sinusoidal motion, producing the cupula frequency response data tabulated below. Determine the transfer function of the angular deflection of the cupula with respect to the skull angular frequency.

ω	$\lvert G(i\omega)\rvert$, db	$\underline{/G(i\omega)}$, deg
0.01	−16.0	81
0.05	−2.5	62
0.10	1.0	45
0.20	2.5	25
0.50	4.0	10
1.0	4.0	3
5.0	4.0	−15
10.0	3.5	−26
20.0	1.0	−45
50.0	−4.5	−68
100.0	−10.0	−78

8. The loop-transfer function for a liquid-level control system is

$$\frac{31.5}{(s+1)(30s+1)(s^2/9+s/3+1)}e^{-s}$$

(a) Draw the Bode diagram for the denominator factors alone and the Bode diagram with the additional phase lag due to the time delay.

(b) Verify that the phase margin without the pure time delay is 40° and with the time delay is −3° and the system is unstable.

(c) Show that in order to obtain a phase margin of 30°, the system gain $K = 31.5$ must be decreased by a factor of 5 db to $K/1.78 = 17.7$.

9. Show that the systems with transfer functions

$$G(s) = \frac{s+3}{(s+1)(s+10)} \qquad H(s) = -\frac{(s-3)}{(s+1)(s+10)}$$

have the same gain curves but different phase curves.

10. A simplified model of the control system for regulating the pupillary aperture in the human eye is shown in the diagram below [19]. The gain $K = 4$ represents the pupillary gain. The pupil time constant = 0.5 sec. The time delay is equal to 0.2 sec.

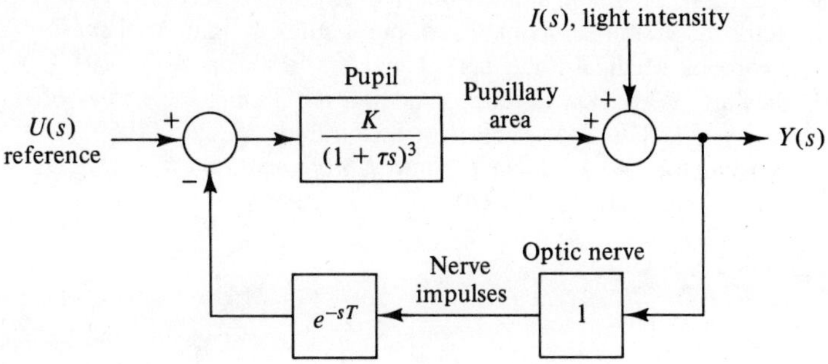

(a) Assuming that the time delay is negligible, draw the Bode diagram for the system. Determine the phase margin for the system.

(b) Include the effect of the time delay by adding the phase shift due to the delay. Determine the phase margin with the time delay included.

REFERENCES

1. Bode, H. W. *Network Analysis and Feedback Amplifer Design*. Van Nostrand, Princeton, NJ, 1945.
2. Cheng, D. K. *Analysis of Linear Systems*. Addison-Wesley, Reading, MA, 1959.
3. Crochetiere, W., Vovovnik, L., and Resnick, J. B. The design of control systems for electronic activation of human limb movements. *Proc. 1967 Joint Automatic Control Conference*, 1967, pp. 51–57.
4. Dorf, R. C. *Introduction to Computers and Computer Science*. Boyd and Fraser, San Francisco, 1972.
5. Durell, C. V., and Robson, A. *Advanced Algebra*, Vols. I–III. Bell and Sons, London, 1952.
6. Evans, W. R. Graphical analysis of control systems. *Trans. AIEE 67*, 547–551, 1948.
7. Fortmann, T. E., and Hitz, K. L. *An Introduction to Linear Control Systems*. Marcel Dekker, New York, 1977.
8. Friedman, B. Note on approximating complex zeros of a polynomial. *Comm. on Pure and Applied Math II*, 195, 1949.
9. Hurwitz, A. On conditions under which an equation has only roots with negative real parts. *Math. Annalen 46*, 273–284, 1895. In *Selected Papers on Mathematical Trends in Control Theory*. Dover, New York, 1964, pp. 70–82.

10. Korta, L. B., Horgan, J. D., and Lange, R. L. Stability analysis of the human respiratory system. *Prac. Natl. Electronics Confer. XXI,* 201–206, 1965.
11. MacFarlane, A. G. J. *Dynamical System Models*. Harrap, London, 1970.
12. Nyquist, H. Regeneration theory. *Bell System Tech. J. 2,* 126–147, 1932.
13. Phillips, E. G. *Functions of a Complex Variable with Applications*. Oliver & Boyd, Edinburgh, 1949.
14. Porter, B. *Stability Criteria for Linear Dynamical Systems*. Oliver & Boyd, Edinburgh, 1967.
15. Ralston, A., and Rabinowitz, P. *A First Course in Numerical Analysis*. McGraw-Hill, New York, 1978.
16. Routh, E. J. *Dynamics of a System of Rigid Bodies*. Macmillan, New York, 1892.
17. Shinners, S. M. *Modern Control System Theory and Application*. Addison-Wesley, Reading, MA, 1972.
18. Smith, W. A. *Elementary Numerical Analysis*. Harper & Row, New York, 1979.
19. Stark, L. Vision-servoanalysis of pupil reflex to light. *Medical Physics Yearbook,* Vol. III. Yearbook Medical Publishers, Chicago, 1960, pp. 702–719.
20. Stewart, W. G. On Lehmer's method for finding the zeroes of a polynomial. *Math. of Comp. 23,* 829–835, 1969. Correction ibid. *25,* 205, 1969.
21. Washington, A. J. *Basic Technical Mathematics with Calculus*. Cummings Publishing Co., Menlo Park, CA, 1970.

Appendix A

Method of Least Squares

Frequently, scientists want to represent some experimental data in an explicit quantified form. Even though it is thought that the observed values y are some function of another variable—for example, $y = mx + b$—all of the observed values will generally not satisfy the proposed relationship because of experimental error. In other words, a plot of the experimental data in the xy plane will not fall on the curve representing the given function. The question arises as to which curve fits the data "best."

The following criterion was proposed by Gauss for the best fit. Let the observed values of y at the point x_i be denoted by y_i^*. Then the error E between the value of y at the point x_i, denoted by $y(x_i)$, and the observed value y_i^* is defined to be

$$E = \sum_{i=1}^{n} [y(x_i) - y_i^*]^2 \tag{1}$$

where n is the number of observations. The reason for the square in the terms in (1) is to avoid the possibility that large negative and positive values of $y(x_i) - y_i^*$ cancel each other out in the sum $\sum_{i=1}^{n} [y(x_i) - y_i^*]$ and give the appearance of a small error. The quantities x_i and y_i^* are given numbers, so that E is a function of the parameters appearing in the function $y = y(x)$. For example, for the straight line $y(x_i) = mx_i + b$, (1) becomes

$$E = \sum_{i=1}^{n} (mx_i + b - y_i^*)^2 \tag{2}$$

Note that E is a function of the parameters m and b. The function E must be minimized with respect to its parameters. The necessary condition for this is that the partial derivatives of E with respect to its parameters are equal to zero. In (2), $E = E(m, b)$

and the required conditions are

$$\frac{\partial E}{\partial m} = 2(m \sum x_i^2 + b \sum x_i - \sum x_i y_i^*) = 0 \tag{3}$$

$$\frac{\partial E}{\partial b} = 2(m \sum x_i + b \sum 1 - \sum y_i^*) = 0 \tag{4}$$

where all the summations extend from $i = 1$ to $i = n$. In general, there will be as many equations as unknown parameters, and these equations can be solved to determine the parameters.

Equations (3) and (4) can be solved for the quantities m and b, yielding

$$m = \frac{n \sum x_i y_i^* - \sum x_i \sum y_i^*}{n \sum x_i^2 - (\sum x_i)^2} \tag{5}$$

$$b = \frac{\sum x_i^2 \sum y_i^* - \sum x_i y_i^* \sum x_i}{n \sum x_i^2 - (\sum x_i)^2} \tag{6}$$

$$= n^{-1}(\sum y_i^* - m \sum x_i)$$

where all the summations extend from $i = 1$ to $i = n$. It can be shown that these values of m and b minimize E.

The above procedure for fitting observations to an assumed functional form is called the *method of least squares*. There must be at least as many observations as parameters to apply this method.

Example 1

Find the best straight line $y = mx + b$ fitting the following data:

x_i	-1	0	1
y_i^*	1.1	3	4.9

Using the $n = 3$ observations: $\sum x_i = 0$, $\sum y_i^* = 9$, $\sum x_i^2 = 2$; $\sum x_i y_i^* = 3.8$. From (5), $m = 3(3.8)/3(2) = 1.9$; and from (6), $b = 3^{-1}[9 - 1.9(0)] = 3$.

The answer to how "good" is the fit and other statistical questions can be found in elementary statistics books. A detailed discussion of these topics appears in [1].

REFERENCES

1. Kendall, M. G., and Stuart, A. *The Advanced Theory of Statistics,* Vols. 2–3, 4th ed. Macmillan, New York, 1977.

Appendix B

The Dirac Delta Function

Many biological and physical problems deal with phenomena of an impulsive nature. For example, consider striking a baseball with a bat. The batter tries to exert as large a force as possible during the short time interval that the bat is in contact with the ball. If the ball approaches the batter with a velocity $v(t_0)$, it leaves the bat with a velocity $v(t_1)$, characteristic of the force exerted by the batter. To discuss these observations from the viewpoint of physics, consider Newton's second law written in the form

$$\frac{d[mv(t)]}{dt} = f(t) \tag{1}$$

where m is the mass of the ball, v is its velocity, and $f(t)$ is the total force acting on the ball. Integrating Eq. (1) between t_0 and t_1 gives

$$mv(t_1) - mv(t_0) = \int_{t_0}^{t_i} f(t)\, dt \tag{2}$$

The quantity mv is called the *momentum* of the ball. Equation (2) states that the change of momentum of the ball from t_0 to t_1 equals $\int_{t_0}^{t_1} f(t)\, dt$. Note that the physically important quantity is the integral of the force or the *impulse,* rather than the force itself.

A mathematical approximation to a large force acting over a small time interval so that its impulse is a constant I can be obtained as follows. Let $t_1 = t_0 + \epsilon$, where $\epsilon > 0$, and define the function

$$\delta_\epsilon(t - t_0) = \begin{cases} 0 & t < t_0 \\ \dfrac{1}{\epsilon} & t_0 < t < t_0 + \epsilon \\ 0 & t > t_0 + \epsilon \end{cases} \tag{3}$$

It is easy to verify that

$$\int_0^\infty \delta_\epsilon(t - t_0)\, dt = 1$$

Then, letting $f(t) = I\delta_\epsilon(t - t_0)$, it follows that

$$\int_{t_0}^{t_0+\epsilon} f(t)\, dt = \frac{I(\int_{t_0}^{t_0+\epsilon} dt)}{\epsilon} = I \tag{4}$$

Moveover, if ϵ is small, then $f(t)$ is large over the interval $t_0 < t < t_0 + \epsilon$. Next, let $g(t)$ be a continuous function; then

$$\int_0^\infty g(t)\, \delta_\epsilon(t - t_0)\, dt = \frac{1}{\epsilon}\int_{t_0}^{t_0+\epsilon} g(t)\, dt = g(\tau) \qquad t_0 < \tau < t_0 + \epsilon \tag{5}$$

where the last equality is a consequence of the mean value theorem for integrals.

Letting ϵ approach zero and denoting the resulting "function" by δ, reasonable conditions that δ must satisfy can be obtained from (4) and (5). Formally taking the limit under the integral sign gives

$$\int_0^\infty \delta(t - t_0) = 1 \qquad (t_0 > 0) \tag{6}$$

$$\int_0^\infty \delta(t - t_0)g(t)\, dt = g(t_0) \tag{7}$$

where the last equality is a consequence of (5) and the fact that $\lim_{\epsilon\to 0} \tau = t_0$. The function δ, known as the *Dirac delta function,* was proposed by the physicist P. A. M. Dirac for dealing wth impulsive functions. Dirac realized that δ was not an ordinary function, since $\delta(t - t_0) = 0$ for $t \neq t_0$ and ∞ for $t = t_0$. Nevertheless, he proposed formally operating with δ as if it really was an ordinary function using properties (6) and (7).

At that time mathematicians ridiculed this method but could not explain why Dirac and his followers always got the right answer. It was not until 1940 that Laurent Schwartz, a French mathematician, succeeded in rigorously justifying the use of the delta function. A brief description of Schwartz's method for placing the δ function on a firm mathematical foundation will be given.

The germ of Schwartz's method was the realization that a function could also be uniquely determined by what it did to other functions through integration. More precisely, consider the set C of functions ϕ, each of which is infinitely differentiable and which vanishes for $|t|$ sufficiently large. Let f be a continuous function. To each function a number $K_f[\phi]$ can be assigned as follows:

$$K_f(\phi) = \int_{-\infty}^{\infty} \phi(t)f(t)\, dt \tag{8}$$

It can be shown that if $K_f[\phi] = 0$ for all ϕ in C, then $f(t) = 0$ for all real t. This last result implies that if f and g are continuous functions and $K_f[\phi] = K_g[\phi]$ for all ϕ in C, then $f = g$. Thus the function K defined on C by (8) can be considered to contain all the continuous functions by identifying f and K_f. Since (8) is also well defined for a piecewise continuous function, the set of all functions K defined on ϕ by (8) is an extension of the set of all continuous functions, just as the complex numbers are an extension of the real numbers.

The function K is an operator acting on functions ϕ in C and associates a real number with each function ϕ. Such functions are called *functionals*. The functional K is *linear,* since

$$\begin{aligned} K_f[c_1\phi_1 + c_2\phi_2] &= \int_{-\infty}^{\infty} (c_1\phi_1 + c_2\phi_2)(t)f(t)\,dt \\ &= c_1 \int_{-\infty}^{\infty} \phi_1(t)f(t)\,dt + c_2 \int_{-\infty}^{\infty} \phi_2(t)f(t)\,dt \\ &= c_1 K_f[\phi_1] + c_2 K_f[\phi_2] \end{aligned}$$

The set of functionals K_f, where f is a piecewise continuous function defined on $-\infty < t < \infty$, can be enlarged by adding all linear functionals defined on C. In particular,

$$K[\phi] = \phi(t_0)$$

is a linear functional defined on C, since

$$K[c_1\phi_1 + c_2\phi_2] = c_1\phi_1(t_0) + c_2\phi_2(t_0) = c_1 K[\phi_1] + c_2 K[\phi_2]$$

To mimic (8), K can be written symbolically as

$$K[\phi] = \int_{-\infty}^{\infty} \phi(t)\delta(t - t_0)\,dt \tag{9}$$

The symbol δ can be regarded as a "generalized function." Note that the value of δ at any time t is not defined, and only the expression $\int_{-\infty}^{\infty} \phi(t)\,\delta(t - t_0)\,dt$ or $K[\phi]$ is meaningful, and the value $\phi(t_0)$ is always assigned to this expression. The enlarged set of linear functionals defined on C is called the set of *distributions* or *generalized functions*.

Although it is conceptually difficult to think of a function in terms of the linear functional (8) it induces, this approach has the advantage of formalizing the delta function. Moreover, a derivative can be assigned to every piecewise continuous function and more generally to every generalized function. This is accomplished by first considering a differentiable function f and defining

$$K'_f[\phi] = \int_{-\infty}^{\infty} \phi(t)f'(t)\,dt \tag{10}$$

in order that the generalized derivative will coincide with the ordinary derivative for a differentiable function f. Integrating the integral appearing in (10) by parts and using the fact that $\phi(t) = 0$ for $|t|$ sufficiently large gives

$$K'_f[\phi] = \int_{-\infty}^{\infty} -\phi'(t)f(t)\,dt = K_f[-\phi'] \tag{11}$$

Notice that the formula

$$K'[\phi] = K[-\phi'] \tag{12}$$

makes sense for any linear functional on C even if K is not induced by a differentiable function f. The linear functional K' is called the *derivative* of K. It is easy to verify that if K is induced by a differentiable function f, then K' is induced by f'. Hence the derivative of f is still f' when a differentiable function is considered a generalized function.

The following example illustrates the definition of the derivative of a distribution given by (12).

Example 1

(a) The Heaviside function or the unit step function u is defined by

$$u(t - t_0) = \begin{cases} 0 & t > t_0 \\ 1 & t > t_0 \end{cases}$$

The corresponding generalized function is

$$K_u[\phi] = \int_{-\infty}^{\infty} u(t - t_0)\phi(t)\, dt$$

According to (12),

$$K_u'[\phi] = \int_{-\infty}^{\infty} u[(t - t_0)(-\phi'(t))]\, dt = -\int_{t_0}^{\infty} d\phi(t) = \phi(t_0)$$

since $\phi(\infty) = 0$. It is evident that the derivative of the Heaviside function is the delta function $\delta(t - t_0)$.

(b) The derivative of the delta function is the linear functional

$$K_\delta'[\phi] = \int_{-\infty}^{\infty} -\phi'\,\delta(t - t_0)\, dt = -\phi'(t_0)$$

For further details on distribution or generalized functions see [1].

REFERENCES

1. Gelfand, I. M., and Shilov, G. I. *Generalized Functions,* Vols. 1–5. Academic Press, New York, 1964–1968.

INDEX

Note: Italicized page numbers indicate diagrams and tables.

A

Acute gouty arthritis, *43*
Adolph, E. F., 4
Aggregation of slime mold amoeba as distributed-parameter system, 30–32
Albumin synthesis, 132–134
  simplified model for, *133*
ALGOL (program), 247
Amplitude of motion, 111
Analog computers
  and nonlinear systems, 212
Analytical modeling, 9–11
  steps of, *9*
Anticipatory systems, 134
Automatic steering of ships, 3

B

Ballism, 118*n*
Bairstow's method, 251–252
BASIC (program), 212
Bayliss, L. E., 3
Bell Laboratories, 3
Bernard, Claude, 2
Bilinear transformation of *z* plane into *s* plane, *258*
Binary fission
  block diagram of, *205*
  described, 11
  flow graph of, *205*
  modeling process for, 11–14
  simplified flow graph of, *205*
Biological systems
  as open systems, 4
Biologists
  use of transfer function by, 191
Biology
  and control theory, 1
Black box, 5
Block diagrams, *5*, 196–201
  of input-output relations, 4–8
  purpose of for transfer functions, 196
  and signal flow graphs, 204–206
  for subsystems, 6
  of transformations, *196, 197, 198, 199*
Block model
  of schistosome life cycle, *83*
Blood glucose regulating system, 235–236
Bode, H. W., 3, 271
Bode diagrams, *273, 274, 276, 277*
  relationship to Nyquist diagrams, 279
Bode plot, the, 271–283
Borelli, Giovanni, 2
Branches
  in signal flow graphs, 202
Brobeck, J. R., 3
Broughton suspension bridge (England), 118
Brownian motion, 32

C

Cannon, W. B., 2, 3
Cauchy's argument principle, 263
Chaotic limit cycle, *86*
Characteristic equation, 72, 108
Characteristic polynomial, 74
Characteristic roots, 72, 108
Chlorinated hydrocarbons, 98–99
Circadian oscillators, 45
Circadian rhythms, 45
Closed-loop negative feedback system
  advantages of, 223
Closed-loop system
  disturbance in, *229*
  with measurement noise, *230*
Closed systems, 4, 40–50
  block diagram for, *41*
Coefficients
  of difference equations, 59
  of differential equations, 93
Cofactor
  of flow graph, 208
Compartment notion, 120
Complex translation
  of Laplace transform, 170–171
Conjugate roots, 110
Constant coefficients, 59
Continuous system, *29,* 33–34
Control engineering, 3
Control systems
  accurate, 227
  classification of, 29–40
  defined, 4
  offset in, 227
  transient response of, 229
Control theory
  beginning of, 1
  brief history of, 1–4
  defined, 4
  and engineering problems, 2–3
  highlights (table), *1*
  history (table), *2*
  military applications of, 3
  motivating factors in development, 1
  in USSR, 3
Convolution
  of Laplace transforms, 71
Convolution representation, 128–138
  origin of term, 130
  and population size as function of immigration, 132
  and rate of albumin synthesis, 132–134
  for time-invariant linear systems, 135–138
  and weighting function, 131–132
Convolution-summation
  of z-transforms, *153*
Convolution theorem, 71
Critically damped unit step response, 220
Crowell, J. W., 46
CSMP (program), 212
Cybernetics, 3

D

Dalton's law of partial pressure, 238
Damped forced vibrations of leg, 116–117
Damped free-swinging leg, 115–116
"Dashpot," 116
DDT, 98–99
Decade, 271
Decomposition of output, 215–217
Delay, *50*
DeMoivre's theorem, 110, 111, 150*n*
Descartes' rule, 247–248
Determinant
  of flow graph, 208
Deterministic system, *29,* 32–33
Diabetes mellitus, 121
Difference equations, 57–91
  definitions concerning, 57–60
  and digital computer, 57
  and discrete systems, 33
  and distributed-parameter systems, 30
  existence and uniqueness of solutions of, 62–66
  linear, 59
  and linear deterministic systems, 50
  and linear systems, 36
  and lumped-parameter systems, 30
  nonlinear, 59
  See also Linear difference equations
Difference operator
  and difference equations, 57
Differential equations, 92-127
  and continuous systems, 33
  definitions concerning, 92–101
  and distributed-parameter systems, 30
  existence and uniqueness of, 101–105
  explicit and implicit solutions for, 93
  initial conditions of, 101–102
  linear, 93
  and linear deterministic systems, 50
  and linear systems, 36

and lumped-parameter systems, 30
nonlinear, 93
order of, 92
and resonance, 117
See also Linear differential equations
Differential-equation model of governors, 2
Differentiation
of Laplace transforms, 171–172
of z-transforms, 156
Digital computers
and nonlinear systems, 212
Dirac, P. A. M., 288
Dirac delta function, 287–290
Direction field, 112
Discrete systems, *29*, 33–34
See also One loop discrete systems
Distributed-parameter systems, *29*, 30–32
Disturbance in system process function, *229*
Dominant poles, 217–218
response property of, 218–219
Drebbel, Cornelis, 2
Driving an automobile system, *7*–8
Driving term
of differential equations, 93
Drosophilia, 96*n*
Dynamical system, 5
DYNAMO (program), 212

E

Elements
of linear systems, *50*
Endogenous variables, 10
Engineering
and control theory, 1
Engineers
and control problems, 4
use of transfer function by, 191
Environment, 4, 10
Error signal, 224, 225
Existence theorem
for linear differential equations, 103
Existence and uniqueness
of difference equations, 62–63
Exogenous variables, 10
Experimental modeling, 9
Explicit solution
of differential equations, 93
Exponential order function, 169
External description, 5
External description of systems, 4–8
first examples of, 35
Extrapyramidal nervous system, 118*n*
Euclid's algorithm, 243–244
Euler's identity, 150, 262

F

Fast responding systems, 142
Feedback
in living organisms, 41
Feedback configuration, 146
Feedback configuration block diagram, *146*
Feedback control systems, 40
characteristics of, 223–233
Fibonacci, Leonardo, 2
Field mouse population growth, 106–107
Filter theory, 215–216
Final value theorem
and Laplace transforms, 173–174
of z-transforms, 155
Flow graph diagram transformations (table), *204*
Flyball governor, 2
Forced free vibrations, 117–118
Forcing term
of difference equations, 59
of differential equations, 93
FORTRAN (program), 212
Fourier's theorem, 261
Forward path
in signal flow theory, 203
Fredericq, L., 2
Free-swinging leg diagram, *113*
Free-swinging limb, 113–114
Frequency analysis methods
analysis of, 280
Frequency-domain, 3
Frequency response data, *281*
Frequency response method, 260–263
Friedman's method, 248, 249–251
Fundamental set of solutions, 69–70, 71–80

G

Gain (system)
defined, 233
Gain change, *50*
Gain of arterial pressure regulating system, 234–235

Gains
  in signal flow graphs, 202
Gauss, C. F., 285
Generating sequence of z-transforms, 147
Glucose tolerance test, 120–121
Goodwin's model of control of protein synthesis, 40
Graphical analysis
  of one-loop discrete systems, 80–91
  of salmon population, 86–91
Gray, J. S., 238
Great Lakes drainage (table), *98*
Greatest common divisor (g.c.d.), 243
Grodin, F. S., 3
Growth of human population
  modeling process for, 15–16
Guyton, A. C., 3, 46

H

Hageman factor, 43
Hazen, H. L., 3
Heaviside function, 137, 290
Hemorrhagic shock, 46–47
Heron of Alexandria, 2
Herring gull population
  block diagram of, *206*
  flow graph of, *206*
  flow graph simplification, *206*
Highest common factor (HCF), 243*n*
Hoffman, K., 45
Homeostasis
  defined, 2
Homeostatic index, 233
Homogeneous system, *29*, 37
Human circulatory system
  block diagram of, *6–7*
Hurwitz, A., 254
Hybrid systems, 33

I

Implicit solution
  of differential equations, 93
Improper fractions, 178–179
Impulse, 287
Index
  and difference equations, 57
Indicator-dilution method, 122–125
Indicator-dilution method diagram, *123*
Infusoria, 96*n*
Initial conditions, 101–102
Initial value theorem
  and Laplace transforms, 173
  of z-transforms, 155
Input-output relations, 4–8
Input signal mode, 214, 215
Input variables, 5
Insulin, 121
Integral equations
  and Laplace transforms, 185
Integration
  of Laplace transforms, 172–173
Integrator, *50*
Integrodifferential equations
  and Laplace transform, 185
Interconnections of linear systems, 144–146
Interval rise time, 222
Inverse Laplace operator, 169
Inverse z-transforms, 158–165
  direct method of finding, 165–166
  partial practice procedure for determining, 163
Inverses
  of series configurations, 145
Isoclines, 102

J

Jokipii, S. G. and Turpeinen, O., 121
Junction, *6*

K

Kalmus, H., 3
Kronecker Delta input, 141–142
Ktesibios, 2

L

Lag phase of growth, 96
Laplace transforms, 3, 167–188
  definition of, 167–170
  of periodic functions, 174
  properties of, 170–175, *175*
  and solutions of differential equations, 184–188
  tables of, 175–178, *176, 177,* 176
  and zero initial conditions, 189
Laplace operator, 171
Lee and Weiner, N., 3
Lehmer's method, 247

Left shifting
  of z-transforms, 154
Li, T. Y., and York, J. A., 86
Liapunov, 3
Life cycle of female herring gull
  example of pictorialization, 205–206
Life cycle of schistosomes, 81–86
  block model of, *83*
  simulation diagram of, 83
Life cycle of salmon
  simulation diagram of, *87*
Limited population growth
  nonlinear model of, 36
Line element diagram, *102*
Line elements, 112
Linear dependence, 208
Linear dependent functions, 106
Linear difference equations, 66–71
  with constant coefficients, 71–80
Linear differential equations, 105–108
  characteristics of, 93
  with constant coefficients, 108–120
  and damped forced vibrations of leg, 116
  and damped free-swinging leg, 115–116
  existence theorem for, 103
  and forced free vibrations, 117–118
  and free-swinging limb, 113–114
  and glucose tolerance tests, 120–121
  homogeneous character of, 93
  and indicator-dilution method, 122–125
  nonhomogeneous character of, 93
Linear operator, 171
Linear stationary systems
  determining stability of, 242
  stability criteria for, 241–283
Linear system diagrams
  basic units of, *50*
Linear systems, *29*, 34–36
  convolution representations of, 128–138
  importance of, 35
  interconnections of, 144–146
  properties of, 138–141
  response of, 214–240
  steady-state response in, 217–223
  and transfer function, 211
  transient response in, 217–223
Linearity, 138–139
  characteristics of, 66–67
  of Laplace transforms, 171
  of z-transforms, 152
Logarithmic plots, 271–283 passim
Logistic growth model, 64
Logistic law of growth, *36*, 95–97
Loop
  in signal flow theory, 203
Loop gain
  in signal flow theory, 203
Lumped-parameter system, *29*, 30–32

M

Machin, K. E., 238
Magnitude asymptotes of poles and zeroes, *277*
Mason's Loop Rule, 208–210
Mathematical models
  described, 8
Mathematical theory of regulators, 2
Mathematicians
  and control problems, 4
Mathematics
  and system analysis, 4
Maxwell, James Clark, 2
Mayr, 0., 4
Measurement sensor, 230
Mechanics of animal motion, 2
Mercury, 99
Method of least squares, 285–286
Method of variation of parameters, 78, 111
Milhorn, H. T., 3
Milsum, J. H., 3
MIMIC (program), 212
Minorsky, N., 3
Modeling process
  examples of, 11–22
Models
  defined, 8
  principal uses of, 22
Momentum, 287
Multiple loop systems
  study of, 47
  symbol-arrow graph for, 45–*46*
Multiplication property
  of z-transforms, 154–155
Multiplier, *50*

N

Natural frequency of motion, 111
Natural period of motion, 111
Negative-feedback closed-loop system, *224*
Negative-feedback loop systems, *42*-43

Negative-feedback systems
disturbances in, 229–230
and error reduction, 224
Negative-feedforward loop systems, *44*-45
Nervous system
modeling process for, 19–22
Newton's equation for angular motion, 114
Newton's equation $m\ddot{y} = u$
simulation diagram for, 52
Newton's second law, 287
Nodes
in signal flow graphs, 202
Noise signal, 230
Nonanticipatory systems, 134
Nonhomogeneous system, *29,* 37
Nonlinear system, *29,* 34–36
and computer investigation, 212
Nonlinearity
and transfer function, 212
Nonstationary system, *29,* 37
Nontouching loops
in signal flow theory, 203
Nyquist, H., 3, 262
Nyquist criterion, 254, 264–269
Nyquist diagram, 261, 262, *265, 267*
relationship to Bode diagram, 279
Nyquist path, 263, *264, 265*

O

Octave, 271
Offset
in control system, 227
Oil lamp and float regulator, 2
One dominant real pole situation, 221
One-loop discrete systems
analysis of dynamics of, *85*
graphical analysis of, 80–91
and life cycle of schistosomes, 81–86
simulation diagram for, *81*
Open- and closed-loop systems, 224–227
Open-loop gain
methods for calculating, 233–238
of the respiratory system, 237–239
Open-loop state of the carotid sinus loop, *234*
Open-loop systems, 40–50, *224*
disturbance in, *229*
Open system, 4
Optimal control theory, 3, 4
Optimal system, 222
Order of differential equations, 92
Output
characteristics of higher order systems, 222
decomposition of, 215–217
Output variables, 5
Overdamped unit step response, 220–221

P

Papin, Dennis, 2
Parallel configuration, 144–145
Parallel systems block diagram, *145*
Parameter variations, 227–229
Partial fractions
expansion into, 178–184
and inverse z-transforms, 158–165
and Laplace transforms, 176
Particular solutions
and beaver population study, 76
methods for finding, 74–75
Past time
in linear discrete systems, 129
Path
in signal flow theory, 203
Path gain
in signal flow theory, 203
Peak time, 222
Percent overshoot, 222
Performance index, 222
Periodic sequences
of z-transforms, 155
Pharmacokinetics, 15–16
Phase angle of motion, 111
Phase curve, *278*
Philon, 2
Phosphorous, 99
Physiologists
and control problems, 3
Piecewise continuous function, 167
Planimeters, 125
Pneumatica (Heron), 2
Pole characteristics of stable system (table), *219*
Polynomial
approximating roots of, 244–246
finding complex roots of, 248–252
finding multiple roots of, 247–248
finding real roots of, 243–248
improving accuracy of approximate root, 246–247
reducing to simple polynomial, 244
Polzunov, I., 2
Population of beavers study, 76

Population size as function of immigration, 132
Portal circulation, 7
Positive-feedback loop systems, 43–44
  uses of, 44
Positive-feedback system, 224
Positive-feedforward loop systems, *45*–47
  examples of, 45
  and multiple loop systems, 45–*46*
  and predictions, 45
Present time
  in linear discrete systems, 129
Pressure regulator (steam), 2
Principle of linearity, 138–139
Probability theory
  and stochastic systems, 32
Proper fractions, 178–179
Protein synthesis
  linear model of, 225–226
  linear model with negative feedback, 226
  and Routh-Hurwitz criteria, 257
Protozoa, 96*n*

R

Rabbit populations, growth of, 2
Radioactive dating of fossils, 94–97
Rainey, R. H., 97, 98
Rapp, P. E., 226
Reaction time
  described, 19
  modeling process for, 19–22
Real roots of polynomial
  steps in locating, 243–248
Real-time translation
  of Laplace transforms, 170
Recursive equations, 15
Region of convergence, 148, *149*
Region of divergence, 148
Regulatory agencies
  according to Fredericq, 2
Relative stability
  of linear systems, *268*
Resonance
  and differential equations, 117
  in mechanical systems, 118
Rieman sums, 125
Right hand side
  of difference equations, 59, 93
Right shifting
  of z-transforms, 152–153
Rise time, 222
  of unit step input, 143
Roots of denominator of system transfer function, 241–242
  with negative real parts, 242
  with positive real parts, 241
  with zero real parts, 242
Rosenbleuth, A., 3
Routh, E. J., 254
Routh-Hurwitz stability criterion, 254–260

S

Safe drug dosages
  modeling process for, 16–19
Salmon population analysis, 86–91
Savageau, M. A., 43
Scale change
  of Laplace transforms, 170
Schwartz, Laurent, 288
Separable equations, 94
Separation of variables, 94–97, 97–99
  and dating of fossils, 94–97
  and logistic law of growth, 95–97
Sequence of numbers
  of difference equations, 60
Series (cascade) configuration, 145
Series systems block diagram, *145*
Servomechanisms, 3
"Set-point theory," 41
Settling time, 222
  of unit step input, 143
Signal flow graphs, 201–211
  and block diagrams, 204–206
  definition of, 202
Signal flow theory terminology, 203
Simple harmonic motion, 111
Simulation, 212
Simulation diagrams, 50–51
  examples of, *51, 52, 53*
  systematic approach to, 51
  uses of, 51
Single-valued function generator, *50*
Sink
  in signal flow theory, 203
Sinusoidal response
  of unit step input, 143
Sluggish systems, 142
Solution
  of difference equation, 59–60
Solution curve, 112
Source
  in signal flow theory, 203

Special test inputs
responses to, 141–144
Splitting point, 50–*51*
Sputnik, 3
Stability of amplifiers, 3
Stability of stationary linear system
defined, 241–242
Stable limit cycle, *85, 88*
Static gain, 233–238
Stationary or equilibrium size of population, 64
Stationary phase of growth, 96
Stationary system, *29,* 37
Steady-state response, 217–223
Stewart Hamilton (indicator-dilution) method, 122–125
Stimulus, 19
Stochastic system, *29,* 32–33
Sturm sequence, 245
Sturm's method, 245
Sturm's theorem, 245–246
Subsystems, 6
Summing junction, *50*
Superposition
principle of, 138
Symbol-arrow graph, 46, *47*
Symbols
of linear system elements, *50*
System frequency response, 260–261
System identification problem, 280
System memory, 141–142
System mode, 214, 215
System sensitivity, 228
Systemic circulation, 6
Systems
and block diagrams, 4–8
boundaries and modeling, 10
defined, 4
examples of (table), *6*
external descriptions of, 4–8
and ideal system, 10
kinds of models for, 9
mathematical models of, 8–22
notations used to describe, 4–5
as part of real world, 4
Systems theory
defined, 4

T

Tacoma Narrows Bridge (Washington), 118
Taylor's series, 102
Temperature control by means of blood flow, *42*
Temperature regulator, 2
Thermoregulation, 41
Thompson, D'Arcy, 2
Threshold level, 19
Time-domain, 3
Time-domain formulation, 3
Time-invariant property, 139–140
Time-invariant systems, 37
Time-varying systems, 37
Total magnitude curve, *278*
Touching loops
in signal flow theory, 203
Transfer function, the, 189–192
black box problem of, 192
and circulatory system example, 190–191
and drug concentration example, 193–194
and leg vibration system example, 191
and system of beavers example, 191
and use of impulse inputs, 192–195
Transient response, 217–223
control of, 229
Transmittance
in signal flow graphs, 202
Trapezoidal rule, 125
Two dominant complex conjugate poles situation, 221–222
Two-path systems (Mason's Rule)
diagram of, 209
flow graph of, 209

U

Undamped motion, *116*
Underdamped unit step response, 219–220
Unforced systems, 37
Unit step function, 218
Unit step input, 142–143
and damping, 142
and special responses, 143
and stability, 142
Unit step response system
characteristics of, *222*
Unknowns
and difference equations, 57
Unstable limit cycle, 85, *86*

V

Vectors, 5
Vicious circle of pathophysiological changes, 43
Vyshnegradskii, I. A., 2

W

Water clock and regulator, 2
Water float regulator, 2
Water pollution of lakes, 97–99
Watt, James, 2
Weighting function, 131–132, 141
Wiener, N., 3

Y

Yamamoto, W. S., 3

Z

z-transforms, the, 147–166
  definition, 147–152
  properties of, 152–157
z-transform pairs
  table of, 157–158